Theory and Applied Macroeconomics

Sadi Uzunoğlu / Adil Akinci (eds.)

Theory and Applied Macroeconomics

PETER LANG

Berlin - Bruxelles - Chennai - Lausanne - New York - Oxford

Library of Congress Cataloging-in-Publication
A CIP catalog record for this book has been applied for at the Library of Congress.

Bibliographic information published by the Deutsche Nationalbibliothek
The Deutsche Nationalbibliothek lists this publication in the Deutsche Nationalbibliografie; detailed bibliographic data is available on the Internet at http://dnb.d-nb.de.

ISBN 978-3-631-92755-7 (Print)
E-ISBN 978-3-631-93526-2 (E-PDF)
E-ISBN 978-3-631-93527-9 (E-PUB)
DOI 10.3726/b22747

Published by Peter Lang GmbH, Berlin, Germany

info@peterlang.com - www.peterlang.com

Foreword

Dear Reader,

Researchers from various universities and institutions provided support for the book of "Theory and Applied Macroeconomics". Studies have been specifically made for Türkiye and generally made in global scale.

All publishing rights of the published book chapters belong to the publisher, and the content responsibility of the published articles belong to the author. The opinions in the book chapters are the personal opinions of the authors; it is not an official opinion of any institution or organization.

I express my gratitude to the authors of the book chapters that contributed to the book "Theory and Applied Macroeconomics".

Best regards,

Professor Sadi UZUNOĞLU
Trakya University

Table of Contents

Bahar Erdal[1] & Murat Ali Yülek[2]

An Evaluation of Monetary Policy Implementations in Türkiye Between 1986 and 2023

1 Introduction

Monetary policy can be defined as the change of money supply by the central bank using various monetary policy tools in order to achieve its ultimate goal. In Türkiye, the Central Bank of the Republic of Turkey (CBRT) is responsible for implementing monetary policy. Türkiye experienced high and volatile inflation rates starting from the 1980s until the implementation of implicit inflation targeting regime in 2002. During this period, different monetary policies are implemented depending on the economic and financial conditions in the Turkish economy. Regarding monetary policy implementation, 1986 was a turning point. Before 1986, since the financing of the public sector budget deficit was provided by the CBRT financial resources, monetary and fiscal policies were carried out interdependently. In 1986, a significant change was made in the implementation of monetary policy. Starting from this year, there was a tendency to control total reserves of the banking sector in order to control the money supply and increase effectiveness of interest rate policy.

In 2002, the CBRT started to implement "implicit inflation targeting" monetary policy regime. The main objective of the implicit inflation targeting regime was to ensure price stability. The CBRT started to implement "inflation targeting regime" at the beginning of 2006. The inflation target is disclosed to the public at three-year periods in order to be in line with the three-year budget plans that have been implemented since 2006.

1 Ph.D., Assoc. Prof., Ostim Technical University, Faculty of Economics and Administrative Sciences Department of Economics, bahar.erdal@ostimteknik.edu.tr, ORCID: 0000-0002-0684-8999.

2 Ph.D., Prof., Ostim Technical University, Faculty of Economics and Administrative Sciences Department of Economics, murat.yulek@ostimteknik.edu.tr, ORCID: 0000-0001-7533-5882.

In this study, a general evaluation of the monetary policies implemented in Türkiye around past 37 years between 1986 and 2023 is done. When doing this evaluation, three important pre-requisites for an effective implementation of monetary policy should be kept in mind: 1- The implementation of fiscal policy and debt finance in an predictable way, 2- Political stability and 3- Efficiently working banking system.

In this framework, following questions are tried to be answered: What were the objectives of the monetary policies during this period? Which monetary policies were implemented? How did the inflation rates evolve? Were the monetary policies successful in reducing inflation? Which monetary policies were successful to support economic growth?

2 Monetary Policy and Central Banking

Monetary policy can be defined as the change of the money supply by the monetary authority, i.e., central bank, using various policy tools in order to achieve its ultimate goal regarding economic or financial variables. Theoretically, the ultimate goal of monetary policy can be one or more of the following: price stability, financial stability, economic growth, full employment, exchange rate stability and interest rate stability. The ultimate goal of the monetary policy has been changed overtime within the framework of changing economic conditions. For example, before the 1970s, the goal of monetary policy was full employment and economic growth. In the 1970s, oil prices increased due to oil crises and then, high inflation has emerged in many countries. As a result, central banks implemented expansionary monetary policies in order to eliminate the negative effects of oil price increases in their economy. Therefore, the goal of monetary policy was shaped as controlling inflation and ensuring monetary stability. Today, the primary objectives of monetary policy implementations are price stability and financial stability.

Price stability is defined by the former Federal Reserve Board (FED) Chairman Alan Greenspan as " Price stability is best thought of as an environment in which inflation is so low and stable over time that it does not materially enter into the decisions of households and firms." (Greenspan, 2002: 6). As can be understood from this definition, in order to ensure price stability, the inflation rate must be low and the low inflation rate must be maintained over time. No numerical value is given for the inflation rate. *Financial stability* is defined as the efficient working of financial system and transfer of funds from savers to investors without interruption. The central banks are institutions that are authorized to issue banknotes and to implement monetary policies of countries. Central banks provide loans or accept deposits to banks as the "ultimate lending authority" or "last lending authority". For this reason, central banks are also described as "banks of banks". Historically, central banks were established as private commercial banks rather than as non-profit public institutions (Capie, Goodhart & Schnadt, 1994: 4–5). The first truly central bank in the world was the Bank of England, founded in 1694. The Bank of England had the three functions that a central bank should have: the power to issue banknotes, being the bank of the state, and being the lender of last resort.

The main reason why countries want to have their own central bank is that they can keep control of issuing money under a single authority. Every independent country wants to have a national currency and central bank as a sign of its independence. Loss of control over money issuance means increased inflation. For example, after the Napoleonic wars, excessive printing of money to cover war expenses caused high inflation, so the idea of giving money printing to an independent organization arose. The second important reason is that central banks obtain seigniorage revenue by printing banknotes.

3 History of Central Banking in Türkiye and the CBRT

The historical development of central banking dates back to the Ottoman Empire (Tekeli and İlkin, 1997). The bank that carried out the first central banking activities during the Ottoman Empire was "Bank-ı Osmani Şahane" which was established in 1863 for the purpose of paying foreign debts between the Ottoman Empire and European countries and providing foreign financing. Bank-ı Osmani Şahane, founded as a British and French partnership, undertook some of the modern central banking functions such as the state treasurer collected the state's revenues and made payments, discounted treasury bills and made principal and interest payments of domestic and foreign debts. The privilege to print banknotes was given to this bank for 30 years. After the establishment of the Republic of Türkiye, the idea of establishing a national central bank was first brought to the agenda at the Turkish Economic Congress held in 1923. Although some attempts were made to transform Bank-ı Osmani Şahane into a state bank, this could not be realized due to unsuitable economic and financial conditions. The works about establishing a national central bank started again in 1927.

The Central Bank of the Republic of Turkey (CBRT), which is responsible for the implementation of monetary policy in Turkey and has the sole authority to issue banknotes, was established as a joint stock company with private law legal personality on June 11, 1930 by Law No. 1715 and started its activities in 1931 (Bakır, 2007). As can be understood from this definition, the CBRT is not a public institution. CBRT is a joint stock company established by its own special law, and in cases where there is no clarity in the founding law, private law provisions will be implemented. The founding capital of the CBRT is 15 million Turkish Lira. The authority to issue banknotes was granted for a period of 30 years, and it was stipulated that this period could be extended five years before its end.

When the CBRT was established, its main purpose was to support economic development. In this respect, it would determine the rediscount rates, regulate the money market and the circulation of money, carry out treasury transactions and take all measures jointly with the government to protect the value of the Turkish Lira. The CBRT Law has been amended many times in order to adapt to changing economic conditions over time. Most of the changes made were to increase the loans given to the Undersecretariat of Treasury and State Economic Enterprises. Thus, the main function of the CBRT has been to finance the public sector deficits rather than carrying out an effective monetary policy.

With the beginning of the planning period in the 1960s, the idea that the existing CBRT Law was not sufficient to carry out the monetary policies necessary for economic development gained importance. In order to meet the changing economic and financial needs, Law No. 1715 was repealed and Law No. 1211 came into force on January 14, 1970. In the new CBRT Law, the main duties of the CBRT are stated as carrying out the monetary and credit policy, taking measures to protect the value of money, and carrying out the printing and lending of money. Amendments were made to Law No. 1211 from time to time. The CBRT Law currently in force is Law No. 1211, amended by Law No. 4651 and dated 25 April 2001.

4 Monetary Policies Implemented Between 1986 and 2001

Türkiye experienced high and volatile inflation rates during the 1980s and 1990s (Figure 1). The main reasons for high and volatile inflation rates were financing high budget deficits by the CBRT funds followed by pricing policies implemented to finance public sector deficits, political instability in the economy and increases in oil prices in the world. High budget deficits prevented implementation of monetary policy and other policies in an effective way as well as structural reforms could not be realized.

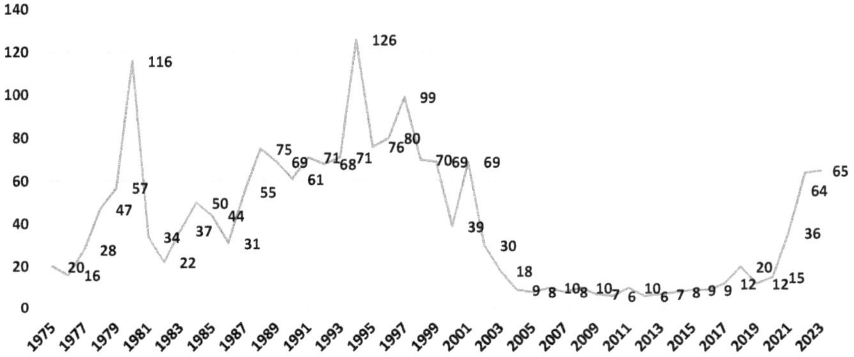

Figure 1: Inflation Rate (%) Between 1975 and 2023
Source: Turkish Statistical Institute

Therefore, the authorities tried to maintain and manage current economic situation without causing bigger problems. The vulnerabilities have increased in the economy and an environment has been created for possible negative developments to create large jumps in inflation. During this period, economic growth rate was also not stable, it had up and downs and had negative values during the crisis years such as 1980, 1994, 1999, 2001, 2008 (Figure 2).

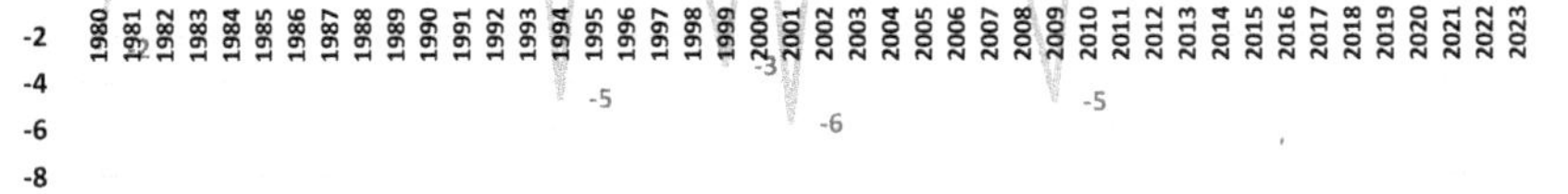

Figure 2: Economic Growth Rate (%) Between 1980 and 2023
Source: Turkish Statistical Institute

1986 was a turning point in monetary policy implementation. Before 1986, since the financing of the public sector was done with the resources of the CBRT, monetary and fiscal policies were carried out interdependently. Since 1986, there has been a tendency to control total reserves of the banking system in order to control the money supply and increase the effectiveness of interest rate policy. In this context, in 1987 and 1988, monetary policies were implemented by the CBRT in a way to carry out these policies. In 1989, taking into account the impact of the size of the Central Bank's balance sheet on monetary developments, monetary policy was implemented to control this size. During this period, domestic credit expansion was stopped and balance sheet growth resulted only from the increase in foreign assets and the change in the valuation account.

In 1990, monetary policy was announced to the public in detail for the first time. The CBRT wanted to control its balance sheet size by placing limits on the total balance sheet size, total domestic liabilities, total foreign assets and central bank money. In 1991, due to impact of the Gulf Crisis, monetary program was not announced and as a result of the rapid increase in public deficits, the targeted monetary aggregates were significantly exceeded.

In 1992, the monetary program was announced, but since the announced targets were largely exceeded as a result of the rapid increase in public deficits, monetary policy was shaped to prevent excessive fluctuations in the exchange rates. In 1993, the announcement of the monetary program was abandoned and it was declared that monetary policy would focus on maintaining stability in financial markets. After the decisions of April 5, 1994, monetary policy was directed towards ensuring stability in financial markets and increasing foreign exchange reserves.

In 1995, a stand-by agreement was signed with the IMF and targets were set for Net Domestic Assets (NDA) and Net International Reserves (NIR) in the CBRT balance sheet. Exchange rate policy as in the previous periods, was used as a nominal anchor in the fight against inflation and an exchange rate increase at or below the inflation rate was envisaged. The stand-by agreement was ended after the general elections held in December of 1995.

In 1996, due to increasing uncertainties in the economy, stability in financial markets was maintained instead of fighting against inflation. Exchange rates were kept under surveillance and balance sheet size of the CBRT was tried to be kept under control.

In 1997, monetary policy continued its implementation in 1996. While attempts were made to ensure the stability of financial markets, exchange rates continued to be used as an intermediate target in the fight against inflation.

In 1998, while stability in financial markets continued to be maintained, the fight against inflation was included among the objectives of monetary policy. In this context, a monetary program targeting the reserve money size in quarterly and six-month periods was prepared and announced to the public.

In the first months of 1999, monetary policy operated as a continuation of the monetary policy of 1998. In December 1999, a stand-by agreement was signed with the IMF and within this framework, the "Exchange Rate Based Inflation Reduction Program" was put into practice. The program in question adopted a tight fiscal policy, increasing the primary surplus, carrying out structural reforms, accelerating privatization and implementing an income policy compatible with the inflation target.

In 2000, the basic element that determined the framework of monetary policy was the "Exchange Rate Based Inflation Reduction Program". Starting from the beginning of 2000, an inflation reduction program that will cover a three-year period has been implemented. The aims of the inflation reduction program were to reduce Consumer Price Index (CPI) inflation to 25 % at the end of 2000, 12 % at the end of 2001 and 7 % at the end of 2002, to reduce real interest rates to reasonable levels, to increase the growth potential of the economy and to ensure an effective and fair distribution of resources in the economy.

Following the financial crises in December 2000 and February 2001, a new economic program was implemented in May 2001 under the name "Transition to a Strong Economy". Within the framework of this program, monetary policy would focus on ensuring stability in the financial markets within a period called the transition period, then its main goal would be to ensure price stability. The main difference of the current program from past programs is that monetary policy is implemented under a floating exchange rate regime. During this period, the monetary base was accepted as an intermediate target (nominal anchor). In order to ensure monetary control, upper limits were set for NDA and lower limits for NIR.

In sum, it could be said that although many different monetary policy programs were implemented to reduce inflation rates between 1986 and 2001, they were unsuccessful to reduce high inflation rates (see Figure 1). The major reasons for failure to reduce high inflation rates were financing budget deficits with the CBRT funds, pricing policies implemented to finance public sector deficits, failure to realize structural reforms such as privatization and banking sector reform, implementation of short-term economic programs instead of long-term programs due to frequent elections, political instability, common distrust to policies by the public, dollarization and inertia in the economy due to high inflation (i.e., economic decisions are taken based on past experiences).

5 Implicit Inflation Targeting Regime

In 2002, a stabilization program was launched, aiming to reduce the inflation rate to 12 % over a three-year period, and then to single digits. The main points of implicit inflation targeting regime can be summarized as follows (Pınar and Erdal, 2016: 164–165):

The ultimate goal of monetary policy was to reduce CPI inflation rate to 35 % by the end of 2002. In order to achieve this goal, the monetary base would continue to be used as an intermediate target and the monetary base would be increased by 40 %, which is the nominal GDP increase. The upper limits on NDA would be monitored as an indicator criteria and the lower limits on NIR would be monitored as a performance criteria. The short-term interest rates would be used to reduce inflation. The CBRT could change short-term interest rates by taking into account the values that inflation could take in the future.

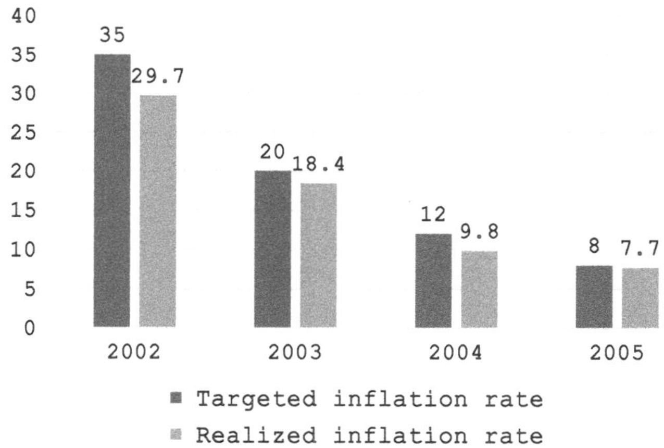

Figure 3: Inflation Rate (%) Between 2002 and 2005
Source: CBRT

Implicit inflation targeting regime was successful in reducing inflation rate from 69 % in 2001 to 29,7 % in 2002, then it decreased to 7.7 % in 2005 (Figure 3).

6 Inflation Targeting Regime

The inflation targeting regime is a monetary policy regime that is widely used in both developed and developing countries. Inflation targeting is carried out as the implementation of monetary policy towards the inflation target in order to ensure and maintain price stability. Inflation targeting firstly implemented in New Zealand in 1990. In the countries where inflation targeting regime is implemented inflation rates and its volatility decreased and economic growth increased. The CBRT started to implement "inflation targeting regime" at the beginning of 2006. The main elements of the inflation targeting regime can be summarized as follows (Pınar ve Erdal, 2016: 167–169):

- The inflation target is determined jointly with the Government, and monetary policy implementation is at the initiative of the CBRT.
- The targeted inflation rate is determined as a "point target", taking into account the advantages of being easily understood by the society and effectiveness in directing expectations.
- The inflation target is determined as the year-end CPI inflation and is calculated as an annual percentage change.
- The inflation target is announced to the public every 3 years in order to be compatible with the three-year budget plans that have been implemented since 2006. In this context, year-end inflation targets for 2006, 2007 and 2008 were determined as 5 %, 4 % and 4 %.
- Within the framework of the CBRT's accountability mechanism, an "uncertainty range" of ± 2 points above and below the inflation target has been created.
- Directing expectations is of great importance in inflation targeting regime. The CBRT announces inflation forecasts covering 2 years, creates a path compatible with the target as of the end of each quarter and declares an uncertainty range of ± 2 points around this path.
- Inflation forecasts show the path that inflation is expected to follow while reaching the final inflation target. For this reason, it would be appropriate for economic units to use inflation forecasts as a reference in the short-term and the inflation target in the medium-term.
- If inflation remains outside the uncertainty range, the CBRT notifies the Government and announces to the public the reasons for this deviation and the measures to be taken to ensure that inflation reaches the target again, within the framework of Article 42 of the CBRT Law.

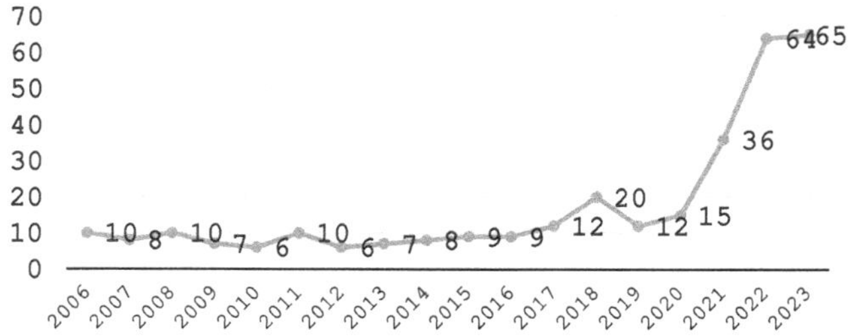

Figure 4: Inflation Rate (%) Between 2006 and 2023
Source: CBRT

In this framework, monetary and exchange rate policies have been successful in reducing inflation rates from two digits to single digits together with the success of measures to reduce government budget deficit and the implementation of structural reforms within the framework of stabilization programs that had been implemented since 2002 (Figure 4). Besides, economic growth rate was higher and stable in the inflation targeting regime (average 5 % between 2002 and 2023) as compared to other monetary policy regimes (average 4 % between 1986 and 2002).

However, there has been an increase in the inflation rates after COVID-19 pandemic starting from 2021. The major reasons for increase in inflation rates can be summarised as follows: the increase of demand due to credit expansion by banks during the COVID-19 pandemic, increases in prices of basic goods groups as a result of increase of import prices due to the depreciation of Turkish lira, increases in the prices of fresh vegetables and fruits, increases in the prices of Services Sector (restaurants and hotels) through food services due to increase in food prices, geopolitical risks due to wars in the neighbor countries and inflation expectations.

7 Conclusion

In this study, the monetary policy implementations by the CBRT to decrease and stabilize the high and volatile inflation rates are evaluated for the period between 1986 and 2023. The high and volatile inflation rates experienced in Türkiye from 1980 to 2001. Then, inflation rates started to decrease since 2002 with the implicit followed by explicit inflation targeting regimes together with the reduction of budget deficits and accomplishment of structural reforms regarding privatization and banking sector. Besides, economic growth rates were higher and more stable in the inflation targeting regimes as compared to those in the other monetary policy regimes. Turkish high inflation experience supports the importance of three pre-requisites necessary for effective implementation of monetary and exchange rate policies that we mentioned at the beginning of the paper: the implementation of fiscal policy and debt finance in a predictable way, political stability and efficiently working banking system.

References

Bakır, C. (2007). *Merkezdeki Banka: Türkiye Cumhuriyet Merkez Bankası ve Uluslararası Bir Karşılaştırma*. İstanbul: Bilgi Üniversitesi Yayınları.

CBRT, internet page (www.tcmb.gov.tr).

Capie, F., Goodhart, C., & Schnadt, N. (1994). *The Development of Central Banking*. Cambridge University Press, pp. 4–5.

Greenspan, A. (2002). "Transparency in Monetary Policy", Federal Reserve of St. Louis Review, 84(4), July/August: 5–6.

Tekeli, İ., İlkin, **S**. (1997). Para ve Kredi Sisteminin Oluşumunda Bir Aşama, Türkiye Cumhuriyet Merkez Bankası, TCMB, Ankara.

Turkish Statistical Institute, internet page page (www.die.gov.tr).

Pınar, A., & Erdal, B. (2016). Para Banka Mali Kuruluşlar ve Uluslararası Mali Sistem, Turhan Kitabevi, 3. Bası, Ankara.

Ayşegül Şahin[1]

Does Monetary Policy Uncertainty Affect Prices?

1 Introduction

> *"The Federal Reserve's experiences over the past two decades make it clear that uncertainty is not just a pervasive feature of the monetary policy landscape, it is the defining characteristic of that landscape." – Alan Greenspan*

In his statement at the American Economic Association meeting held in California in 2004, Greenspan demonstrated the importance of monetary policy uncertainty with these words. Baker et al. (2016) initiated the policy uncertainty calculation methods and later guided the calculation of different uncertainties. Husted et al. (2020) used this methodology to create a monetary policy uncertainty index using monetary policy-related news in newspapers and magazines. Accordingly, the increase in news may affect economic activities by causing an increase in uncertainty. Uncertainties in the economy can be one of the factors explaining economic fluctuations (Caldara et al., 2016). Cascaldi-Garcia et al. (2020) note, "Researchers, policymakers, and market participants are increasingly focusing on the effects of uncertainty and risk on financial market and economic outcomes." Increased uncertainty typically leads to declines in real GDP, consumption, investment, employment, and interest rates – see, for example, Elder & Serletis (2010), Jurado et al. (2015), and Caldara et al. (2016), among others.

The widespread view in the literature is that economic uncertainties can have a significant impact on the inflation rate, mainly by affecting consumer decisions. Consumers who cannot predict their future income in the economy may tend to save by consuming less in the future. The contraction in consumption expenditures may also lead to a decline in the general level of prices. Similarly, increases in uncertainty affect investors' decisions, causing investment expenditures to decrease. Therefore, there is an increase in the general level of prices.

---

1 Ph.D., Anadolu University, Faculty of Economics and Administrative Sciences Department of Economics, aysegulsahin@anadolu.edu.tr, ORCID: 0000-0002-4278-0266.

There are many studies explaining the relationship between prices and uncertainties (Bloom, 2009; Colombo, 2013; Baker et al., 2016; Caldara et al., 2016; Alam & Istiak, 2020; Karagöl, 2023). Although interest in economic policy uncertainty has increased in recent years, studies on monetary policy uncertainty remain limited (Kurov & Stan, 2018; Ugurlu-Yildirim et al., 2020; Azad & Serletis, 2022).

Schaling & Nolan (1998) based the most important way of reducing both monetary policy uncertainty and inflation on the solution of the central bank accountability problem. Yellen (2017) suggested that the 2 % inflation target could be achieved over time by making gradual adjustments in monetary policy and reducing internal uncertainties. Azad & Serletis (2022) proved that US monetary policy uncertainty has negative effects on the macroeconomic and financial fundamentals of emerging markets.

This paper contributes to two literatures. Firstly, it contributes to the uncertainty literature. Higher uncertainty, whether resulting from trade restrictions, political unrest, or the Covid-19 crisis, often has a negative impact on the level of economic activity, and the study proves this. Second, it contributes to the growing international literature on monetary policy uncertainty and time-varying causality.

The article is organized as follows. In Section 2, the Methodology used in the study will be mentioned, and in Section 3, the Data, Model, and Empirical Findings presented. The last section provides a brief conclusion

2 Methodology

The existence and detection of causal relationships are important in determining the interaction mechanism between variables in the economy (Wang & Fu, 2022). If the inclusion of a variable in the model does not improve the prediction of the second variable, this variable is said to be non-Granger causal for the second variable (Balcilar et al., 2010). Granger causality analysis is analyzed using Wald, Lagrange multiplier (LM), and likelihood ratio (LR). Here, the test statistics must be stationary, otherwise, it will cause difficulties in the level predictions of VAR models (Toda & Phillips, 1994). To overcome these difficulties, Toda and Yamamoto (1995) proposed an extended VAR where the coefficients are non-stationary or a modified Wald test to obtain long-term causality estimation of VAR(p) coefficients. If we consider a two-variable VAR(p) process:

$$y_t = \theta_0 + \theta_1 y_{t-1} + \ldots + \theta_P y_{t-p} + \varepsilon_t, t = 1, 2, \ldots, T \quad (1)$$

$$\begin{bmatrix} y_{1t} \\ y_{2t} \end{bmatrix} = \begin{bmatrix} \theta_{10} \\ \theta_{20} \end{bmatrix} + \begin{bmatrix} \theta_{11}(L) & \theta_{12}(L) \\ \theta_{21}(L) & \theta_{22}(L) \end{bmatrix} \begin{bmatrix} y_{1t} \\ y_{2t} \end{bmatrix} + \begin{bmatrix} \varepsilon_{1t} \\ \varepsilon_{2t} \end{bmatrix}$$

where p is the lag order, $\theta_{ij}(L) = \sum_{k=1}^{p} \theta_{ij,k} L^k, i, j = 1,2$ and L is the lag operator.

In these tests, the assumption that the parameters of the VAR model remain constant over time prevails. However, structural changes often violate this assumption. Studies often use techniques such as dummy variables and sample splitting and can identify the presence of structural changes. For this reason, rolling window (RO) bootstrap estimation was applied, which allows structural breaks caused by economic and political changes in the United States economy during the review period to change the coefficients over time. The rolling window Granger causality test takes into account the shifts that structural changes can cause in parameters and the time-changing pattern of the causal relationships (Balcilar et al., 2010). Balcilar et al. (2010), utilizing the above methodology developed by Hacker & Hatemi-J (2006), developed a causality method to investigate bootstrap causality in rolling window subsamples. This method uses the bootstrap test $t = \tau - l + 1, \tau - l, \ldots, \tau$ and $\tau = l, l + 1, \ldots, T$ where l is the size of the rolling window. The rolling window technique uses a moving window of fixed length sequentially from the beginning to the end of the sample by adding an observation at the beginning and subtracting an observation from the end, and each rolling window subsample containing l observations. At each step, apply the causality test to each subsample, providing a $(T - l)$ set of causality tests rather than just one. The most important advantage of rolling window estimation is that it recognizes that the relationship between variables changes over time and takes into account instabilities between different subsamples due to structural changes.

3 Data, Model and Empirical Findings

3.1 Data

This study aims to determine the impact of monetary policy uncertainty on inflation in the United States. To measure monetary policy uncertainty, the US monetary policy uncertainty index (*mpu*) developed by Husted et al. (2020) is used. Monetary policy uncertainty is calculated using Baker et al. (2016) suggested methodology for measuring policy uncertainty based on newspaper reports and articles. When considering monetary policy uncertainty, the inflation rate (*inf*) is the most important indicator of stability in monetary policy. Due to availability, the monthly data interval of this research sample ranged from January 2002 to April 2023. The *mpu* data was obtained from www.policy uncertainty.com (Accessed on: 01 May 2024) and the *inf* data was obtained from Worldbank Databank (Accessed on: 01 May 2024).

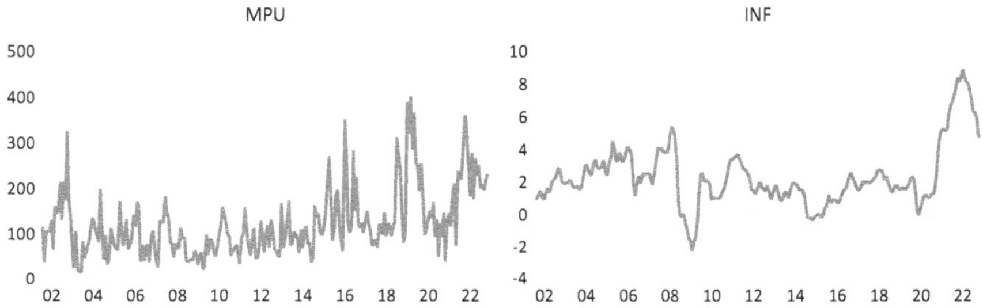

Figure 1: The trend of monetary policy uncertainty and inflation rate

Figure 1 shows the time trends of the monetary policy uncertainty index and inflation rate between 2002 and 2023. The most obvious fluctuation in the monetary policy uncertainty index begins with the 2003 Iraq invasion, continues with the 2008 Global Financial Crisis, and the Quantitative Easing policies implemented thereafter. In recent years, there has been the Brexit process, the US Presidential election, and the Covid-19 pandemic. The right panel of the figure shows the approximately 20-year change in the inflation rate. Considering that the long-term inflation target is 2 % on average during this period, it is seen that the rate has remained under the target for many years. Negative inflation rates were observed especially with the recession during the 2008 Global Financial Crisis. With the Covid-19 pandemic, there is a serious increase in the inflation rate in the US economy, as in the rest of the world. Do the uncertainties in monetary policy have an impact on the rising inflation, especially in 2008, and the extreme increase that has occurred recently? It is not possible to interpret this directly from the figure. Therefore, the causality relationship between the two variables needs to be examined.

3.2 Model and Empirical Findings

Following Balcilar et al. (2010), this study analyzes the Granger causality between monetary policy uncertainty (*mpu*) and inflation rate (*inf*) using the vector autoregressive (VAR) framework. For this purpose, the stationarity levels in the time series must be determined through unit root tests. The stationarity levels of the variables are determined by Augmented Dickey-Fuller (ADF) unit root tests developed by Said and Dickey (1984) and Phillips-Perron (PP) unit root tests developed by Phillips and Perron (1988) and by Lee and Strazicich (2004) unit root test with structural breaks. The null hypothesis of the ADF and PP unit root test is that the variables are not stationary. In the Lee & Strazicich unit root test, the null hypothesis is that the variables are not stationary with a single break. If the calculated test statistics are greater than the critical values or the probability values of these statistics are less than the significance levels such as 10 %, 5 %, and 1 %, these basic hypotheses are rejected. Unit root test results are presented in Table 1.

Table 1: ADF, PP and Lee & Strazicich Unit Root Tests Results

With Constant							
Test	**ADF**		**PP**		**Lee & Strazicich**		
Variables	**Test Statistics**	**Prob.**	**Test Statistics**	**Prob.**	**Test Statistics**	**Critical Value**	**Break Date**
mpu	-5.2699	0.0000*	-6.9733	0.0000*	-5.1042*	-4.0820	2006M01
inf	-2.3099	0.1697	-2.5398	0.1073	-3.0338	-4.0820	2008M10
Δinf	-5.9517	0.0000*	-9.3791	0.0000*	-10.2792*	-4.0814	2007M09
With Constant and Trend							
Test	**ADF**		**PP**		**Lee & Strazicich**		
Variables	**Test Statistics**	**Prob.**	**Test Statistics**	**Prob.**	**Test Statistics**	**Critical Value**	**Break Date**
mpu	-8.1695	0.0000*	-8.2970	0.0000*	-5.5313*	-4.8345	2008M06
inf	-2.3286	0.4164	-2.6424	0.2619	-4.6677	-4.1233	2020M05
Δinf	-6.0358	0.0000*	-9.3567	0.0000*	-10.5746*	-4.6471	2020M12
*Notes: * denote significance at 1 %, respectively. Δ is the difference operator.*							

According to ADF and PP test results, while monetary policy uncertainty is stationary at level one of the variables, inflation data is stationary at the first difference. When we look at the Lee-Strazicich unit root test results, which take into account structural breaks, the result does not change. Once the stationarity levels of the variables are determined, time-varying causal relationships between the variables at different window siess will be determined. Since the study aims to determine whether monetary policy uncertainty causes the inflation rate, the relationship between the variables will be modelled as seen in Equation (2):

$$inf_t = \delta_0 + \delta_1 mpu_t + \varepsilon_t \tag{2}$$

In rolling window regression, the model does not represent the entire subsample period. For a better representation, the window size should also be chosen as small. Representativeness increases with a smaller window because as the size of the subsample decreases, the likelihood of structural change occurring within the subsample decreases (Nyakabawo et al., 2015). A small window size reduces the risk of including multiple shifts in subsamples. However, the smaller subsample size means that the accuracy of parameter estimates decreases as standard errors increase. On the other hand, a large window size may increase the accuracy of predictions, but it reduces the representativeness of the model, especially in the presence of heterogeneity (Peseran & Timmermann, 2005). Although there is no consensus in the literature on how to determine the window size, Caspi (2017) suggested in his study that the window size should be determined with the formula $m = T\left(0{,}01 + \left(1.8/\sqrt{T}\right)\right)$. In this formula, T represents the number of observations. Considering the number of observations of the study, it turns out that the window size can be m≈30. Additionally, Balcilar et al. (2010) stated that there is no criterion for determining window size. There are studies in the literature that prove the robustness of the results using alternative window sizes (Nyakabawo et al., 2015; Tümtürk & Kırca, 2024; Zhu et al., 2021). Considering that different window sizes may give different findings in this study, alternative window sizes are used and the results are generalized. Figure 2 reports time-varying causal relationships from monetary policy uncertainty to inflation at different window sizes (20, 30, 40, 50).

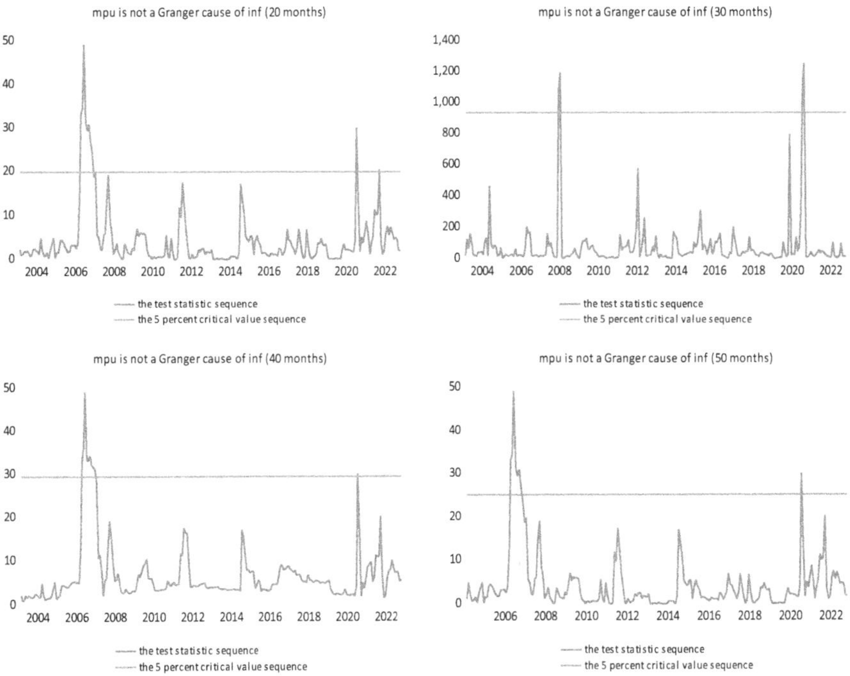

Figure 2: Rolling Window Causality Analysis Results

When all window sizes are examined, the existence of a causal relationship that starts in 2006 and covers the whole of 2007 is proven. It is seen that the causality relationship revealed by calculating the number of windows as suggested by Caspi (2017), emerged in 2008. When the US economy is examined for the period 2006–2008, the constant reduction of the overnight borrowing interest in the interbank market to keep the inflation rate under control and increase growth rates created uncertainty in monetary policy. Increasing mortgage credits along with falling interest rates also caused serious fluctuations in housing prices. It is said that the foundations of the Global Financial Crisis were laid during this period. Although monetary policy uncertainty was not the only reason for the increases in the inflation rate in that period, it can be said that it had a significant impact. When the graphs are examined, it is seen that a second causality relationship emerged towards the end of 2020. The Federal Reserve's tightened monetary policy in the face of contracting economic activities due to the Covid-19 pandemic has likewise increased monetary policy uncertainty, causing inflation. In addition, failure to achieve the previously stated 2 % inflation target may weaken the credibility of the Federal Open Market Committee (FOMC), causing inflation expectations to deviate and actual inflation and economic activity to become more volatile.

4 Conclusion

This study aimed to determine whether the monetary policy uncertainty created by economic news in newspapers and magazines affects the inflation rate. For this purpose, data from the 2002:01-2023:04 sample period are used for the US economy. To determine the causality relationship that changes over time, the window sizes are first determined. Since there are examples in the literature showing that results may vary with different window sizes, the results were strengthened by using alternative window sizes.

The first conclusion from the study is that uncertainties in monetary policy are the Granger cause of inflation. This result is robust to different window sizes in the time-varying causality test. Increases in monetary policy uncertainty seen in the US economy may increase inflation. The existence of this causality relationship is solid in the review period, 2006–2007 and 2020. When the data is examined, it is observed that the inflation rate in the US is quite high during these periods. Monetary policy uncertainty has an impact on inflation by affecting both consumer and investor decisions.

The findings of the study show that policies that reduce monetary policy uncertainty also have a reducing effect on inflation. Economic decision-makers and policymakers avoiding discourses that will create uncertainty will contribute to reducing the inflation rate in the economy. The findings of the study produce similar results to Schaling and Nolan (1998), Yellen (2017), and Tümtürk and Kırca (2024).

Actually, this study has some limits. First of all, calculating monetary policy uncertainty only for the US and Japan limits the study. Calculating data for new economies will be guiding for many current studies. It may be possible to say more about economic conditions and outlook when new data arrives. As a result, considering that the relationship between monetary policy uncertainty and the inflation rate changes over time, reducing uncertainties will be an important step for the targeted inflation rate.

References

Alam, M. R., & Istiak, K. (2020). Impact of US Policy Uncertainty on Mexico: Evidence from Linear and Nonlinear Tests. *Quarterly Review of Economics & Finance*, *77*, 355–366.

Azad, N. F., & Serletis, A. (2022). Spillovers of US Monetary Policy Uncertainty on Inflation Targeting Emerging Economies. *Emerging Markets Review*, *51*, 100875.

Baker, S. R., Bloom, N., & Davis, S. J. (2016). Measuring Economic Policy Uncertainty. *The Quarterly Journal of Economics*, *131*(4), 1593–1636.

Balcilar, M., Ozdemir, Z. A., & Arslanturk, Y., (2010). Economic Growth and Energy Consumption Causal Nexus Viewed Through a Bootstrap Rolling Window. *Energy Economics*, *32*(6), 1398–1410.

Bloom, N. (2009). The Impact of Uncertainty Shocks. *Econometrica*, *77*(3), 623–685.

Caldara, D., Fuentes-Albero, C., Gilchrist, S., & Zakrajšek E. (2016). The Macroeconomic Impact of Financial and Uncertainty Shocks. *European Economic Review, 88,* 185–207.

Cascaldi-Garcia, D., Sarisoy, C., Londono, J.-M., Rogers, J., Datta, D., Ferreira, T., Grishchenko, O., Jahan-Parvar, M.R., Loria, F., Ma, S., Rodriguez, M., & Zer, I. (2020). What Is Certain About Uncertainty? *Board of Governors of the Federal Reserve System International Finance Discussion Paper,* Number F 1294.

Caspi, I. (2017). Rtadf: Testing for Bubbles with EViews. *Journal of Statistical Software*, *81*(1), 1–16.

Colombo, V. (2013). Economic Policy Uncertainty in the US: Does It Matter for the Euro Area?. *Economics Letters*, *121*(1), 39–42.

Elder, J., & Serletis, A. (2010). Oil Price Uncertainty. *Journal of Money, Credit and Banking*, *42*(6), 1137–1159.

Hacker, R. S., & Hatemi-J. A. (2006). Tests for Causality between Integrated Variables Using Asymptotic and Bootstrap Distributions: Theory and Application. *Applied Economics,* 38, 1489–1500.

Husted, L., Rogers, J., & Sun, B. (2020). Monetary Policy Uncertainty. *Journal of Monetary Economics*, 115, 20–36.

Inflation Data, https://databank.worldbank.org/ (Accessed on: 01 May 2024).

Jurado, K., Ludvigson, S. C., & Ng, S. (2015). Measuring Uncertainty. *American Economic Review*, *105*(3), 1177–1216.

Karagöl, V. (2023). Ekonomik Politika Belirsizliğinin Gıda Fiyatlarına Etkisi: Seçilmiş Ülkeler İçin Zamanla Değişen Nedensellik Analizi. *İktisat Politikası Araştırmaları Dergisi*, *10*(2), 409–433.

Kurov, A., & Stan, R. (2018). Monetary Policy Uncertainty and the Market Reaction to Macroeconomic News. *Journal of Banking & Finance*, *86*, 127–142.

Lee, J., & Strazicich, M. C. (2004). Minimum LM Unit Root Test with One Structural Break. *Appalachian State University Working Papers, 04*(17), 1–15.

Monetary Policy Uncertainty, www.policyuncertainty.com (Accessed on: 01 May 2024).

Nyakabawo, W., Miller, S. M., Balcilar, M., Das, S., & Gupta, R. (2015). Temporal Causality between House Prices and Output in the US: A Bootstrap Rolling-Window Approach. *The North American Journal of Economics and Finance*, *33*, 55–73.

Pesaran, M. H., & Timmermann, A. (2005). Small Sample Properties of Forecasts from Autoregressive Models under Structural Breaks. *Journal of Econometrics*, *129*(1–2), 183–217.

Phillips, P. C. B., & Perron, P. (1988). Testing for a Unit Root in Time Series Regression. *Biometrika, 75*(2), 335.

Said, S. E., & Dickey, D. A. (1984). Testing for Unit Roots in Autoregressive-Moving Average Models of Unknown Order. *Biometrika*, *71*(3), 599.

Schaling, E., & Nolan, C. (1998). Monetary Policy Uncertainty and Inflation: The Role of Central Bank Accountability. *De Economist, 146*, 585–602.

Toda, H. Y., & Phillips, P. C. (1994). Vector Autoregression and Causality: A Theoretical Overview and Simulation Study. *Econometric Reviews*, *13*(2), 259–285.

Toda, H. Y., & T. Yamamoto., (1995). Statistical Inference in Vector Autoregressions with Possibly Integrated Processes. *Journal of Econometrics, 66*, 225–250.

Tümtürk, O., & Kırca, M. (2024). Belirsizliklerin Enflasyon Üzerindeki Etkisinin Zamanla Değişen Nedensellik Yöntemiyle Analizi: Türkiye Örneği. *Fiscaoeconomia*, *8*(1), 222–243.

Ugurlu-Yildirim, E., Kocaarslan B., & Ordu-Akkaya B. M. (2020). Monetary Policy Uncertainty, Investor Sentiment, and US Stock Market Performance: New Evidence from Nonlinear Cointegration Analysis. *International Journal of Finance & Economics*, *26*, 1724–1738.

Wang, M., & Fu, Z. (2022). A New Method of Nonlinear Causality Detection: Reservoir Computing Granger Causality. *Chaos, Solitons & Fractals*, *154*, 111675.

Yellen, J. L. (2017). Inflation, Uncertainty, and Monetary Policy. *Business Economics, 52*, 194–207.

Zhu, H., Chen, W., Hau, L., & Chen, Q. (2021). Time-Frequency Connectedness of Crude Oil, Economic Policy Uncertainty and Chinese Commodity Markets: Evidence from Rolling Window Analysis, *The North American Journal of Economics and Finance, 57*, 101447.

Serap Ece Karazincir[1]

Endogeneity Analysis of Money Supply: OECD Country Examples

1 Introduction

Beyond being a tool that allows the exchange of goods and services, money constitutes the foundation of economic activities. From the early periods when humanity began forming commercial relationships due to its endless needs to the present day, money has undergone significant changes in form and content. The study of money has explored not only political and institutional developments but also the reactions and tendencies of individuals and banks. In this context, the way money is approached has led to the formation of different schools of thought in economic literature, each with its own theoretical and methodological assumptions. While the Classical School assumptions came to the fore in the literature, later on, due to changing economic conditions, the Monetarist and Keynesian School assumptions began to be discussed. The views of these schools, which have been around for many years, have been joined by the Post Keynesian School, which is a different interpretation of Keynesian economics and brings a new interpretation to the recently emerging and developing financial markets and banking sector. With the Post Keynesian school, the concept of endogeneity of money supply has also gained importance. According to the concept of endogenous money supply, the central bank has no control over the money supply and the money supply is determined by the demand for credit. In this context, the endogeneity of money supply will be investigated with the PCSE estimator for 18 OECD countries using annual data for the period 2010–2022.

2 Approaches to Endogeneity and Exogeneity of Money Supply

Differences of opinion regarding money have brought about the importance of the concept of money supply and the debates on whether this concept is endogenous or exogenous in economic processes. The basis of these discussions is the

1 Ph.Dc., YÖK 100/2000, Çukurova University, Faculty of Economics and Administrative Sciences, Department of Economics, eceerkk@gmail.com, ORCID: 0000-0003-2655-1719.

Bullionist-Anti-Bullionist views that emerged between 1797 and 1821 and the differences of opinion between the Cash School, the Banking School and the Free Banking School, which were its continuation. The understanding of the endogeneity of money supply is based on the fact that the central bank does not have control over the money supply and that the money supply is determined by the demand for credit. Here, credit demand is affected by variables that affect the production level. The exogenous money supply concept is the situation in which the central bank directly affects the money supply, regardless of the developments in the market. According to the Classical, Neoclassical, and Monetarist schools of thought, money supply is considered exogenous. As distinguished by Schumpeter (1954: 277–278), they are included in the tradition of Real Analysis, asserting that as long as money serves as a medium of exchange, it cannot affect economic processes and remains neutral. However, in the approach called the Monetary Analysis tradition, which includes Marx, Keynes and Post Keynesians, money is the basis of the economic system and is biased. Post Keynesians, who adopt the Monetary Analysis approach and argue that the money supply is created internally by the credit mechanism, thus reflect an attitude against the Orthodox Approach. This understanding expresses the basic understanding of money of Post Keynesians, especially those gathered around the Journal of Post Keynesian Economics.

Discussions between Classical, Neo-Classical and Monetarists in the real analysis tradition and Keynes, Marx and Post Keynesian economists in the monetary analysis tradition about whether it is possible to control the money supply and whether the money supply is endogenous or exogenous; It has been the subject of research with the Bullionist and Anti-Bullionist understanding and later between groups called the Cash School and the Banking – Free Banking School.

The Bullionist and Anti-Bullionist debate emerged with the policies of the Central Bank in England between 1797 and 1821. During these years, with the suspension of full convertibility of banknotes to gold, an extreme increase in gold and foreign currency prices was observed and sterling lost value. The debate takes place between the Bullionist approach, which criticizes the policies of the Bank of England and advocates convertibility into gold, and the Anti-Bullionist approach, which defends the currently implemented policies, that is, the policies of the Central Bank (Humphrey, 1993: 144). Paper money in use in England during these years was issued by the Bank of England and some provincial banks. However, since it was a state bank, the Central Bank had a larger share in the use of banknotes. As a result of England entering a war with France between 1792 and 1815 and the expenditures made to finance this war, the foreign exchange reserves of the Central Bank were exhausted. As a result, in 1797,

England abolished the convertibility between gold and banknotes for a while due to the difficulty of gold payments and began to experience inflationary effects. Thus, rising inflation and appreciating exchange rates led to the questioning of the Central Bank's policies (Laidler, 1989: 61). According to Bullionists, who act from this point, the factor that causes the increase in inflation and the depreciation of money is the excessive banknote issue by the Central Bank. While the Central Bank determines the amount of non-convertible paper money, the amount of money determines domestic prices through total expenditures, and the exchange rate between non-convertible money and gold standard money determines the premium on gold. For this reason, excessive money export is the basis of the depreciation of money both inside and outside (Işık, 2010: 25). On the other hand, Anti Bullionists attribute the rise in the exchange rate to the war environment, extraordinary payments and real shocks. The reason for the price increases is the increase in exchange rates due to product shortages due to inefficiency in agricultural products and the supply of these through imports, and as a result, the currency loses value. For this reason, the inflation that occurs for Anti Bullionists is due to real shock (Laidler, 1989: 64). Although Anti-Bullionists did not accept the excessive banknote issuance of the central bank, they adopted the view that later became known as the Real Bills Theory in the literature. Accepting the Real Bills Theory also means accepting the endogeneity of the money supply. According to the real bills theory, the banking system can provide short-term loans to economic decision-making units and finance real production, and this does not result in excessive banknote issuance that will lead to inflation.

With the Bank of England suspending convertibility in 1821, the new form of the Bullionist – Anti-Bullionist debate changed to the Cash School, Banking School and Free Banking School. Price increases as a result of financial crises and monetary expansions that occurred between 1836 and 1839 showed that convertibility alone did not provide assurance in terms of stability. During this period, the continuous decrease in gold reserves poses a serious problem for the progress of convertibility. And the issues of whether the Central Bank follows rule-based or discretionary policies have begun to be discussed (Işık, 2010: 27). As a result of these discussions, the Cash School, like the Bullionists, Monetarists and Classical economists, accepted the money supply as an exogenous indicator under the control of the central bank, while the Banking School and the Free Banking School, like the Anti-Bullionists and Post Keynesians, argued that the money supply was endogenous.

The Cash School attributes the main reason for the crises and convertibility threats that occurred in England to the fact that the reserve money could not be narrowed as in the pure coin system. Adjustments in the monetary system are insufficient due to the Central Bank's inability to take the necessary precautions against high gold outflows during crises or because the measures are insufficient. Cash School adopts an understanding they call the Cash Principle. Accordingly, in mixed currency systems, if banknotes change as in the pure coin system, banknote issuance will not pose any problem. Gold exports in the system reduce the volume of the domestic currency and increase its value. With the increase in value, there will be a decrease in the general level of prices, and as a result, exports will increase and imports will decrease (Cramp, 1962: 9).

The Banking School, on the other hand, opposed and criticized the views of the Cash School. In fact, although they agree on the continuation of convertibility, there are disagreements about what the most appropriate institutional arrangements would be to maintain convertibility. According to the Banking School, which proceeds in line with the understanding of endogenous money supply, it is not possible to prevent the excessive export of money. According to the approach they adopt, called the Reflux Law, over-issued banknotes or loans return to the system through different ways (Langus & Lipatov, 2001: 8). According to Fullarton, even if the banking system exports an asset in excessive amounts, it will ensure its return to the system in the following ways; First, excess issuances can be replaced with assets, and secondly, excess issuances will turn into liabilities such as deposits. The views of the Banking School, its emphasis on expenditures financed by providing credit and its defense that the size of money in circulation will be shaped by commercial activities, are an approach related to the acceptance of the endogeneity of the money supply.

According to the Free Banking School, which is the other side of the debate, banks should be competitive in all their services, including the issuance of banknotes, and also argued that the central bank should not have monopoly power in the issue of banknotes. Accordingly, the Bank of England's excessive banknote issuance is due to the absence of another institution that can compete with the bank. Again, for this reason, they think that the Reflux Law is not effective. Because if there is competition in banknote issuance, excessive issuance will not occur. They thought that when individuals realized that a bank's banknotes had increased too much, they would turn to other banks' banknotes and excessive money export would be prevented. With these approaches, the Free Banking School also adopted the endogenous money supply approach (Langus & Lipatov, 2001: 10).

To summarize the approaches of the Cash School, Banking School and Free Banking School, regarding excessive banknote issuance, the Cash School interprets this situation as a situation that every bank can realize, while the Banking School argues that no bank can carry out excessive issuance due to the Reflux

Law. The Free Banking School, on the other hand, thought that it could only issue excessive banknotes because the Central Bank had monopoly power. While the Cash School and Banking School support banknote monopoly power, the Free Banking School opposes this with a competitive approach. The main factor on which the Cash School and the Banking School disagree is whether the monetary authority should exercise limited or unlimited authority. While the Cash School advocates limited power, the Banking School supports unlimited power. The Free Banking School, on the other hand, advocates a system of free and competitive banknote issuance instead of these two options (Işık, 2010: 38).

3 Endogenous Money Supply and Post Keynesians

The debates on the endogeneity-exogeneity of money supply in modern monetary theory date back to the periods when Classical monetary theory emerged. The movement, which started with the Bullionists and continued with the Cash School, is continued by the Classical School and Monetarists today. Their general assumption is that if the credit system works as in the pure monetary system, there will be no inflation due to excessive inflationary banknote issuance. On the other hand, according to the Anti-Bullionist Approach, the Banking School and its current representatives, the Post Keynesians, the necessary monetary system is a monetary system that can be shaped endogenously according to the needs of commercial activities (Işık, 2010: 49).

The Radcliffe Committee, which met in Britain in 1959, has a very important place in the emergence of modern endogenous monetary theory. II. After the World War, high levels of liquidity emerged in England, leading to concerns that this would lead to high spending rates and inflation. Due to these developments, the implementation of contractionary monetary policies came to the fore and it was investigated whether the amount of liquidity could be taken under control. Nicholas Kaldor, who participated in the committee and influenced the report with his approaches, asserted that money supply is endogenous rather than exogenous, contrary to the views advocated by the Monetarist School. According to Kaldor, the increase in money supply is due to the increase in credit expenditures, and the concept that becomes important here is the circulation speed of money. According to him, money supply is determined by credit demand, and the stable relationship between expenditures and money supply is a proof of this. Kaldor argued that money supply is determined endogenously by credit demand, whereas interest rates are determined exogenously through the central bank (Wray, 1990: 132–133). This report is considered an important step for the money supply endogeneity approach because the authors of the report find the monetary policies implemented in the period in question faulty and reveal that the policies were mainly based on the acceptance of the exogeneity of the money supply. Accordingly, the policies implemented are based on the money supply accepted under the control of the Central Bank and ignore the financial institutions and financial innovations in the system. The reason why the report was taken as a reference by Post Keynesian economists was that the assumptions accepted in the Classical Quantity Theory were seen as largely flawed and the Quantity Theory was found to be unsuccessful in practice (Tokucu, 2008: 10–13).

The understanding of endogenous money supply and the Post Keynesians who have adopted this understanding express a criticism and opposition to the Classical Approach and Monetarists, generally referred to as the Orthodox view, and their monetary theory, the Quantity Theory. In this approach, which rejects the assumptions of the mainstream and is called the Heterodox view, Post Keynesians have a very important place in the economic literature with their formation and general acceptance.

Post Keynesian Economics is a school that developed in England and the USA in the late 1960s and early 1970s. Economists such as Nicholas Kaldor, Richard Kahn, Michal Kalecki, Joan Robinson, Piero Sraffa, Victoria Chick, Alfred Eichner, Basil Moore, Sidney Weintraub, and Paul Davidson are pioneering Post Keynesians. While these economists form the theoretical foundations of Post Keynesian economics, they argue, like Keynes, that the models created should be realistic and relevant to the world in which we live (Davidson, 1980: 152). Post Keynesian economics is accepted as a special interpretation of Keynesian economics with its general features. They accept the principle of effective demand, supported by Keynes, as the basis of macroeconomic theory. In this context, the Post Keynesian approach, which includes a reaction to the mainstream Neoclassical synthesis, does not reveal a homogeneous structure within itself, but exhibits pluralism. This pluralism is considered an asset for the Post Keynesian approach. Although economists who specialize in different subjects do not put forward a common and general theory determined by all of them, they have common features that can be called Post Keynesian (Wrıghtsman, 1983).

Although Post Keynesian economists accept that the money supply is endogenous, they examine this hypothesis through credit demand and therefore pioneer the hypothesis. According to them, the main assumptions about the endogeneity of the money supply are; The entire money supply is not determined by monetary authorities and most of the money emerges through credit markets. And since money is formed by monetary policy implemented by the central bank, it is predominantly subject to shaping by firms and banks and is therefore considered endogenous.

4 Econometric Approach

In this study, annual data for the period between 2010 and 2022 and 18 countries (Turkey, Australia, Denmark, Iceland, Norway, England, USA, Sweden, Japan, Mexico, Czechia, Hungary, Poland, South Korea, Chile, Israel, Colombia, Costa Rica) the endogeneity of money supply for is evaluated within the scope of panel data analysis.

The data for the variables used in the analysis were obtained from the World Bank database and the model in equation 1 was created.

$$M2_{i,t} = \beta_0 + \beta_1 Credit_{i,t} + \beta_2 GDP_{i,t} + \beta_3 CPI_{i.t} + \varepsilon_{i,t} \quad (1)$$

The "M2" dependent variable in the model represents the broad money supply, the "Credit" independent variable represents the loans given to the private sector, the "GDP" independent variable represents growth, and the "CPI" independent variable represents the inflation rate. "e" indicates the error term.

Panel data regression models are based on the assumption of cross-sectional independence between units. Baltagi (2005) states that in panel data analysis, the cross-sectional dependence that may occur in the series in the long term causes the standard fixed effect (FE) and random effect (RE) estimates to be consistent but ineffective and the estimated standard errors to be deviated. For this reason, different tests that test cross-sectional dependence are used in panel data analyses. Among these tests, the Pesaran (2004) test, which examines the existence of correlation between units, the non-parametric Friedman (1937) test, which calculates the average value based on Spearman's rank correlation coefficient, and the Frees (1995) test, which is based on the sum of the squares of the rank correlation coefficients, were used in the study. The results are in Table 1.

Table 1: Cross Section Dependency CD Test Results

Test	Test Statistic	P value
Peseran CD Test	*20.813*	*0.000*
Friedman CD Test	*92.681*	*0.000*
Frees' CD Test	*5.433*	***Q Distribution Critical Values*** *0.10 for 0.198* *0.05 for 0.262* *0.01 for 0.390*

As a result of the three tests performed, cross-sectional dependence was detected between the variables. In the continuing part of the study, Wooldridge Auto-Correlation Test was applied to detect the autocorrelation problem and Wald heteroskedasticity test was applied to detect the heteroscedasticity problem and the results are presented in Table 2.

Table 2: Auto Correlation and Heteroscedasticity Test Results

Test	The calculated test statistic	Probability Value	Conclusion
Wooldridge Autocorrelation Test	*240.000*	*0.000*	There is a first-order autocorrelation problem.
Wald Heteroscedasticity Test	*1489.67*	*0.000*	There is a heteroscedasticity problem

According to the test results in Table 2, both autocorrelation problems and heteroskedasticity problems were determined in the series. This problem affects the validity of variances and standard errors, t and F statistics, and confidence intervals of R2. BeckKatz (1995) recommends the "Panel Adjusted Standard Errors (PCSE)" method for such problems. Panel Adjusted Standard Errors estimation results obtained with the Beck Katz estimator, which is resistant to the CD tests, auto-correlation problem and heteroskedasticity problem within the scope of the analysis, are shown in Table 3.

Table 3: PCSE Test Results

M2	Coef.	Std. Err.	z	P>|z|
Credit	*0.557*	*0.013*	*39.97*	*0.000**
GDP	*-0.723*	*0.580*	*-1.25*	*0.212*
CPI	*-0.464*	*0.249*	*-1.86*	*0.063**
cons	*34.917*	*3.016*	*11.57*	*0.000**

According to the results in Table 3, the probability values of the "Credit" and "CPI" variables were found to be statistically significant, while the "GDP" variable was statistically insignificant. Accordingly, a 1 unit increase in the "Credit" variable increases the "M2" dependent variable by approximately 0.56 units. A 1 unit increase in the "CPI" variable reduces the "M2" dependent variable by

0.46 units. Therefore, for the country group discussed in the study, there can be a positive relationship between the "Credit" and "M2" variables, and a negative relationship between the "CPI" and "M2" variables.

7 Conclusion

Although people met their needs with their own production in the early periods, the emergence of social life and the increasing needs over time started the monetary process and gave rise to commercial activities. The activities in question appeared as barter transactions and various objects used to carry out these transactions were included in the use. However, later, with the weakness of barter activities, a new commercial life emerged and precious metals were used as money, laying the foundation of the concept of money known today. But over time, due to the difficulty in using money obtained from precious metals and the increasing volume of commercial activity, another stage was passed in the monetary system and the use of paper money emerged. With increasing trade activities and the evolving concept of money, structural transformations have occurred in economies, banking activities have developed and financial markets have progressed. As a result of these processes, the concept of money has gained more importance and the issues of money demand and money supply have come to the fore. Approaches to these issues have begun to be explained with the views of very important schools in the economic literature. Issues such as how money is defined, why it is demanded and how the money supply is formed have become the main subjects of study, and while mainstream economic assumptions emerged with the views of the Classical School and Monetarists at first, Keynesian economics came to the fore as a result of changing economic conjunctures and crises. While comparisons between mainstream economics and Keynesian economics have been made for many years, Post Keynesian economics, which is a heterogeneous formation, has recently come to the fore. Post Keynesians have created a paradigm different from the mainstream, based on Keynes' monetary production economics. One of the main differences between mainstream and Post Keynesian economics, in terms of the way they deal with the concept of money and economic processes, has been their approach to money supply. While mainstream economics accepts that the money supply is determined exogenously under the control of the Central Bank, Post Keynesian economists argue that the money supply emerges endogenously through the credit channel. In this context, in order to test the endogeneity theory of money supply, PCSE test was conducted for 18 OECD countries with annual data for the period 2010–2022. According to the analysis results, there is a positive relationship between money supply and loans in 18 OECD countries. While a negative relationship is detected between inflation and money supply, GDP is statistically insignificant. Thus, it has been determined that the increase in loans given to the private sector has an increasing effect on the amount of money supply. This shows that the theory of money supply endogeneity is supported in the relationship between money supply and loans given to the private sector in the countries within the scope of the study.

References

Baltagi, B. (2005). *Econometric Analysis of Panel Data* (3rd ed.). John Wiley & Son Press.

Beck, N., & Katz, J. N. (1995). What To Do (And Not To Do) With Time-Series Cross-Section Data. *American Political Science Review*, *89*(3), 634–647.

Cramp, A. B. (1962). Two Views on Money. *Lloyds Bank Review*, July, *65*, 1–16.

Davıdson, P. (1980). "Post Keynesian Economics: Solving the Crisis in Economic Theory". The Public Interest-Special Education.

Frees, E. W. (1995). Assessing Cross-Sectional Correlation in Panel Data. *Journal of Econometrics*, *69*(2), 393–414.

Friedman, M. (1937). The Use of Ranks to Avoid the Assumption of Normality Implicit in the Analysis of Variance. *Journal of the American Statistical Association*, *32*(200), 675–701.

Humphrey, T. (1993). "Kaldor Versus Friedman in Historical Perspective". In H.Thomas (Ed.), *Money Banking and Inflation: Essays in the History of Monetary Thought*. Aldershot: Edward Elgar, pp. 144–157.

Işık, S. (2010). Para, Finans ve Kriz Post Keynesyen Yaklaşım, Palme Yayıncılık, 2.Baskı, Ankara.

Laidler, D. (1989). "The Bullionist Controvers". J. Eatwell, M. Milgate & P. Newman, New Palgrave: A Dictionary of Economics, London.

Langus, G., & Lipatov, W. (2001). "Assess The 'Real-Bills Doctrine' And The 'Principle of Reflux' Which Figured Prominently In The Three Cornered Debates Between The Currency School, The Banking School, and The Free Banking School In Mid Nineteenth Century England", Central Europan University Department of Economics.

Pesaran, M. H. (2004). General Diagnostic Tests For Cross Section Dependence In Panels. Cesifo Working Paper, No. 1229.

Schumpeter, J. (1954). *History of Economic Analysis*. New York: Oxford University Press.

Tokucu, E. (2008). Yapısalcı Post Keynesyen İçsel Para Arzı Yaklaşımı Ve Türkiye Ekonomisinde Geçerliliği: 1986–2005. Marmara Üniversitesi Sosyal Bilimler Enstitüsü İktisat Ana Bilim Dalı İktisat Teorisi Bilim Dalı, Doktora Tezi.

Wray, L. R. (1990). *"Money and Credit in Capitalist Economies: The Endogenous Money Approach"*. Aldershot: Edward Elgar.

Wrıghtsman, D. (1983). *An Introduction to Monetary Theory and Policy* (3. bs.). London: The Free Press.

Ufuk Serdar Akalın[1]

Central Bank Communication[*]

1 Introduction

Before the 1990s, central banks had a structure that kept all information to themselves and limited communication as much as possible. In terms of monetary policy communication, the general opinion was that monetary policy makers spoke very little and followed a cryptic way while doing so (Blinder et al., 2008). The most important basis supporting the positions of central banks against the background of their communication attitudes is that the opinion that the applied monetary policy should be a policy that the public does not expect in order to achieve its goal comes to the fore.

In 1990, a significant development occurred in the field of monetary policy and New Zealand adopted an inflation targeting regime as its monetary policy strategy. Following New Zealand, Canada in 1991 and the United Kingdom in 1992 also adopted inflation targeting as their main monetary policy strategy. Along with the adoption of the inflation targeting regime, another notable policy change has occurred. Within this change, central banks began to shift away from their previously secretive nature and adopted a more transparent communication strategy. This change in strategy marked a significant shift towards increased communication by central banks.

Taking the example of the United Kingdom, one of the early adopters of the inflation targeting regime, the Bank of England began presenting inflation reports containing its economic views on a quarterly basis starting from 1993. In addition to these inflation reports, the Financial Stability Review was introduced biannually from 1996. During this period, the number of speeches made and published by members of the Monetary Policy Committee (MPC) increased significantly, from 13 in the 1990s to 34 in the 2000s. Furthermore, since 1997,

---

1 Ph.D., Research Assistant., Karamanoğlu Mehmetbey University, Faculty of Economics and Administrative Sciences Department of Economics, userdarakalin@kmu.edu.tr, 0000-0002-0590-8419.

* This study was developed based on the PhD thesis titled "The impact of the Central Bank of the Republic of Turkey communication instruments on macroeconomic indicators", written by the author.

minutes of the MPC meetings have been published following each meeting (Haldane, 2017). In this context, the inflation targeting regime served as a catalyst for increased communication.

Monetary policy authorities implementing an inflation targeting regime must ensure that economic agents act in accordance with the set targets to achieve price stability objectives. Therefore, it is essential to communicate the framework of the policy to economic agents, including the additional measures that will be taken if potential risks arise or the extent to which deviations from the intended targets will be perceived as risk indicators by the authority. This constitutes an effective communication strategy. In this context, with proper communication, the extent of changes needed in policy implementation can be reduced. In other words, through an effective communication strategy, the central bank can also minimize potential costs associated with policy implementation.

Central banks implement policies using specific instruments aimed at achieving certain objectives. Similarly, central banks also use specific tools through communication to reach their objectives. The following sections of the study will focus on these objectives and tools.

2 Goals of Central Bank Communication

The change in the communication approach of central banks stems from their desire to achieve specific objectives through communication. Central banks use communication to guide expectations and provide information about potential future policy changes. These practices are related to enhancing the effectiveness of monetary policy.

In addition to enhancing the effectiveness of monetary policy, the success of a central bank's monetary policy implementation is possible through establishing public trust in the central bank. This trust requires central banks to be transparent and accountable. Therefore, we will consider the goals that central banks aim to achieve through communication in terms of both enhancing the effectiveness of monetary policy and increasing transparency and accountability.

2.1 Increasing the Effectiveness of Monetary Policy

Enhancing the effectiveness of monetary policy is a fundamental objective of central bank communication. The philosophy is that a better-informed public will make more accurate and appropriate decisions. Providing the public with information about the central bank's future actions and anchoring expectations form the core of communication (Issing, 2019). In this context, effective communication plays a crucial supporting role in the success of the policies implemented by the central bank. As illustrated by Gürkaynak et al. (2004) in the Federal Open Market Committee (FOMC) case, the ability to manipulate financial market expectations without hindering policy execution can be achieved by central banks through strategic communication.

Stabilizing or guiding inflation expectations within the desired policy framework is a crucial aspect of central bank communication, whether considering the expectations of market experts or the general public. Central banks aim for long-term inflation expectations to align closely with their long-term inflation targets. Additionally, they may desire short-term expectations to decrease during periods of monetary tightening and increase during periods of monetary easing. One of the most significant challenges central banks face is predicting the expectations of economic actors. While the expectations of financial market participants can be gauged through asset prices, estimating and influencing the expectations of the general public, non-financial institutions, and wage earners require more effort (Blinder et al., 2022).

Predicting and guiding the future behavior of economic agents is crucial for central banks to decrease the negative effects of incomplete information on the success of their policies. Communicating information about the set inflation target allows for more desirable outcomes in economic stability and price stability compared to situations where economic agents have less information about inflation targeting (Orphanides & Williams, 2004).

To enhance the success of monetary policy implementation, central banks use forward guidance to align expectations with policy actions. First, by communicating their policy path to economic agents, central banks aim to reduce uncertainties they may face. Additionally, when interest rates approach the zero lower bound, central banks use forward guidance as an alternative tool to facilitate monetary easing. Forward guidance at the zero lower bound requires a sacrifice. Central banks must forgo some future policy flexibility since they cannot achieve monetary easing through traditional means (Campbell, 2013)

The problem of interest rates' reaching the zero lower bound posed a challenge for the Bank of Japan (BOJ), which had been struggling with deflationary pressures for many years. At the stage of implementing monetary expansion, traditional methods were limited. It could be argued that the assurance by the BOJ in April 1999 that overnight interest rates would be kept at zero until deflationary concerns were resolved represented perhaps the first application of forward guidance (Ueda, 2013).

However, by the year 2000, during the implementation phase, the BOJ abandoned forward guidance and opted to raise interest rates. The reason for this policy change was the diminishing deflationary pressure caused by weakness in demand, alongside improvements in the economic outlook. Due to a timing inconsistency, the expected impact did not emerge. Along with the global impact, the deflation pressure has deepened further in Japan. As a result, from 2001 onwards, the BOJ returned to using forward guidance, stating that it would continue to do so "until consumer price inflation stably exceeds zero." (Ueda, 2013). As can be seen from the example of Japan, timing and consistency stand out as an important element in the implementation of forward guidance.

2.2 Increasing the Transparency and Accountability

Central banks have taken numerous steps in guiding the public to act towards targets and shaping their expectations with the period when the fight against inflation began to be considered as the main objective of monetary policy. A crucial precondition for effectively guiding expectations is the public's confidence in the central bank. To enhance this confidence, it is essential to increase the transparency and accountability of the central bank.

Until 30 years ago, Central Banks preferred a mysterious structure to transparency. at the end of the 1980s, Federal Reserve (FED) chairman Greenspan mentioned that he did not make a statement about the future of interest rates, and those who thought he was expressing himself clearly misunderstood what he was saying. Today, Central Banks provide indicators about their current policies and future development to the press after policy decisions (Barthélemy et al., 2019). Informing about future policies is one dimension of central bank transparency. In addition to policy transparency, central bank transparency encompasses dimensions such as transparency in governance, transparency in operations, transparency in outcomes, and transparency in official relationships (International Monetary Found (IMF, 2020).

Under each of these five dimensions, there are subcomponents related to those dimensions. Within this, the subcomponents of governance transparency include legal framework, mandate, independence, decision-making arrangements, risk management, internal accountability, communication, and confidentiality. The subcomponents of policy transparency encompass policy framework, policy decisions, and supporting analyses. Regarding transparency in operations, subcomponents include transparency in the use of policy tools, as well as scope and accessibility. Transparency in outcomes is constituted by final reports provided by central banks. Lastly, in relationships with governments, and other institutions, international relations, and transparency in official relationships form subcomponents of the dimension of transparency in official relationships.

The independence of central banks in implementing monetary policy and their choice of instruments is crucial in keeping the implemented policies free from political influence. Alongside policy implementation, the independence of central banks in institutional practices also entails accountability. Central banks are responsible towards the public for the decisions they make and the outcomes of those decisions. In this regard, accountability necessitates effective communication, in addition to the transparency previously emphasized. Fulfilling accountability reinforces the emphasis on the independence of the central bank and ensures societal support for policy implementation (TCMB, 2011). An independent Central Bank may be accountable not only to the public but also directly to the legislative or executive body. Even in cases where a separate oversight board exists, the oversight authority can be conferred to the legislative or executive body by law (Lybek, 2008).

3 Tools of Central Bank Communication

Central banks aim to influence the behavior and expectations of market participants, either separately or collectively, to align with their objectives by utilizing the communication factor. To achieve this goal, they use various communication channels. With technological advancements, the speed and form of information dissemination are increasing, thereby expanding the range of tools available for central banks to communicate. The main tools used by central banks for communication include press releases and conferences, reports, MPC decisions and meeting summaries, websites, social media, and speeches by officials.

3.1 Press Releases and Meetings

Press releases are among the most frequently used communication tools by central banks. They are used to share with the public the reports presented by the bank, the decisions made in meetings, the measures to be taken in extraordinary situations, and the meetings held by officials. Along with press releases, central banks also present their analyses of the current situation and the steps that have been or will be taken to relevant economic circles. Central banks share these press releases via their websites.

Press conferences are informational meetings organized for the economic press. These conferences, held with the participation of senior management, include a question-and-answer session following the official's speech. Considering the media's power to influence the behavior and expectations of economic actors, the Q&A session is important for addressing public concerns and information gaps about policies. The transparency and clarity of the responses given by officials are crucial in achieving the desired effect.

It is important for the language used in both press conferences and press releases to be simple and understandable. Economic actors prefer press releases on the same subject to maintain a similar structure over time. A different press release following similar ones can cause significant levels of high volatility in the market (Ehrmann & Talmi, 2020).

3.2 Reports

Similar to press releases and conferences, another tool frequently used by central banks to achieve their objectives is reports. Reports are used for specific purposes such as inflation and financial stability, as well as to provide a general overview. Given that the inflation targeting regime is used by the majority of countries worldwide, inflation reports are among the most frequently utilized types of reports by central banks in this context.

Although inflation reports are not used by all countries that have adopted an inflation targeting regime, they are utilized as a valuable communication tool. For an inflation report to have the desired impact, it should clearly outline the policy objectives, the decision-making process, and how conflicts between targets will be addressed. Additionally, the analytical framework and information on which the policies are based should be assessable. Finally, the report should include information on past performance along with the presentation of inflation forecasts (Fracasso et al., 2003). Inflation reports are typically presented by central banks on a quarterly basis shortly after adopting an inflation targeting regime. In many countries that have adopted this regime, inflation reports, often referred to as monetary policy reports, should not be confused with the summaries of MPC meetings.

Similar to inflation reports, another specific type of report used by central banks is financial stability reports. Financial stability is important for preventing price fluctuations in the market under an inflation targeting regime. Financial stability reports are significant as they include the structure of the financial system, potential risks, and policy changes to be implemented in case these risks materialize. Starting in 1996, financial stability reports began to be presented by central banks and are now issued by many developed and developing countries. These reports, typically published by central banks, sometimes feature the central bank as a participant rather than the main publisher. For example, as of 2011, in the United States, financial stability reports are presented by the Financial

Stability Oversight Council, with the FED being represented on this council (Čihák et al., 2012).

3.3 MPC Decisions and Meeting Summaries

Although the decision-making body in monetary policy varies by country, it is evident that since the late 1990s, MPC's have emerged as the authoritative bodies for making decisions. In situations where the central bank governor makes decisions individually, the decisions may vary according to the governor's views. The MPC is important in terms of including a plurality in the decision-making process according to the situation in which the central bank makes decisions alone. Developments in the decision-making process of monetary policy will positively affect transparency and predictability, which are important conditions for central banking success (Maier, 2010). This, in turn, will lead to inflation expectations being anchored at better or lower levels.

In the types of central banking where monetary policy is determined by the PPC, the ways in which policy is determined may also differ. This diversity can be understood through the distinction between equal responsibility and individualism. Under equal responsibility, central banks (such as the FOMC and the Governing Council of the European Central Bank (ECB)) strive to make decisions by consensus, regardless of whether there is a formal voting process in their decision-making bodies. In such cases, individual members are expected to stand behind the group decision, and the principle of accountability resides within the group. In decision-making bodies where individualism is prominent (such as the Bank of England and the Riksbank of Sweden decision-making bodies), members are expected to articulate and present their own policy preferences for voting. In decision-making mechanisms dominated by individualism, the concept of accountability is reduced to individual votes(Blinder, 2008). Regardless of differences in decision-making processes and responsibilities within these decision-making bodies, the outputs resulting from decisions made by the MPC directly relate to central banks' communication objectives in guiding output expectations.

In economies where MPC holds decision-making authority regarding monetary policy, communication processes follow similar procedures, albeit with variations in dates and frequencies. MPC's typically announce in advance the dates of their upcoming meetings to the public. Additionally, in light of extraordinary developments, the MPC may convene meetings outside of the scheduled dates to make monetary policy decisions. The general framework of decisions taken during these meetings is communicated to the public at the specified time on the day of the meeting. A summary of the decisions made in the meeting is usually shared with the public approximately one week after the meeting.

3.4 Web Sites and Social Media

With the development and proliferation of technology, new tools are being added to the means of communication used by central banks. Websites, in this regard, offer conveniences in delivering desired information to even the most remote areas with the widespread geographic accessibility of the internet. Central banks' websites serve as an important bridge in easily conveying communication tools such as reports, presentations, speeches, and most notably, MPC decisions and meeting summaries, where they present their main policy actions. These communication tools are organized under specific headings on central banks' websites and are accessible to relevant individuals and institutions.

Another tool added to the means of communication used by central banks as a result of technological advancements is social media. Social media stands out as a platform where individuals, institutions, and organizations share their information and thoughts. In this context, it serves as a tool that facilitates reaching market actors in terms of central bank communication. Central banks typically share information on the market concisely through their social media accounts and direct those seeking detailed information to resources available on the institution's website.

3.5 Speeches Made by Authorities

The most important aspect that distinguishes speeches made by officials from official policy announcements is the flexibility they contain. Unlike the communication carried out within a cautious and well-defined framework in official policy announcements, officials allow for more flexible communication to market participants regarding past and ongoing economic activities and upcoming policies (Ranaldo & Rossi, 2010). In addition to providing information, speeches, when compared to meeting decisions, offer the advantage of not being bound to a predetermined schedule. This allows for oral communication to be carried out on desired topics at desired times (Ehrmann & Fratzscher, 2005). Considering the flexibility that speeches provide, central banks have the power to guide the behavior of relevant market participants in alignment with policy goals through the expectations channel, without the need to use traditional policy tools.

Speeches also provide central bank officials with flexibility in terms of audience selection. By choosing a target audience, officials can, for example, organize a speech specifically aimed at the officials of participation banks when there is a need for communication on the subject of participation banking. These speeches can be conducted exclusively with their participation.

4 Conclusion

Central banks have been intensively using communication as a policy tool, especially in the last 30 years. During the implementation of monetary policy, communication ensures that necessary information is conveyed to economic actors through press releases, press conferences, reports, MPC decisions, MPC meeting summaries, websites, social media, and speeches by officials. These mentioned communication channels constitute the tools of the communication policy.

In addition to enhancing the effectiveness of monetary policy implementation, central banks aim to increase transparency and accountability through the use of these tools. For the success of monetary policy implementation, it is of great importance that central banks are transparent and accountable. One of the prerequisites for the successful implementation of monetary policy is the confidence that market participants have in the monetary policy. This confidence can be established through the transparent and accountable structures of the central banks implementing the monetary policy.

A well-structured communication framework allows central banks to achieve their ultimate goals without the need to use monetary policy tools that entail certain costs. A sound monetary policy framework supported by communication can align the expectations of economic actors with the objectives. Consequently, the market behaviors of economic actors will also be shaped in line with the objectives. Developments in technology indicate that the importance of communication in terms of monetary policy implementation will continue to increase in the near future.

References

Barthélemy, J., Dupraz, S., Gaballo, G., & Istref, K. (2019). *Trends in Central Bank Communication: From Secrecy to Transparency* (Bulletin de La Banque de France 226/2).

Blinder, A. S. (2008). *The Quiet Revolution: Central Banking Goes Modern*. Yale University Press.

Blinder, A. S., Ehrmann, M., de Haan, J., & Jansen, D.-J. (2022). *Central Bank Communication with the General Public: Promise or False Hope?* Publications Office.

Blinder, A. S., Ehrmann, M., Fratzscher, M., De Haan, J., & Jansen, D.-J. (2008). Central Bank Communication and Monetary Policy: A Survey of Theory and Evidence. *Journal of Economic Literature*, *46*(4), 910–945.

Campbell, J. R. (2013). Odyssean Forward Guidance in Monetary Policy: A Primer. In *Forward Guidance Perspectives from Central Bankers, Scholars and Market Participants* (pp. 87–105). Centre for Economic Policy Research (CEPR).

Ehrmann, M., & Fratzscher, M. (2005). How Should Central Banks Communicate? *SSRN Electronic Journal*.

Ehrmann, M., & Talmi, J. (2020). Starting from a Blank Page? Semantic Similarity in Central Bank Communication and Market Volatility. *Journal of Monetary Economics*, *111*, 48–62.

Fracasso, A., Genberg, H., & Wyplosz, C. (2003). *How Do Central Banks Write?: An Evaluation of Inflation Reports by Inflation Targeting Central Banks*. Centre for Economic Policy Research.

Gürkaynak, R. S., Sack, B. P., & Swanson, E. T. (2004). *Do Actions Speak Louder than Words? The Response of Asset Prices to Monetary Policy Actions and Statements* (SSRN Scholarly Paper 633281).

Haldane, A. (2017). *A Little More Conversation A Little Less Action*. Macroeconomics and Monetary Policy Conference.

IMF. (2020). *The Central Bank Transparency Code* (Press Release Pr 20/258; IMF Policy Paper). International Monetary Fund (IMF).

Issing, O. (2019). *The Long Journey of Central Bank Communication*. MIT Press.

Lybek, T. (2008). Central Bank Autonomy, Accountability, and Governance: Conceptual Framework. In *Current Developments in Monetary and Financial Law, Vol. 4* (pp. 133–157). International Monetary Fund.

Maier, P. (2010). How Central Banks Take Decisions: An Analysis of Monetary Policy Meetings. In M. E. Wohar, M. T. Bohl, & P. L. Siklos (Eds.), *Challenges in Central Banking: The Current Institutional Environment and Forces Affecting Monetary Policy* (pp. 320–356). Cambridge University Press.

Orphanides, A., & Williams, J. (2004). Imperfect Knowledge, Inflation Expectations, and Monetary Policy. In *The Inflation-Targeting Debate*. University of Chicago Press.

Ranaldo, A., & Rossi, E. (2010). The Reaction of Asset Markets to Swiss National Bank Communication. *Journal of International Money and Finance*, *29*(3), 486–503.

TCMB. (2011). *Merkez Bankaları ve İletişim Türkiye Cumhuriyet Merkez Bankasında İletişim Politikalarının Gelişimi*. Türkiye Cumhuriyet Merkez Bankası.

Ueda, K. (2013). A Long and Bumpy Journey ahead for Forward Guidance. In den H. Wouter (Ed.), *Forward Guidance Perspectives from Central Bankers, Scholars and Market Participants* (pp. 165–170). Centre for Economic Policy Research (CEPR).

Duygu Kalkay[1]

The Relationship Between Interest Rate Policy and Inflation in Periods of Narrowing Traditional Monetary Policy Space

1 Introduction

With the 2008 global financial crisis, unconventional monetary policies were implemented, especially by developing countries, due to the ineffectiveness of conventional monetary policy in response to economic activities coming to a standstill. The results of the global crisis have revealed that monetary policies should pursue financial stability as well as price stability. In general judgment, unconventional monetary policies are implemented in times of crisis, and despite many contradictions, they have helped revitalize countries' economies. In this context, during the COVID-19 pandemic, many countries prevented severe economic recessions by implementing unconventional monetary policies. As of 2009, the most significant criticism of the expansionary monetary policies was that they would cause severe inflation after ten years.[2] In this direction and due to the expansionary monetary policies implemented within the scope of pandemic measures, rising inflation was faced worldwide.

One factor that significantly affects the direction of economic activity in economies is the interest rate policy conducted within the scope of monetary policy. Central banks take many factors into account when setting the policy interest rate. Inflation is one of the most critical indicators affecting the direction of interest rate policy in advanced and emerging economies. Since 1970, fighting inflation has become the most important task of central banks. Inflation slows down economic activity as it reduces the value of money, which is very important for the economy and, thus, the purchasing power. Fighting inflation is the most fundamental task of central banks for sustainable growth.

This study evaluates the relationship between the effectiveness of monetary policy, macroeconomic indicators and the transmission mechanism since 2020,

1 Ph.D., Lecturer, Bandırma Onyedi Eylül University, Faculty of Economics and Administrative Sciences, Balikesir, Turkiye. dkalkay@bandirma.edu.tr, ORCID: 0000-0002-8729-9177.

2 Bernanke (2022).

when the effects of the Covid-19 pandemic were felt. In this context, rising global inflation and the direction of monetary policy in Turkey will be discussed.

2 Monetary Policy and Its Effectiveness in Extraordinary Circumstances

As of the 1980s, capital mobility in financial markets and the tightening of economic relations in financial markets, together with globalization and developing technology, have changed the form and the nature of economic policies in general and monetary policies in particular. Monetary policy is one of the most critical macroeconomic instruments for national economies. Therefore, monetary policy must be implemented within the framework of the conditions of the period. Monetary policy is one of the most critical macroeconomic instruments for national economies. According to the Taylor rule, macroeconomic developments only affect short-term interest rates. Monetary aggregates are assumed to respond endogenously to the policy rate (Turner, 2014: 3). However, this expected relationship did not materialize during the financial crisis. Although short-term interest rates were close to the lower bound, they did not affect long-term interest rates, and economic activity slowed down. Accordingly, central banks of advanced economies have implemented unconventional monetary policies to keep long-term interest rates low. With the decline in long-term interest rates, investments increased, and the economy was revitalized (El-Erian, 2016) because the interest rates set by central banks are also crucial for sustainable economic growth and employment (Değirmen, 2008).

The institutional details of monetary policymaking in emerging economies and the complexity of financial markets due to increased global integration also limit the effectiveness of the monetary transmission mechanism. The monetary transmission mechanism should work effectively for monetary policies to affect markets and economic activity in line with expectations. The effectiveness of monetary policy on macro variables through the transmission mechanism depends on dollarization, confidence in central bank policy, liquidity conditions, the depth of financial markets and the exchange rate regime. Countries like Turkey, which are exposed to commodity markets with relatively low elasticities, can partially reduce price volatility by increasing the demand and supply elasticities of many commodities through fiscal measures and dynamic monetary policy to mitigate the effects of potential shocks (Albrizio et al., 2022, 2023; ve Alvarez et al., 2023).

The Central Bank activates the macroprudential toolkit to support financial stability and strengthen the monetary transmission mechanism. In June 2022, by establishing Turkish lira (TL)-denominated securities over foreign currency deposit/participation fund liabilities, banks established securities at rates determined according to their conversion targets to Turkish lira deposit/participation accounts. In late 2022, the policy-credit rate spread widened significantly, reducing the effectiveness of monetary transmission (CBRT, 2022). The CBRT started a robust monetary tightening in June 2023, raising the policy rate from 8.5 % to 50 %. In this period, macroprudential policies were also implemented to enhance the efficiency of the market mechanism and ensure financial stability. For the monetary policy rate to have a tightening effect consistent with the inflation target, it should be supported by complementary and effective macroprudential tools. Figure 1 shows the transmission of changes in the policy rate to financial conditions (CBRT, 2024a).

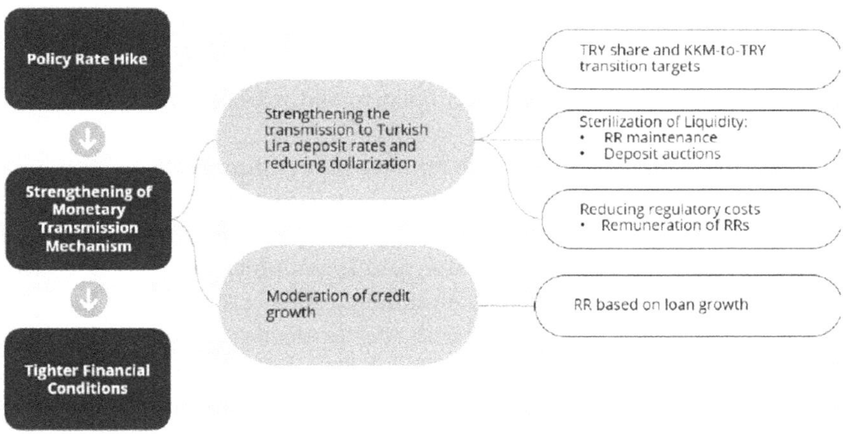

Figure 1: Enhancing Effectiveness of the Monetary Transmission Mechanism
Source: CBRT, 2024a

Central banks use the volume of money, the exchange rate level and the inflation rate as anchors for monetary policy (Karahan ve Gencür, 2019). Unconventional monetary policies had contributed to easing financial conditions when the effectiveness of conventional monetary policies weakened and the equilibrium real interest rate was close to the lower bound, such as during the fight against the Covid-19 pandemic and the financial crisis. Financial conditions have been made more supportive through various channels that interact intensively with each other and reinforce and complement each other's transmission through the economic system.

3 The Relationship Between Monetary Policy Effectiveness and Inflation

Regarding financial stability in national economies, the relationship between the policy and inflation rates and the correct analysis of this relationship is meaningful, especially for developing countries. Central banks affect inflation, international capital flows and exchange rates by raising or lowering the policy interest rate depending on the course of inflation. Since foreign capital is attracted to high interest rates, it strengthens the exchange rate for the country in question and helps control inflation's cost side. Otherwise, capital outflows and exchange rate-driven cost inflation will increase. Due to structural vulnerabilities, managing the relationship between monetary policy, interest rates and inflation is more difficult in emerging economies than in advanced economies. In these countries, an economic shock can rapidly affect inflation and weaken the effectiveness of monetary policy (IMF, 2024).

Global inflation rates, which had been increasing since the 1970s, were partially brought down to low levels with many measures taken, particularly by developed countries. As of 2021, when the results of the measures taken during the COVID-19 pandemic were monitored, inflation, which had been at low levels for a long time, started to rise. Accordingly, many central banks have significantly increased their policy interest rates to combat inflation. While high interest rates have been essential to fight global inflation, rising interest rates have led to a partial economic slowdown. On the one hand, high interest rates have been the most critical tool used to combat global inflation, while on the other hand, rising interest rates have led to a partial slowdown in economies (IMF, 2024).

The direction of the relationship between inflation and interest rates is constantly debated in the context of country dynamics. What matters here is whether inflation in a country is supply-side or cost-side. If there is demand and cost inflation, it is essential to determine how much of the inflation is demand-driven and cost-driven. If half of inflation is demand-driven and half cost-driven, raising interest rates will increase cost inflation and will not be beneficial. On the other hand, lowering interest rates to reduce cost inflation would increase demand inflation (Eğilmez, 2014).

Since late 2021, central banks worldwide have raised policy interest rates at a pace not seen in the last 40 years to achieve price stability. Economies have experienced a series of supply shocks due to supply chain disruptions caused by measures taken in the context of the COVID-19 pandemic and the war in Ukraine. These shocks, combined with the extraordinarily accommodative monetary and fiscal policies implemented by countries in response to these shocks, pushed inflation to levels not seen for many years. On the other hand, there has been a sharp rise in interest rates worldwide. This has led to a sharp decline in growth rates by 2023 (IMF, 2024).

Table 1: World Central Banks Policy Interest Rates

Central Bank	Interest rate	Last update	Inflation Rate
CBRT	50 %	21.03.2024(500bp)	71.60 %
FED	5.50 %	26.07.2023 (25bp)	3 %
ECB	4,25 %	06.06.2024(-25bp)	2.5 %
BOE	5,25 %	03.08.2023(25 bp)	2 %
SNB	1,25 %	20.06.2024 (-25bp)	1.3 %
RBA	4,35 %	07.11.2023 (25bp)	3.6 %
BOC	4,50 %	24.07. 2024 (-25bp)	2.7 %
BOJ	0,10 %	19.03.2023(20bp)	2.8 %
CBR	18,00 %	26.07.2024(200bp)	8.6 %
RBI	6.50 %	08.02.2023(25bp)	5.08 %
PBOC	3,35 %	22.07.2024 (-10bp)	0.2 %
BCB	10,50 %	08.05.2024 (-25 bp)	4.23 %
BCRA	40 %	06.2024	272 %

Source: https://tr.tradingeconomics.com

4 The Relationship Between Monetary Policy and Inflation after the Covid-19 Pandemic in Turkey

During the COVID-19 pandemic in March 2020, when the first cases started in Turkey, the Turkish economy was fragile. As in the 2008 crisis, Turkey implemented expansionary monetary policies at the onset of the pandemic to prevent economic activity from falling into recession, as in other countries. In this framework, the CBRT adopted a tight monetary policy as of June 2023, taking into account the effects of the exchange rate on costs, domestic demand conditions, the rise in commodity and food prices and their negative impact on inflation expectations (Özkök, 2021).

The global risk appetite, which contracted significantly in early 2020 due to the pandemic, gradually recovered towards the end of the year thanks to central banks' expansionary monetary policies and governments' comprehensive fiscal stimulus packages. As risk appetite increased, stocks in developed countries appreciated, and commodity prices soared. Although risk premiums in Turkey and other emerging economies declined, they did not fall to pre-pandemic levels. Moreover, capital outflows from these countries continued to decline. In the third quarter of 2020, emerging market currencies appreciated more against the US dollar than the Turkish lira (CBRT, 2020)

The CBRT uses the short-term policy rate to control market Turkish lira interest rates and money supply/demand. However, the increase in the share of foreign currency in the system complicates the effective use of the policy rate. Moreover, high dollarization weakens the monetary transmission mechanism by affecting the economy through the exchange rate. Increases in exchange rates and gold prices increase money supply in nominal terms. This creates demand inflation, pushes up asset prices and adversely affects price stability.

Turkey's pro-growth policies have reduced reserves and increased vulnerabilities. In 2021, a new policy model aimed to reduce the current account deficit, support growth and reduce inflation. The most important pillar of the model was low interest rates. It was also based on measures to reduce dollarization, credit growth and financial volatility. Successive interest rate cuts increased dollarization. Exchange Rate Protected Deposits were introduced as part of the Monetization strategy to combat the emerging vulnerabilities (IMF, 2024).

Interest rate hikes in monetary policy and macroprudential measures on credit growth pushed TL loan rates higher, credit costs increased, and financial conditions tightened (Figure 2). In 2024, credit card interest rates were also adjusted; in March, the maximum interest rate for cash withdrawals and overdraft accounts was raised to 5 %, and in April, the maximum interest rate for credit card shopping transactions was raised to 4.25 % (Figure 3). Improved expectations regarding the depreciation of the TL and reduced exchange rate volatility led to an increase in foreign currency loan demand. This led to lower foreign currency loan rates compared to TL loan rates. Moreover, banks' reduced need for swaps and depositors' tendency to switch to TL deposits had a downward impact on foreign currency loan rates (CBRT, 2024).

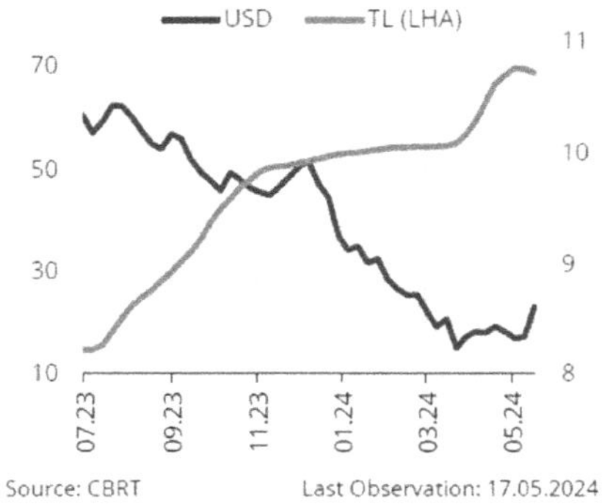

Figure 2: Commercial Loan Rates (Flow, 4WMA, %)

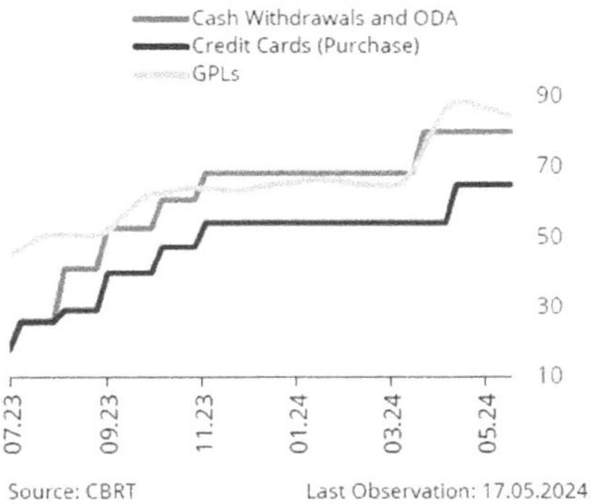

Figure 3: General-purpose Loan and PCC Interest Rates (Flow, 4WMA, %)
Source: CBRT, Last Observation: 17/05/2024

In the CBRT's May 2024 Financial Stability Report, the tight monetary policies implemented by central banks of advanced economies to fight inflation and global financial conditions restrain global growth. Interest rate cuts by central banks of advanced economies and uncertainties in the global risk appetite cause portfolio flows to err in emerging economies. Recently, the contribution of consumption to growth rates has been decreasing under tight policies, while the contribution of net exports has been increasing.

While the effects of the pandemic were still ongoing, expansionary fiscal policies were implemented due to the earthquake in February 2023. As a result, inflation in Turkey reached very high levels compared to other countries. Annual inflation increased notably in the services sector. As a result of the policy rate hike and macroprudential measures under the tight monetary policy, expectations of a depreciation in the TL weakened, exchange rate volatility decreased, and foreign currency loan utilization increased. Moreover, these measures make the TL more attractive, and banks' foreign exchange and TL liquidity increase. These developments improved the risk perception towards Turkey and led to a decline in the risk premium (CBRT, 2024).

The central bank raised interest rates above expectations in Turkey, but inflation remains high (Figure 4). Economists emphasize that interest rate hikes should be gradual. Because this gradual hike delays foreign capital inflows, it fails to improve inflation expectations.

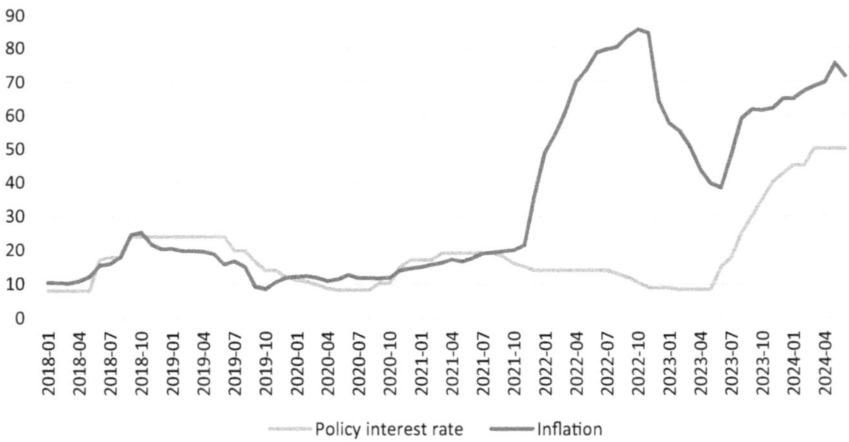

Figure 4: The Relationship between Policy Rate and Inflation (2018–2024)
Source: *CBRT*

As Figure 4 shows, the gap between interest rates and inflation widened after the second half of 2021. Despite rising inflation, a low interest rate policy was implemented between the second quarter of 2021 and the first quarter of 2023. After the 2023 general elections, it was stated that a tight fight against inflation would be launched, and interest rates would be increased accordingly. Despite the increase in interest rates, inflation remains high.

6. Conclusion

The first objective of the world's central banks, struggling with high inflation since the 1970s, is to achieve price stability. The success of monetary policies is critical in determining interest rates with rational reality, especially in times of crisis. In the fight against the 2008 financial crisis, Turkey implemented economic policies to control the large amounts of capital flowing into the country due to the expansionary monetary policies implemented by advanced central banks. During the Covid-19 pandemic, policies were implemented to prevent the depreciation of the TL due to capital outflowing from the country. In this context, monetary policies should evaluate domestic and external dynamics together.

This study analyzes the direction of the relationship between monetary policy interest rates and inflation in extraordinary times. As a result of the expansionary monetary and fiscal policies implemented within the scope of the Covid-19 pandemic measures, the gap between the interest rate and the inflation rate widened. In particular, the tight monetary policies implemented by central banks of advanced economies to combat rising inflation and global financial conditions restrain global growth. At this point, the relationship between the policy and the inflation rates and the correct analysis of this relationship is essential for economic activities.

References

Albrizio & others. (2023). "Sectoral Shocks and the Role of Market Integration: The Role of Natural Gas." *American Economic Review Papers and Proceedings, 113*, 43–46.

Albrizio and others. (2022). "Market Size and Supply Disruptions: Sharing the Pain of a Potential Russian Gas Shut-Off to the European Union." IMF Working Paper 22/143, International Monetary Fund, Washington, DC.

Bernanke, B. S. (2022). 21. Yüzyıl Para Politikaları, çev. Kılınç, D., Scala Yayıncılık.

CBRT (2022). "Makroihtiyati Tedbirler Hakkında Basın Duyurusu". Available At: 20.07.2024, https://www.tcmb.gov.tr/wps/wcm/connect/TR/TCMB+TR/Main+Menu/Duyurular/Basin/2022/DUY2022-36

CBRT (2024a). "Finansal İstikrar Raporu, Finansal Kesim". Available At: 15.01.2024. https://www.tcmb.gov.tr/wps/wcm/connect/TR/TCMB+TR/Main+Menu/Yayinlar/Raporlar/Finansal+Istikrar+Raporu/2024/Sayi+38

Čihák, M., Muñoz, S., Sharifuddin, S., and Tintchev, K. (2012), Financial Stability Reports: What Are They Good For? IMF Working Paper 12/1, Washington, DC, USA.

Değirmen, S. (2008). "Türkiye'de Ortodoks İktisat Politikası ve IMF: Enflasyon Raporları". *Toplum ve Demokrasi Dergisi*, *2*(2), 133–148.

El-Erian, M. A. (2016). Şehirdeki Tek Oyun, çev. Feyyat, C., İstanbul, Scala Yayıncılık.

Karahan, Ö., & Gencür, A. S. (2019). "Döviz kuru geçişkenliği ve merkez bankası politikaları". *Balkan Sosyal Bilimler Dergisi*, *8*, 381–387.

Özkök, Y. (2021). COVİD-19 Pandemisinde Türkiye'de İzlenen Para Politikasının Parasal Göstergeler Üzerinden Analizi. IV. Internatıonal Covıd-19 And Current Issues Congress.

Turner, P. (2014). The exit from the non-conventional monetary policy: What challenges? BIS Working Papers, No 448.

Eğilmez, M. (2014). "Enflasyon ile faiz ilişkisi". Available At:15.07.2024. https://www.mahfiegilmez.com/2014/05/enflasyon-ile-faiz-iliskisi.html

IMF. (2024). " Steady But Slow: Resılıence Amıd Dıvergence," Available At:08.07.2024, https://www.imf.org/en/Publications/WEO/Issues/2024/04/16/world-economic-outlook-april-2024

Trading Economics, Available At:15.07.2024, https://tr.tradingeconomics.com

TCMB. (2024). "Finansal İstikrar Raporu, Finansal Kesim". Available At: 15.01.2024, https://www.tcmb.gov.tr/wps/wcm/connect/tr/tcmb+tr/main+menu/yayinlar/raporlar/finansal+istikrar+raporu/2024/sayi+38

TCMB. (2022). "Makroihtiyati Tedbirler Hakkında Basın Duyurusu". Available At: 20.07.2024, https://www.tcmb.gov.tr/wps/wcm/connect/TR/TCMB+TR/Main+Menu/Duyurular/Basin/2022/DUY2022-36

TCMB. (2020). "Finansal Koşullar ve Para Politikası". Available At:10.07.2024, https://www.tcmb.gov.tr/wps/wcm/connect/95958d62-854f-4a38-ba4e-3109bc2d1fb5/5b20-3.pdf?MOD=AJPERES&CACHEID=ROOTWORKSPACE-95958d62-854f-4a38-ba4e-3109bc2d1fb5-nepI3eZ

TCMB. (2024). "Finansal istikrar Raporu". Available At:10.07.2024 https://www.tcmb.gov.tr/wps/wcm/connect/6305e9ca-dd2c-4f7e-b855-33e3077122d0/Genel+De%C4%9Ferlendirme.pdf?MOD=AJPERES&CACHEID=ROOTWORKSPACE-6305e9ca-dd2c-4f7e-b855-33e3077122d0-o.dA2J4

TCMB. (2024a). "Finansal İstikrar Raporu/Makroekonomik Görünüm". Available At 12.07.2024 https://www.tcmb.gov.tr/wps/wcm/connect/tr/tcmb+tr/main+menu/yayinlar/raporlar/finansal+istikrar+raporu/2024/sayi+38

Ayşe Özge Artekin[1]

The Impact of Technological Development Indicators on Economic Growth: A Panel Data Analysis on Emerging Economies

1 Introduction

Today, the close relationship between technological development and economic growth is a generally accepted fact. Technology is seen as a fundamental element that leads the development of nations and at the same time increases the living standards and welfare of countries. This close relationship between economic growth and technological development enables countries to increase their competitiveness and become more effective in global markets. Therefore, it is inevitable to take into account key technological development indicators such as research and development (R&D) expenditures and patent numbers when formulating economic growth strategies.

R&D can be defined as activities that are carried out by utilizing human capital and existing knowledge stock in order to develop new products, services and processes and to meet the needs of the economy by offering them to the market, and whose outputs have original, scientific and technical content (Karagöl & Karahan, 2014). In another definition, R&D is express as "the process of transformation of knowledge into tangible products with high added value" (Yaylalı et al., 2010). This process is at the center of the strategic plans of countries that want to achieve sustainable economic growth figures, and it is known that this importance given to R&D, particularly since the 1990s, has reaped its fruits in the medium and long term. As a matter of fact, countries that have made notable progress in technology, industry and science by making important investments in R&D have visibly enhanced their global competitiveness in the this time (Karagöl & Karahan, 2014).

A patent is the legal right of an inventor to prevent others from making or using a particular invention. This right is sometimes referred to as "intellectual property right" and is seen as an incentive for innovation (Hall, 2007).

1 Ph.D., Asst. Prof., Selçuk University, aoartekin@selcuk.edu.tr, ORCID: 0000-0001-7658-416X.

Patents are also an indicator of a new invention/product. Many research have determined a positive linkage between patents and innovation (Mercan et al., 2011: 33). Patents aim to encourage the development of new products and processes. According to standard patent theory, they do so in two ways. First, they provide profits associated with market power by giving inventors a limited right to exclude others; these profits create incentives for R&D activities. Second, patents encourage the disclosure of information to be kept secret. Sharing this information in exchange for exclusive rights can be useful for follow-up research and for copying inventions after the patent expires, thus promoting innovation (Sampat, 2018).

Technological development indicators such as R&D expenditures and the number of patents are important factors affecting economic growth. An analysis of R&D and patenting activities in emerging economies reveals that these countries have the potential to increase their competitiveness. The concept of emerging economy or emerging market has its origins in the term "emerging economy", which was used in the 1970s to describe underdeveloped markets. This term was not fully adequate as it was based on both objective and subjective assessments. Hence the need for a new term that more appropriately characterizes economies that fall into this category. In 1980, World Bank economist Antoine van Agtamael introduced the term "emerging economy" to describe an intermediate status between developing and developed countries. The term refers to countries that have the potential for rapid economic growth, are in the process of transitioning to a market economy and are beginning to assume important roles in the global economy. Emerging economies are generally defined as countries that are rapidly industrializing, have a high capacity to attract investment, and are well on their way to integrating into global trade networks. This new classification goes beyond the category of developing countries and more accurately reflects their dynamic and growing markets (Daniela-Neonila & Roxana-Manuela, 2014).

The importance of being an emerging economy country plays a critical role both for the internal dynamics of these countries and for the sustainability and balance of the global economic system. Emerging economy countries contribute significantly to the growth, diversification and stability of the world economy. These countries, which often have rapid economic growth rates, become attractive for foreign investment. In turn, this leads countries to improve their infrastructure, promote technology transfer and increase production capacity, as well as increase the volume and diversity of international trade.

Leading emerging economy countries such as China, India, Mexico and Türkiye have adopted economic liberalization policies and started to play important roles in the global economy. These countries have opened their doors to foreign investments, removed legal and invisible barriers to trade, reduced import quotas and opened their markets to the final products of western firms (Çivi and Çavuşgil, 2001). These steps accelerated global economic integration and increased the growth potential of countries.

As one of the world's fastest growing and largest markets, China is rapidly increasing its income level. In addition, its strong investments in high-tech production areas support the country's long-term growth. India, notable for its recent increase in growth rates and rapidly developing service sector, has strengthened its economic activities within its own borders, increasing local production-consumption shares and contributing to economic growth. Mexico, one of the largest economies in North America, stands out with its strong infrastructure in the manufacturing industry. However, although it has attempted technological development, it has not yet reached its potential. Türkiye, with its strategic location and various incentives and investments in technological development in recent years, is among the countries that continue to grow rapidly. These mentioned emerging economies are making significant contributions to the reshaping of global supply chains and the growth of international trade.

On the other hand, Russia is another emerging economy country that has an influential position in the international arena by utilizing its natural resources effectively. Russia's contribution to global economic integration is through energy trade and natural resources. Russia, which makes more specific investments in technology by focusing on specific sectors, strengthens its investments with the income generated by energy resources. At the same time, Russia's active role in various international economic organizations allows it to have a say in global economic balances.

The purpose of this study is to analyze the effects of patent numbers and R&D expenditures on economic growth for China, India, Mexico, Russia and Türkiye, which are among the emerging economy countries. The study consists of five main parts including the introduction. The introductory part includes explanations on technological development indicators and emerging economies, which are the main points of the study. In the second part, various studies in the literature analyzing the linkage among these variables and economic growth are presented. In the third part, the methodology and data analysis for the study are explained and empirical results are presented in the fourth part. The research, is concluded with a conclusion and evaluation part.

2 Literature Review

Recently, as the significance of the effects of technological developments on economic growth increases, a significant enhance has been noticed in the number of studies analyzing the relationship among R&D expenditures and the number of patents, which are indicators of technological development, and economic growth. It is generally observed in the literature that technological development indicators such as R&D expenditures and the number of patents are analyzed in association with economic growth.

Table 1: Literature Review

Yazar(lar)	Ülke(ler)	Zaman	Metodoloji	Sonuç
Ülkü (2004)	20 OECD and 10 non-OECD countries	1981–1997	Panel Data Analysis	It is concluded that there is a positive linkage between innovation and economic growth in all countries under analysis.
Wu and Zhou (2007)	China	1953-2004	Johansen Cointegration and Granger Causality Analysis	It has been determined that there is a long-term cointegration and bidirectional causality linkage between R&D expenditures and economic growth.
Sinha (2008)	Japan and South Korea	1963–2005	Cointegration and Causality	For Japan, cointegration and bidirectional causality were found between real GDP growth and growth in the number of patents. A cointegration is also found for South Korea. For panel data, real GDP growth Granger-wise causes growth in the number of patents.
Weng at al. (2012)	Shanghai	1991–2009	VAR Analysis	In conclusion of the study, a positive correlation was found among economic growth and other technological indicators.
Işık (2014)	Türkiye	1990:1-2010:4	Granger Causality Analysis	It was concluded that there is a unidirectional causality relationship from patent expenditures to economic growth.
Song and Nan (2014)	China	1993–2012	VAR Analysis	The study concluded that there is a mutual relationship between patent applications and economic growth in the long term.

Table 1: Continued

Yazar(lar)	Ülke(ler)	Zaman	Metodoloji	Sonuç
Sohag at al. (2015)	Malaysia	1985–2012	Cointegration Analysis and ARDL Model	The authors determined that an enhance in technology patents led to an enhance in economic growth in the short and long term.
Ustabaş and Ersin (2016)	South Korea and Türkiye	1989–2014	Engle-Granger and Johansen Cointegration Tests	It has been concluded that economic growth in South Korea, in the long and short term, and in Türkiye, in the short term, is positively affected by high-tech product exports.
Algan at al. (2017)	Türkiye	1996–2015	Granger Causality Analysis	It was concluded that there is a short-term unidirectional causality linkage from R&D expenditures to GDP per capita, and a unidirectional causality linkage from GDP per capita to the number of patent applications. Moreover, in the long run, R&D expenditures and patent applications have positively affected GDP per capita.
Hong (2017)	Korea	1988–2013	Granger Causality Analysis	It has been proven that there is a bidirectional linkage among economic growth and technological development.
Köse and Şentürk (2017)	Türkiye	1989–2012	Granger Causality Analysis	In the findings, there is a two-way positive linkage among R&D expenditures and economic growth. Nevertheless, no important linkage was gained among economic growth and patent expenditures.
n	10 Country	1996–2017	Granger Causality Analysis	Causality from R&D to growth was found in China, India and countries with the average Upper Middle Income group.
Dağlı ve Ezanoğlu (2021)	36 OECD Countries	2007–2017	Dynamic Panel Data Analysis	R&D expenditures and patent numbers have a positive and statistically significant relationship with economic growth.
Bakari (2022)	Romania	1990–2020	Cointegration Analysis and ARDL Model	In conclusion of the analysis, it was found that patents have a positive effect on growth in the long term.

(*continued on next page*)

Table 1: Continued

Yazar(lar)	Ülke(ler)	Zaman	Metodoloji	Sonuç
Karataş and Bozkurt (2022)	22 OECD Countries	1980–2019	Panel Causality Analysis	The research decided that there is a causal linkage between technological developments and economic growth.
Ben Yedder at al. (2023)	MENA Countries	1998–2022	Panel Data Analysis	The authors found that patents had no effect on economic growth.
Var (2023)	Türkiye and 30 OECD Countries	2008–2020	Dumitrescu-Hurlin Panel Causality Test	The study found a bidirectional linkage among economic growth and public/private sector R&D expenditures.

When Table 1 is analysed, it is possible to say that R&D expenditures positively affect economic growth in most of the empirical studies (See: Wu & Zhou, 2007; Ustabaş & Ersin, 2016; Algan et al., 2017; Köse & Şentürk, 2017; Das & Mukherjee, 2020; Var, 2023). Studies such as Sinha (2008), Işık (2014); Song & Nan (2014), Sohag et al. (2015), Algan et al., (2017), Dağlı ve Ezanoğlu (2021) and Bakari (2022) have gained that patents have a positive effect on economic growth. Köse and Şentürk (2017) and Ben Yedder et al. (2023), contrary to the majority, gained that patents do not have any effect on economic growth.

3 Methodology and Data Analysis

In this study, it is analysed whether the independent variables R&D expenditures, number of patents by residents and number of patents by non-residents affect the dependent variable, economic growth, within the scope of technological development indicators for China, India, Mexico, Russia and Türkiye, which are among the emerging economies, for the period among 1996 and 2021 using panel data analysis. The limitation of the number of countries within the scope of emerging economies to the aforementioned 5 countries was made in order to ensure the most comprehensive representation of the data sets. In this respect, the data obtained for all variables are taken from the official website of the World Bank.

$$(\boldsymbol{GDP})_t = \alpha_0 + \alpha_1 (\boldsymbol{RDE})_t + \alpha_2 (\boldsymbol{PAT.RESD}) + \alpha_3 (\boldsymbol{PAT.NRES})_t + \boldsymbol{e}_t$$

In the equation, ***RDE***; stands for R&D expenditures, ***PAT. RESD***; stands for resident patents and ***PAT. NRES*** stands for non-resident patents. α_0, α_1, α_2 and α_3 are the estimated parameters, ***t*** is the time index and ***e*** is the error term.

4 Empirical Findings

For China, India, Mexico, Russia and Türkiye, variables such as R&D expenditures, number of patents, number of patents by residents, number of patents by non-residents and GDP for the period between 1996 and 2021 were taken into consideration and Hausman test was performed first in order to perform panel data analysis. Hausman test is a test used to evaluate the accuracy of the models used in panel data analysis. This test is applied to decide whether to use a model with fixed effects or a model with random effects. Hausman statistical test should be performed to determine the type of effect. The Hausman test results are presented in Table 2 below.

Table 2: Hausman Test Results

Correlated Random Effects – Hausman Test
Test cross-section random effects

Test Summary	Chi-Sq. Statistic	Chi-Sq. d.f.	Prob.
Cross-section random	22.684999	3	0.0000

Cross-section random effects test comparisons:

Variable	Fixed	Random	Var(Diff.)	Prob.
RDE	137596332650.8	46235215304.2	51959711897491608000	0.6886
PAT.RESD	6278642.4	7003363.7	47846654044.7	0.0009
PAT.NRES	42950261.1	29775820.4	16195382348800.3	0.0011

Cross-section random effects test equation:
Dependent Variable: GDP
Method: Panel Least Squares
Sample: 1996 2021
Periods included: 26
Cross-sections included: 5
Total panel (balanced) observations: 130

Variable	Coefficient	Std. Error	t-Statistic	Prob.
C	1.56E+11	2.05E+11	0.761408	0.4479
RDE	1.38E+11	2.80E+11	0.491096	0.6242
PAT.RESD	6278642.	420742.1	14.92278	0.0000

(continued on next page)

Table 2: Continued

PAT.NRES	42950261	4908916.	8.749439	0.0000
		Effects Specification		
Cross-section fixed (dummy variables)				
Root MSE	5.18E+11		R-squared	0.974372
Mean dependent var	2.13E+12		Adjusted R-squared	0.972901
S.D. dependent var	3.25E+12		S.E. of regression	5.34E+11
Akaike info criterion	56.90563		Sum squared resid	3.48E+25
Schwarz criterion	57.08210		Log likelihood	-3690.866
Hannan-Quinn criter.	56.97733		F-statistic	662.6272
Durbin-Watson stat	0.429906		Prob(F-statistic)	0.000000

In Table 2, for the research model, GDP is determined as the dependent variable and R&D expenditures, number of patents by residents and number of patents by non-residents are determined as independent variables. According to the results of the test for emerging economies in Table 2 for the period between 2000 and 2021, the Hausman test rejected the H_0 hypothesis with a p-value of 0.000 and the fixed effect model was found appropriate. As a result of the p-value obtained according to the results of Table 2, the fixed effect model was selected and the effect coefficients of the independent variables on the dependent variable were tested in Table 3 below.

Table 3: Panel Data Analysis Results

Dependent Variable: GDP
Method: Panel Least Squares
Sample: 1996 2021
Periods included: 26
Cross-sections included: 5
Total panel (balanced) observations: 130

Table 3: Continued

Variable	Coefficient	Std. Error	t-Statistic	Prob.
RDE	1.38E+11	2.80E+11	0.491096	0.6242
PAT.RESD	6278642.	420742.1	14.92278	0.0000
PAT.NRES	42950261	4908916.	8.749439	0.0000
C	1.56E+11	2.05E+11	0.761408	0.4479
		Effects Specification		
Cross-section fixed (dummy variables)				
Root MSE	5.18E+11	R-squared		0.974372
Mean dependent var	2.13E+12	Adjusted R-squared		0.972901
S.D. dependent var	3.25E+12	S.E. of regression		5.34E+11
Akaike info criterion	56.90563	Sum squared resid		3.48E+25
Schwarz criterion	57.08210	Log likelihood		-3690.866
Hannan-Quinn criter.	56.97733	F-statistic		662.6272
Durbin-Watson stat	0.429906	Prob(F-statistic)		0.000000

According to the research model, GDP, which is the dependent variable for China, India, Mexico, Russia and Türkiye in the period between 1996 and 2021, is affected by the number of patents of residents and non-residents variables. However, there is no effect of R&D expenditures on GDP. In the analysis, it is observed that there is a positive relationship between the number of patents of residents and non-residents and GDP. As a result of the analysis, the p-values of the number of patents residents and number of patents non-residents variables were found to be "0.0000", which is an important finding. These values indicate the strength of the impact on the relevant dependent variable. In addition, the p-value of the R&D expenditures variable was found to be "0.6242", indicating that this variable does not statistically affect economic growth. Moreover, it is noted that the p value of the F statistic is less than 0.05 and the regression model is generally significant.

5 Conclusion and Evaluation

In the study, the effects of patent applications (resident and non-resident) and R&D expenditures, which are the main indicators of technological development, on economic growth were examined in the context of emerging country economies for the period 1996–2021. The findings of the study are important in terms of understanding the relationship between technological development indicators and economic growth in China, India, Mexico, Russia and Türkiye and determining policies accordingly. According to the inferences obtained from the

Hausman test results, in the analysis made with the fixed effect model, it was found that the number of patents had a significant and positive impact on the GDP of the countries. On the other hand, it has been determined that R&D expenditures do not have a statistically important impact on economic growth.

In recent years, emerging economy countries have gained momentum in technological development in order to be in a stronger position and maintain their existing place in the globalizing world. China and India, in particular, have become technology leaders by showing large increases in patent applications and R&D expenditures. Similarly, Mexico, Russia and Türkiye are also striving to strengthen their economies by increasing their technological development capacity. The importance that countries attach to technological development plays an significant role in increasing their sustainable economic growth and competitiveness. The results of the tests and observations clearly reveal that technological development is necessary for increasing the competitiveness of countries and economic growth. From this perspective, decision makers need to increase their investments in technological development to create new jobs and opportunities.

References

Algan, N., Manga, M., & Tekeoğlu, M. (2017). "Teknolojik Gelişme Göstergeleri ile Ekonomik Büyüme Arasındaki Nedensellik İlişkisi: Türkiye Örneği". In *International Conference on Eurasian Economies*, 332–338.

Bakari, S. (2022). "The Impact of Digitalization and Patent on Economic Growth in Romania". *Journal of Research, Innovation and Technologies*, *1*(1(1)), 49–61.

Ben Yedder, N., El Weriemmi, M., & Bakari, S. (2023). "The Nexus Between Domestic İnvestment and Economic Growth in MENA Countries; Do Patents Matter?". University Library of Munich, Germany.

Çivi, E., & Çavuşgil, S. T. (2001). "Yeni Dünya Düzeninde Güç Kazanan Ülkeler: Yükselen Ekonomiler". *Yönetim ve Ekonomi Dergisi*, *8*(1), 113–128.

Dağlı, İ., & Ezanoğlu, Z. (2021). Ar-Ge, patent ve ileri teknoloji ihracatının ekonomik büyümeye etkileri: OECD ülkeleri için dinamik panel veri analizi. *İnsan ve Toplum Bilimleri Araştırmaları Dergisi*, 10(1), 438–460.

Daniela-Neonila, M., & Roxana-Manuela, D. (2014). "The Emerging Economies Classification in Terms of Their Defining, Grouping Criteria and Acronyms Used for this Purpose". *Management Strategies Journal*, *26*(4), 311–319.

Das, R. C., & Mukherjee, S. (2020). "Do Spending on R&D Influence Income? An Enquiry on the World's Leading Economies and Groups". *Journal of the Knowledge Economy*, *11*(4), 1295–1315.

Hall, B. H. (2007). “Patents and Patent Policy”. *Oxford Review of Economic Policy*, *23*(4).

Hong, J. P. (2017). “Causal Relationship Between ICT R&D Investment and Economic Growth in Korea”. *Technological Forecasting and Social Change*, *116*, 70–75.

Isik, C. 2014. “Patent Harcamalari ve Iktisadi Büyüme Arasindaki Iliski: Türkiye Örnegi”. *Sosyoekonomi*, *1*, 69.

Karagöl, E. T., & Karahan, H. (2014, February). Yeni Ekonomi: Ar-ge ve İnovasyon. Seta.

Karataş, A., & Bozkurt, K. (2022). “Ekonomik Büyüme ve Teknolojik Gelişmeler: Bir Nedensellik Analizi”. *Yönetim ve Ekonomi Dergisi*, *29*(2), 209–219.

Köse, Z., & Şentürk, M. (2017). “Ar-Ge – Patent Harcamaları ve Teknolojik İlerlemenin Ekonomik Büyüme Üzerindeki Etkisi: Ampirik Bir Uygulama”. *Akademik Araştırmalar Ve Çalışmalar Dergisi (Akad)*, *9*(17), 215–221.

Mercan, B., Göktaş, D., & Gömleksiz, M. (2011). “AR&GE Faaliyetleri ve Girişimcilerin İnovasyon Üzerindeki Etkileri: Patent Verileri Üzerine Bir Uygulama”. *Paradoks Ekonomi Sosyoloji ve Politika Dergisi*, 7(2), 27–44.

Sampat, B. N. (2018). “A Survey of Empirical Evidence on Patents and Innovation”. 25383 http://www.nber.org/papers/w25383

Sinha, D. (2008). “Patents, Innovations and Economic Growth in Japan and South Korea: Evidence From Individual Country and Panel Data”. *Applied Econometrics and International Development*, *8*(1), 1–23.

Sohag, K., Begum, R. A., Abdullah, S. M. S., & Jaafar, M. (2015). “Dynamics of Energy Use, Technological Innovation. *Economic Growth and Trade Openness in Malaysia*”. *Energy*, 90, 1497–1507. DOI: 10.1016/j.energy.2015.06.101

Song, Z. S., & Nan, G. J. (2014). “Empirical Studies of The Relationship Between Technological Innovation and Economic Growth-Shanghai-Based Panel Data Analysis”. *BioTechnology: An Indian Journal*, *10*(9).

Ulku, H. (2004). “R&D, Innovation, and Economic Growth: An Empirical Analysis”. *IMF Working Paper*, No.04/185.

Ustabaş, A., & Ersin, Ö. (2016). “The Effects of R&D and High Technology Exports on Economic Growth: A Comparative Cointegration Analysis for Turkey and South Korea”, *International Conference On Eurasian Economies*, Kaposvar, pp. 44–55.

Var, Ö. F. (2023). “Ar-Ge Teşvik ve Harcamalarının Ekonomik Büyüme Üzerindeki Etkisi: Panel Veri Analizi”, Yüksek Lisans Tezi, Sakarya Üniversitesi.

Weng, L., Song,W., & Sheng, S.(2012)."Empirical Research on Scientific and Technical Innovation and Economic Growth in Shanghai". *American Journal of Operations Research.*

Wu, Y., & Zhou, L. (2007). "Cointegration and Causality Between R&D Expenditure and Economic Growth in China: 1953–2004". *In International Conference on Public Administration*, Vol. 76.

Yaylalı, M., Akan, Y., & Işık C. (2010). "Türkiye'de AR&GE Yatırım Harcamaları ve Ekonomik Büyüme Arasındaki Eşbütünleşme ve Nedensellik ilişkisi: 1990–2009". *The Journal of Knowledge Economy&Knowledge Management*, 7(2): 13–26.

Yaşar Turna[1]

The Relationship Between Energy Consumption and Economic Growth in Turkey Based on the Solow Growth Model: The Case of Prais-Winsten Regression

1 Introduction

Energy is important as a factor that has existed throughout human history and used by human beings in many processes. Energy, which is an important input especially in the realization of economic activities, is also considered an important factor in terms of influencing and directing political and economic relations between countries in the world. The manner in which energy is used across the world have started to change, particularly with the advent of the industrial revolution. The requirement for energy and the amount of energy used worldwide have increased as a result of the industrialization phenomena that followed this process. Increasing energy use has become an indispensable element of industrial production in the future (Yıldırım, 2019: 120).

Especially in the 1970s, energy supply shocks have an irreversible effect on the economy, making energy consumption essential. As a matter of fact, the oil crisis in 1973 revealed how effective the energy factor is throughout the process of producing products and services. Therefore, the disruptions caused by the oil crisis in the production process have further highlighted the idea of considering energy as a production factor (Aydın, 2021: 2).

2 Energy Consumption in Growth Theories

Energy input is necessary for a sizable portion of the economy's output operations. Nonetheless, it is noted that conventional economic models fail to consider the direct impact of energy on economic growth (Çetin and Şeker, 2012: 86). In this context, it can be said that the energy factor is indirectly taken into account in traditional economic approaches. Put differently, while conventional

1 Ph.D., Lecturer Pamukkale University, Çivril Atasay Kamer Vocational School, Department of Banking and Insurance,yturna@pau.edu.tr, ORCID: 0000-0002-3972-9099.

methods consider how energy usage affects economic activities, they do not include energy as a component that might spur economic expansion. However, there are several drawbacks to defining economic growth only in terms of the availability of energy (Stern, 2011: 26). Starting from this point, the classical approach is explained by factors such as price mechanism, capital accumulation, division of labor, specialization and population growth. The price mechanism here basically includes energy prices. The inclusion of energy prices in the price mechanism can be explained as energy prices are also effective on the economic growth explained in the classical model. In addition, in the classical approach, it can be considered that the investment factor in the economic growth process also includes energy investments. According to the socialist growth model, It is argued that rising energy consumption per person will raise labor's surplus value and raise per capita income. According to Schumpeter, energy is considered as a produced factor of production. In the Harrod-Domar growth model, it is claimed that the energy factor comes to the fore as the determinant of the upper limit of economic growth. It is claimed that the indirect effects of the energy factor on economic growth can be defined as a direct effect with the Solow growth model. Given this, the energy element in the Solow growth model may be explicitly incorporated and seen as a predictor of economic growth. Accordingly, when the energy factor is included in the production function in the Solow model, the function is;

$$Y = f\left(K,\ L,\ E\right) \tag{1}$$

It is defined as (Gökçe, 2007: 8–18). In the Solow growth model defined in this way, when the energy factor with low substitution elasticity value is added in addition to capital and labor, a new production function that includes energy can be defined. Note that the resultant production function still makes the assumption of constant returns to scaling. It is suggested that the energy factor may affect economic growth depending on the level of use of the energy factor and the development of technical progress in this scale-based constant return model, where the substitution elasticity coefficient is one. In this context, the Solow growth model expanded by including the energy factor;

$$\mathrm{Y} = \left[\left(1-\gamma\right)\left(A_L^{\beta} L^{\beta} K^{1-\beta}\right)^{\varnothing} + \gamma\left(A_E E\right)^{\varnothing}\right]^{\frac{1}{\varnothing}} \tag{2}$$

$$\Delta K = (\gamma - p_E E) - \delta K \quad (3)$$

It is expressed as . In Equation 2, it is defined as the CES production function with constant substitution elasticity that includes the Cobb-Douglas production function. Here, the value $\varnothing = \frac{\sigma - 1}{\sigma}$ included in the production function showing the gross output level is considered as the coefficient of substitution elasticity. Equation 3 is expressed as the capital depreciation equation, which shows that capital loses value at a constant rate under the assumption that the level of gross output saved is constant. The main reason for using the CES production function to show the relationship between variables is to distinguish between technical progress that increases labor and energy consumption. As a matter of fact, in the Cobb-Douglas production function, a distinction cannot be made between innovations that increase labor and innovations that increase energy consumption (Stern, 2011: 34–35).

3 The Relationship Between Energy Consumption and Economic Growth in Turkey

Due to Turkey's fast urbanisation and economic expansion, the country's population is consuming more energy. The increasing trend in energy consumption, especially after the 1980s, attracts attention. These developments have gradually increased the energy demand with the effect of liberalization in foreign trade and overcoming the foreign exchange bottleneck. Given that Turkey currently depends on foreign energy sources for 70 % of its energy, it can be said that it has a significant impact on realizing industrial production and increasing economic growth. In this framework, Turkey's energy strategy can be expressed as increasing the contribution of energy to economic growth, supporting demographic developments, and providing energy at effective, efficient and competitive prices (Aydın, 2010: 328).

The transportation and industrial sectors stand out when looking at Turkey's energy usage by sector. Industry and transportation sectors rank first in terms of energy intensity and energy consumption. In this context, the contribution of energy used especially in industry to economic growth cannot be ignored (Yanıktepe et al., 2021: 453). Therefore, energy consumption data that differs according to sectors has significant effects on economic growth and current account deficits. (Yanar & Kerimoğlu, 2011: 193–194).

4 Literature

There are significant studies of how energy use affects global and Turkish economic growth. It has been noted that the rise in these research is a result of the global recognition of energy's growing significance, particularly following the 1973 oil crisis. Recently, it has been observed that energy efficiency and green energy issues have come to the fore because of the depletion of energy supplies and in this context, the ecological economic approach has developed. Therefore, several pioneering and contemporary research on energy use and economic growth are listed here.

Table 1: Literature

Author	Period/ Country	Method	Results
Kraft and Kraft (1978)	1947–1974 (USA)	*Sims causality test	Energy consumption and economic growth are correlated in a one-way manner.
Sengul and Tuncer (2006)	1960–2000 (Turkiye)	*Toda Yamamoto causality test	There is a one-way causality between GDP and energy use.
Mucuk and Uysal (2009)	1960–2006 (Turkiye)	*Johansen cointegration test * Granger causality test	It has been determined that there are causal and cointegration links between energy use and economic development.
Altıntas (2013)	1970–2008 (Turkiye)	*ARDL test *TYDL causality test	The relationship between economic growth and energy consumption has been found to be cointegrated. Furthermore, there is a one-way causal link between the amount of per capita income and energy use.
Erdogan and Gurbuz (2014)	1970–2009 (Turkiye)	*Gregory-Hansen cointegration test	It has been shown that energy consumption and economic development are cointegrated.
Govdere and Can (2015)	1970–2014 (Turkiye)	*Engle Granger cointegration test	It has been established that energy consumption and economic growth are correlated over the long run.
Dogan and Deger (2016)	1970–2013 (India)	*Johansen cointegration test *Granger causality test	There is a one-way causal link between energy use and economic growth. A cointegration link between the series was found at the same time.

Table 1: Continued

Author	Period/ Country	Method	Results
Acaravcı and Erdogan (2018)	1992–2013 (5 Country)	*Panel data analysis	There is a relation between per capita income and renewable energy.
Syzdykova (2018)	1991–2016 (Middle Asia)	* Panel data analysis	In Central Asian nations, there is a bidirectional causal link between energy use and economic growth.
Altıner (2019)	1971–2014 (MINT)	*Emirmahmutoğlu ve Köse Panel causality test	Only in Mexico was a unidirectional causal link discovered between energy use and economic growth. There has been no evidence of causation in other nations.

According to the literature review, included are several studies that examine the connection between energy usage and global and Turkish economic growth. In the studies examined, it has been observed that there are generally relationships between economic growth and energy usage. Although the methods and times used in the studies vary, the relationships between these two variables support the theoretical literature. However, it seems that the methods used in the studies are causality and co-integration methods, which can be considered classical. In this context, in this study, unlike the analysis methods in the literature, the Prais-Winsten Regression approach is used to analyse the connection between Turkey's energy usage and economic development.

5 Dataset and Model

In this study, where the Solow growth model is used to analyse the link between energy use and economic growth in Turkey, the relationship between the series is autocorrelation by using total factor productivity data expressing GDP, energy consumption, capital accumulation, employment and technology level in Turkey between 1965 and 2023. and it is estimated by prais-winsten regression, which gives stable and reliable results even in the case of heteroscedasticity. The data referenced in the research derived from the Total Economy Database (TED, 2024) and Statistical Review of World Energy Data database (WED, 2024). The linear model defined for the data used in the analysis is;

$$GDP = \beta_0 + \beta_1 TFP + \beta_2 Capital + \beta_3 Labour + \beta_4 Energy + u_t \quad (4)$$

It is expressed as. Descriptive statistics for the data included in the linear model and used in the study is listed in the following table.

Table 2: Descriptive Statistics

	GDP	TFP	Capital	Labor	Energy
Mean	4.7501	0.9040	6.6444	1.4050	6.0946
Median	5.1600	1.5700	6.9100	1.1500	6.3329
Maximum	11.0600	8.6700	15.1400	12.9300	6.9488
Minimum	-5.9200	-11.2900	0.3300	-14.6400	4.2395
Std. Dev.	3.8183	3.6921	2.5196	3.3248	0.6825
Skewness	-0.9686	-0.8485	0.3853	-1.0890	-0.9533
Kurtosis	3.8819	4.1047	4.6229	12.7197	3.1935

6 Prais-Winsten Regression

Prais-Winsten regression was developed by Prais and Winsten in 1954. The Preis-Winsten method is considered an effective method in correcting this autocorrelation problem, especially when there is an autocorrelation problem between the series (Albayrak, 2014: 8). Therefore, this model is expressed as a special form of the generalized least squares method. In the Prais-Winsten method developed within this framework, the least squares method stated;

$$y_t = \alpha + X_t \beta + \varepsilon_t \quad (5)$$

errors obtained from the linear regression model (ε_t);

$$\varepsilon_t = \rho \varepsilon_{t-1} + e_t \quad (6)$$

Prais-Winsten regression, which was developed because the coefficients (ρ) are sequentially dependent, that is, autocorrelation;

$$y_t - \rho y_{t-1} = \alpha(1-\rho) + (X_t - \rho X_{t-1})\beta + e_t \quad (7)$$

It is defined as . This method, which is considered an improved form of the Cochrane-Orcutt method, is frequently preferred in correcting autocorrelation (Kadiyala, 1968: 93–94).

7 Empirical Application

In order to choose the appropriate time series model in analyzing economic variables and to make more accurate and stable predictions about the variables, initially, the stationarity levels of the variables must be ascertained. Unit root tests are used to determine stationarity levels. In this study, Augmented Dickey Fuller (ADF) and Philips Perron (PP) unit root tests were used to determine the stationarity levels of the variables. The results of the ADF and PP unit root tests conducted within this framework are given in the table below.

Table 3: ADF and PP unit root test results

Tests **Variables**	**ADF** **T Statistics**	**PP** **T Statistics**	**Unit Root**
GDP	-7.4907***	-7.4892***	I(0)
TFP	-7.9491***	-7.9499***	I(0)
Capital	-4.1189***	-4.1229***	I(0)
Labor	-7.9897***	-9.9663***	I(0)
Energy	-5.0171***	-5.1113***	I(0)

***,**,* expressions indicate 1 %, 5 % and 10 % significance levels, respectively.

The series are seen to be stationary at level based on the outcomes of the PP and ADF unit root tests. For this reason, the least squares approach (OLS) can be used to estimate series. However, in order for EKK to provide reliable results, there should be no problems such as autocorrelation, heteroskedasticity and normal distribution. Given this, the results of the tests for autocorrelation, heteroskedasticity, and normal distribution among the variables under investigation, it is observed that there is an autocorrelation and heteroscedasticity problem in the series. For this reason, the Prais-Winsten Regression method, which can be used in case of problems such as autocorrelation and heteroscedasticity and is known to give stable and reliable results, was employed to examine the correlation between the variables. The results of the Prais-Winsten method applied in this context are given in the table below.

Table 4: Prais-Winsten Regression results

Variables	Coefficients	T Statistics	Probability
TFP	1.0110	52.24	0.0000
Capital	0.3042	7.94	0.0000
Labor	0.4901	25.18	0.0000
Energy	0.9564	3.56	0.0010
R^2	0.98	-----	------
Model F	-------	969.27	0.0000

Based on the Prais-Winsten regression findings displayed in the table, Economic growth is increased by 1.01 for every unit rise in TFP. Economic growth is increased by 0.30 for every unit increase in capital accumulation. An increase of one employment unit results in 0.49 rises in economic growth. Economic growth is increased by 0.95 for every unit increase in energy use.

8 Result and Conclusion

In this study, where the relationship between energy consumption and economic growth is analyzed based on the Solow growth model in Turkey, the relationship between the series is analyzed by using total factor productivity data expressing GDP, energy consumption, capital accumulation, employment and technology level in Turkey between 1965 and 2023. -Estimated by Winsten regression model. According to the Prais-Winsten regression results, Economic growth is increased by 1.01 for every unit rise in TFP. Economic growth is increased by 0.30 for every unit increase in capital accumulation. An increase of one employment unit results in 0.49 rises in economic growth. Economic growth is increased by 0.95 for every unit increase in energy use.

The analyses conducted in this context show that Turkey's economic development is influenced by capital accumulation, employment levels, technical advancements, and energy consumption. At the same time, when looking at the impact levels of the variables, it is clear that the level of technology and energy consumption in Turkey affect economic growth more than the level of capital accumulation and employment. This situation can be evaluated as the level of capital accumulation and employment within the economic growth dynamics in Turkey has now matured, but investments in technological development and energy consumption can contribute to more stable and higher growth rates in the economy. For this reason, ensuring efficiency in technological progress and energy consumption, which are especially important in industrial production, and reviewing the composition of investments in these areas can be considered as an important factor in ensuring sustainable economic growth.

References

Acaravcı, A., & Erdoğan, S. (2018). "Yenilenebilir Enerji, Çevre ve Ekonomik Büyüme İlişkisi: Seçilmiş Ülkeler İçin Ampirik Bir Analiz". *Eskişehir Osmangazi Üniversitesi İktisadi ve İdari Bilimler Dergisi*, *13*(1), 53–64.

Albayrak, A. S. (2014). "Otokorelasyon Durumunda En Küçük Kareler Tekniğinin Alternatifi Otoregresyon Teknikleri ve Bir Uygulama". *Suleyman Demirel University Journal of Faculty of Economics & Administrative Sciences*, *19*(1).

Altıner, A. (2019). "MINT Ülkelerinde Enerji Tüketimi ve Ekonomik Büyüme İlişkisi: Panel Nedensellik Analizi". *Gümüşhane Üniversitesi Sosyal Bilimler Dergisi*, *10*(2), 369–378.

Altıntaş, H. (2013). "Türkiye'de Birincil Enerji Tüketimi, Karbondioksit Emisyonu ve Ekonomik Büyüme İlişkisi: Eşbütünleşme ve Nedensellik Analizi". *Eskişehir Osmangazi Üniversitesi İktisadi ve İdari Bilimler Dergisi*, *8*(1), 263–294.

Aydın, F. (2010). "Enerji Tüketimi ve Ekonomik Büyüme". *Erciyes Üniversitesi İktisadi ve İdari Bilimler Fakültesi Dergisi*, *35*, 317–340.

Aydın, B. (2021). "Kalkınma Literatüründen Hareketle Enerji-Büyüme İlişkisi". *Finans Ekonomi ve Sosyal Araştırmalar Dergisi*, *6*(1), 1–13.

Çetin, M., & Seker, F. (2012). Enerji tüketiminin ekonomik büyüme üzerindeki etkisi: Türkiye örneği. Uludağ Üniversitesi İktisadi ve İdari Bilimler Fakültesi Dergisi, 31(1), 85–106.

Doğan, B., & Değer, O. (2016). "Enerji Tüketimi, Finansal Gelişme ve Ekonomik Büyüme İlişkisi: Hindistan Örneklemi". *Yaşar Üniversitesi E-Dergisi*, *11*(44), 326–338.

Energy Institute Statistical Review of World Energy. (2024). https://www.energyinst.org/ statistical-review. (Access Date: 05.05.2024).

Erdoğan, S., & Gürbüz, S. (2014). "Türkiye'de Enerji Tüketimi ve Ekonomik Büyüme İlişkisi: Yapisal Kirilmali Zaman Serisi Analizi". *Selçuk Üniversitesi Sosyal Bilimler Enstitüsü Dergisi*, *32*, 79–87.

Gökçe, C. (2007). *"Ekonomik Büyüme Sürecinde Enerjinin Değişen Rolü: Türkiye Örneği"* (Master's thesis, Sosyal Bilimler Enstitüsü).

Gövdere, B., & Can, M. (2015). "Enerji Tüketimi ve Ekonomik Büyüme İlişkisi: Türkiye Örnekleminde Eşbütünleşme Analizi". *Uluslararası İktisadi ve İdari Bilimler Dergisi*, *1*(2), 101–114.

Kadiyala, K. R. (1968). "A Transformation Used to Circumvent the Problem of Autocorrelation". *Econometrica: Journal of the Econometric Society*, 93–96.

Kraft J., & Kraft A. (1978). "On the Relationship Between Energy and GNP". *Journal of Energy and Development*, *3*(2), 401–403.

Mucuk, M., & Uysal, D. (2009). "Türkiye Ekonomisinde Enerji Tüketimi ve Ekonomik Büyüme". *Maliye Dergisi*, *157*(1), 105–115.

Stern, D. I. (2011). "The Role of Energy in Economic Growth". *Annals of the New York Academy of Sciences*, *1219*(1), 26–51.

Syzdykova, A. (2018). "Orta Asya Ülkelerinde Enerji Tüketimi ve Ekonomik Büyüme İlişkisi: Panel Veri Analizi". *Afyon Kocatepe Üniversitesi İktisadi ve İdari Bilimler Fakültesi Dergisi*, *20*(1), 87–99.

Şengül, S., & Tuncer, İ. (2006). "Türkiye'de Enerji Tüketimi ve Ekonomik Büyüme: 1960–2000". *İktisat İşletme ve Finans*, *21*(242), 69–80.

Total Economy Database. (2024). https://www.conference-board.org/data/economydatabase (Access Date: 05.05.2024).

Yanar, R., & Kerimoğlu, G. (2011). "Türkiye'de Enerji Tüketimi, Ekonomik Büyüme ve Cari Açık İlişkisi". *Ekonomi Bilimleri Dergisi*, *3*(2), 191–201.

Yanıktepe, B., Kara, O., & Parlak, T. K. (2021). "Enerji Tüketimi ve Ekonomik Büyüme İlişkisi: Türkiye". *Osmaniye Korkut Ata Üniversitesi Fen Bilimleri Enstitüsü Dergisi*, *4*(3), 452–465.

Yıldırım, C. Y. (2019). "Türkiye'de Enerji Tüketimi ve Ekonomik Büyüme: Granger Nedensellik Yaklaşımı". *İktisadi İdari ve Siyasal Araştırmalar Dergisi (İKTİSAD)*, *4*(9), 119–145.

Halit Yalçın[1]

Effects of Money Supply on Economic Growth and Inflation Rate: The Case of Türkiye

1 Introduction

Among the main objectives of monetary and fiscal policies in an economy is to ensure full employment and price stability. In an economy where price stability is achieved, ensuring stability in interest rates can contribute to economic growth. The monetary authority has the power to change the quantity of money in circulation. Central banks can change the money supply, leading to effects on macroeconomic indicators. Increasing (decreasing) the money supply can increase (decrease) liquidity in the economy, causing aggregate demand to rise (decrease) and thus demand-side inflation to increase (decrease).

One of the fundamental components of the financial system is the monetary base, which is defined as the total of reserve money and open market activities. In this context, money supply may be related to economic growth as well as inflation (Okşak & Özen, 2020: 242). Investment expenditures and therefore savings are very important in ensuring sustainable economic growth. In this situation, the developed structure of financial markets can support the provision of the necessary funds for the financing of economic growth.

The literature has a variety of opinions regarding how the money supply affects inflation. The classical economic theory holds that changes in the money supply have a one-to-one effect on the overall level of prices. The overall level of pricing is impacted by changes in the money supply, but not the level of actual product. Therefore, according to the classical view, money is neutral (Ünsal, 2013: 531–532) while according to the Keynesian economic view, money is not neutral.

The monetarist economic perspective states that, Since inflation is a monetary event, monetary policy should be implemented according to a certain rule. This rule is that the money supply is increased at a fixed rate every year (Friedman, 1988: 95). According to the monetarist view, changes in the money supply have an impact on employment, prices in general level, and production in the

1 Lect, Bilecik Şeyh Edebali University, Department of Foreign Trade, halit.yalcin@bilecik.edu.tr, ORCID: 0000-0002-5035-1525, 11500, Bilecik, Türkiye.

economy. Changes in the money supply primarily affect in the near future, production and the overall level of prices, while they only affect the general level of prices in the long run. For this reason, money is not neutral in the short term and neutral in the long run. For the New Classical view, changes in the money supply have an impact only on the general level of prices if they are made public, while unforeseen policies only have an impact on the level of output and the general level of prices until they are foreseen in the short term. For the New Keynesian view, since wages and prices are stickiness, predicted or unforeseen changes in the money supply affect both the level of output and the general level of prices in the short run (Bulut & Ulusoy, 2019: 145–156).

In order to reduce the inflationary pressures that have increased in the Turkish economy in recent years, contractionary monetary and fiscal policies have been implemented.The study's objective is to analyze how the money supply effects inflation and economic growth in this particular scenario. In the study, monthly data for the period 2006:01-2024:03 (Autoregressive Distributed Lag Bound Test) is analyzed by ARDL boundary test. By employing an up-to-date data set and an up-to-date econometric technique, this study aims to contribute to the literature by evaluating the impact of money supply on economic growth and inflation rate in distinct models. In the study, which was prepared in five sections, the empirical literature review in the second part, the explanations about the data set and method used in the analysis in the third part, the empirical analysis and results in the fourth part, and finally the evaluation of the results obtained in the conclusion section are included.

2 Literature Review

Changes in the money supply can have effects on macroeconomic variables. In this perspective, there is no complete consensus in the economics literature regarding the macroeconomic effects of the money supply. From this point of view, empirical studies analyzing the effects of money supply on macroeconomic variables for other countries, especially Türkiye, are included in historical order.

In the Pakistani economy, Qayyum (2006) looked into the validity of the monetarist theory that claims inflation is a monetary occurrences. Thanks to the research's results obtained by correlation analysis, it has been determined that there is a positive relationship between money supply growth and inflation, that the increase in money supply affects the real product growth at first, affects the inflation in the country in the second stage, and that the contribution of money supply growth to the inflation increase in the country is significant.

Chaker, Inguva and Çetin (2012) analyzed the effect of supply of money on inflation rate and economic growth at the United Arab Emirates in the 2000–2010 period with regression equations. According to the results of the regression analysis, it was stated that effect of supply of money and inflation rate on GDP changes was significant.

By applying the Johansen cointegration and causality test, Masnan, Shaari and Hussain (2013) examined the connections between the supply of money, inflation, and increase in real GDP in the economies of Malaysia, Singapore, and Indonesia. The study's conclusions included the following: there is a one-way correlation between supply of money and inflation in Singapore and Malaysia; the money supply in the Malaysian economy is not the Granger cause of increase in real GDP; it appears that there is an ongoing connection among the series in mentioned countries.

In his analysis of the Turkish economy from 1987 to 2007, Adak (2014) looked at the relationship between money supply and increase in real GDP. There is a causality relationship between money supply and increase in real GDP, based on the research.

In the Kenyan economy, Kiganda (2014) analyzed the relationship between supply of money as well as inflation for the years 1984–2012. The study came to the conclusion that the money supply is an essential factor in determining inflation in the nation over the long term and that there is a long-term, positive, significant relationship between the two.

Şahin and Karanfil (2015) investigated the effect of the budget deficit and real exchange rate on inflation in the Turkish economy for the period 1980–2013. It has been stated that the series are related over an extended period of time and that the money supply and inflation are not causally related directly.

Aslam (2016) analyzed the effects of the supply of on Sri Lankan economy for the period 1959–2013. In the study, in which the multivariate econometrics method is utilized to test the impact of supply of money on increase in real GDP, it is concluded that the positive money Supply impact on increase in real GDP was significant.

Su et al. (2016) investigated the relationship between money supply growth as well as inflation in China with causality test and the rolling-window estimation test for the period 1997–2014. According to the research, there is a one-way relationship between the expansion of the money supply and inflation, that according to the timevarying rolling-window approach, money supply growth has positive and negative effects on inflation in some sub-periods, and that inflation has the same effects on money supply growth.

Dingela and Khobai (2017) analyzed the impact of the broad supply of money on per capita increase in real GDP in South Africa for the period 1980–2016 using the ARDL bounds test method. The analysis conclusion indicated that there is a positive and significant relationship between the supply of money and both short- and long-term increase in real GDP.

George et al. (2018) studied on the impact of supply of money on macroeconomic variables in the Nigerian economy during the period 1985–2016. In the study, it was stated that the money supply in the narrow sense has a positive and significant effect on real GDP and inflation, while the broad supply of money does not have a significant effect on real GDP and inflation, and that inflation and economic growth are a function of the narrow money supply and exchange rate.

Gnawali (2019) applied causality testing and the Vector Error Correction Model to examine

the association between Nepal's supply of money and increase in real GDP 1975–2016. In the study, it was concluded that the supply of money has a significant and positive impact on increase in real GDP. In this direction, it is proposed to increase the money supply in order to increase economic growth in the country.

Okşak and Özen (2020) investigated the association between the growth rate of the monetary base and increase in real GDP in the Turkish economy. Data for the period 1969–2017 is analyzed by ARDL method and causality tests. In the study, it was determined that the growth rate of the monetary base and the economic growth rate were related and there was no causality between the series according to the results of the causality tests.

Kılavuz and Altınöz (2020) analyzed the association between supply of money, exchange rate,interest rate and inflation in the Turkish economy for the period 2006–2018 by using the ARDL bounds test method. Thus, it was stated that there is a positive relationship between money supply and inflation.

Uğur and Atılgan (2021) analyzed how the money supply affected inflation in five selected developing country economies between 1995 and 2017. Based on the panel data analysis findings, it was concluded that the increase in the growth rate of the supply of money in the countries in the sample group increased inflation and inflation was a real phenomenon compared to monetary factors.

Razia and Omarya (2022) analyzed the effect of broad money supply on per capita increase in real GDP in Palestine for the period 2000–2020 with the ARDL method. In the study, it was determined that money supply, inflation rate and total capital formation had a positive effect on increase in real GDP in the short run, yet relevant variables didn't affect increase in real GDP in the long run.

3. Methodology

By this research, the effect of supply of money (MS) on increase in real GDP (G) and inflation rate (IR) is analyzed with monthly data for the period 2006:01-2024:03 by ARDL bounds test method. All of the series analyzed in the study were acquired through the Electronic Data Distribution System of the Central Bank of the Republic of Türkiye (CBRT). The fact that the inflation targeting strategy started to be adopted was effective in determining the beginning of the analysis period as 2006. The variables to be studied in the study were chosen after considering pertinent literature. In this context, since the inflation rate is announced on a monthly and quarterly basis to represent economic growth, based on the consumer price index, the industrial production index data, which provides a leading information about the economic growth trend, and the M3 money supply variable as money supply are included in the analysis.

In order to choose the method to be used in determining the relationships between the variables, it is necessary to determine the stationary states of the series. From this,unit root tests developed by Dickey-Fuller (1981) and Phillips-Perron (1988) is used in the study. Although there are three models in both unit root tests: constant, unconstant and non-trending, and constant and trending, According to the null hypothesis,the relevant series has a unit root, while the alternative hypothesis states that the series doesn't contain a unit root.

The ARDL bounds test, developed by Pesaran et al. (2001), can be used when the series is stationary at the level or first difference, except for the second difference. In addition, with the bounds test, efficient results can be obtained in small-sized samples compared to classical cointegration tests, and unbiased estimates of the long-term model can be revealed. Engle and Granger's (1987) analysis with a single equation can be used to distinguish between dependent and independent variables in the ARDL bounds test. Due to the stated advantages, the analyzes were carried out by ARDL bounds test method in the study. The unconstrained error correction models created in the study are included in equation 1 and equation 2.

$$\Delta G_t = \alpha + \sum_{i=1}^{k} \delta_{1i} \Delta (G)_{t-i} \sum_{i=0}^{k} \phi_{2i} \Delta (\mathrm{MS})_{t-i} + \beta_0 (G)_{t-1} + \beta_1 (\mathrm{MS})_{t-1} + u_t \quad (1)$$

$$\Delta IR_t = \alpha + \sum_{i=1}^{k} \delta_{1i} \Delta (IR)_{t-i} \sum_{i=0}^{k} \phi_{2i} \Delta (\mathrm{MS})_{t-i} + \beta_0 (IR)_{t-1} + \beta_1 (\mathrm{MS})_{t-1} + u_t \quad (2)$$

The u_t in the equations represent the error term, the Δ represents the first difference of the series, α constant term, k represents the lag length, δ, and φ short-term β0 and β1 represent the long-run coefficients.

The cointegration relationship is decided by comparing the F-statistic value acquired by the ARDL boundary test with the lower and upper limit critical values. If the statistical value is above (below) the upper (lower) limit critical value, it indicates that there is (not) a cointegration relationship. If the statistical value is between the critical values, an evaluation of cointegration cannot be made. The error correction model obtained for the model is included in equations 3 and 4.

$$G_t = \alpha + \sum_{i=1}^{k} \delta_{1i} \Delta(G)_{t-i} + \sum_{i=0}^{k} \phi_{2i} \Delta(\mathrm{MS})_{t-i} + \pi ECM_{t-1} + u_t \quad (3)$$

$$IR_t = \alpha + \sum_{i=1}^{k} \delta_{1i} \Delta(IR)_{t-i} + \sum_{i=0}^{k} \phi_{2i} \Delta(\mathrm{MS})_{t-i} + \pi ECM_{t-1} + u_t \quad (4)$$

ECM_{t-1} in equation 5 and equation 6 refers to the error correction term, δ_{1i}, and ϕ_{2i} short-term coefficients, and π refers to the rate at which the model returns to long-term equilibrium and adjusts in the event of a short-term shock. Although the Π coefficient is negative, it should be statistically significant.

4 Empirical Findings

The analysis carried out in the study is first started by determining the unit root states of the series. Accordingly, the findings of Phillips-Perron (PP) and Augmented Dickey-Fuller (ADF) unit root tests are given in Table 1.

Table 1: Tests Results

	ADF		PP	
Variables	**Constant**	**Constant/ Trend**	**Constant**	**Constant/ Trend**
G_t	-0.528 (0.882)	-4.162 (0.01)	0.136 (0.97)	-3.817 (0.017)
ΔG_t	-16.801 (0.0)***	-16.786 (0.0)***	-19.600 (0.0)***	-20.145 (0.0)***
IR_t	0.887 (0.995)	-0.555 (0.98)	-0.378 (0.9092)	-1.571 (0.80)

Table 2: Continued

	ADF		PP	
Variables	**Constant**	**Constant/ Trend**	**Constant**	**Constant/ Trend**
ΔIR_t	-4.445 (0.0)***	-5.382 (0.0)***	-7.842 (0.0)***	-7.958 (0.0)***
MS_t	3.274 (1.0)	-2.152 (0.5132)	-9.306 (0.0)***	-12.023 (0.0)***
ΔMS_t	-11.791 (0.0)***	-6.151 (0.0)***	-77.826 (0.0)***	-123.861 (0.0)***

Note: *The symbols ***, **, and * denote significance levels at 1 %, 5 %, and 10 %, respectively. Probability values are given in parentheses*

According to the findings in Table 1, the results of ADF and PP unit root tests are parallel except for the MS variable. The G and IR variables are stationary in the first difference according to the results of both unit root tests. The MS variable, on the other hand, is stationary at the first difference according to the ADF unit root test result, while it is stationary at the level according to the PP unit root test result. To ascertain the stationarity status of the series, analysis was completed with the ARDL boundary test. In the study, two separate models are created in which money supply is the independent variable and economic growth and inflation rate are dependent variables. The situation in which economic growth is the dependent variable is determined as Model A and the situation in which the inflation rate is the dependent variable is determined as Model B. The optimal delay lengths determined according to the Akaike information criterion are ARDL(1, 0) for Model A and ARDL(6, 2) for Model B. In Table 2, F-statistic, lower and upper limit critical values of the models and diagnostic tests related to the models are included.

Table 2: Bounds and Diagnostic Test Results

F-statistic Model A (10.024**) Model B (43.774**)				
Critical Values	**Model A**		**Model B**	
Significance Level	**I (0)**	**I (1)**	**I (0)**	**I (1)**
%10	5.725	6.45	5.725	6.45
%5	6.82	7.67	6.82	7.67
%1	9.17	10.24	9.17	10.24

(continued on next page)

Table 3: Continued

F-statistic Model A (10.024**) Model B (43.774**)				
Critical Values	**Model A**		**Model B**	
Significance Level	**I (0)**	**I (1)**	**I (0)**	**I (1)**
Diagnostic Tests		Model A		Model B
BG Serial Correlation LM Test		0.449 (0.639)		0.303(0.739)
ARCH Heteroscedasticity Test		1.609 (0.206)		0.517 (0.473)
Ramsey RESET test		0.184 (0.668)		3.470 (0.064)
Model A, B		Cusum Test (S) Cusum Q Test (U)		

Not: Probability values are given in parentheses. U: Unstable S: Stable

The F-statistic value (10.024) in Table 2 is above the lower and upper limit critical values at the 5 % significance level for Model A. The F-statistic value for Model B (43,774) is above the lower and upper bound critical values at the 1 % significance level. This shows that the variables are cointegrated in both models. Considering the probe values of the tests performed to detect any autocorrelation, model construction error and changing variance problems in the models, it is seen that the problems listed are not found in the models. In addition, Cusum test findings indicates the stability of the models.

Table 3: Long Run Results

Variables	Coefficient	Std. Error	t-stat	Prob
Model A MS_t	8.738	5.029	1.737	0.084
Model B MS_t	1.416	1.180	11.996	0.000

According to Table 3, which includes the results of the long-term analysis, although the coefficients of the money supply on the dependent variables are positive in both models, they are statistically significant at the level of 10 % for Model A and 1 % for Model B. A 1 % change in the money supply causes an 8.74 % effect on economic growth and a 1.42 % effect on inflation. According to the results of the analysis, the impact of supply of money on increase in real GDP is higher than inflation. The short-term coefficients of the analysis are given in Table 4.

Table 4: Short Run Results

Model A				
Variables	Coefficient	Std. Error	t-Statistic	Prob
Constant	7.157	1.604	4.462	0.000
TREND	0.048	0.011	4.469	0.000
ECT_{t-1}	-0.172	0.038	-4.488	0.000
Model B				
Variables	Coefficient	Std. Error	t-Statistic	Prob
Constant	0.792	0.288	2.749	0.007
TREND	0.002	0.002	0.770	0.442
ΔIR_{t-1}	0.436	0.063	6.886	0.000
ΔIR_{t-2}	-0.058	0.068	-0.862	0.390
ΔIR_{t-3}	0.201	0.067	3.021	0.003
ΔIR_{t-4}	0.153	0.067	2.287	0.023
ΔIR_{t-5}	0.146	0.060	2.438	0.016
ΔMS_{t}	0.000	0.000	1.934	0.055
ΔMS_{t-1}	0.000	0.000	-3.338	0.001
ECT_{t-1}	-0.123	0.013	-9.380	0.000

As the outcomes in Table 4 demonstrate, the error correction parameter coefficients for both models are between 0 and 1 in negative, significant and absolute value. This supports the long-term relationship. For Model B, the short-run coefficients are positive, but only negative in the second lagging value of inflation, but not significant.

5 Conclusion

Since high inflation and economic contraction are not desirable by economic decision-makers, policymakers use policy tools to eliminate the relevant problems. Which policy tool will be used in line with the objectives, how and in what way is closely related to the effects it will have on the economy. In the theoretical literature, there is a consensus that the increase in supply of money increases inflation in the long term and does not have an effect on the level of output. However, as the relevant empirical literatüre is analysed, it can be observed that there is no complete consensus on the effect of money supply on inflation and increase in real GDP.

High inflation has remained on the agenda as a major economic problem in the Turkish economy for many years. In the period from the 1970s to the early 2000s, the inflation rate was at least double-digit. With the Transition to a Strong Economy Program implemented after the economic crisis in 2001, the inflation rate was below 10 % in 2004. In the following years, the inflation rate did not show major changes, and the per capita income level increased. With the reduction of inflation, CBRT adopted an explicit inflation targeting strategy in 2006. However, economic indicators have deteriorated and the inflation rate has increased in recent years due to terrorist incidents, cross-border operations, the covid-19, exchange rate shocks, changes in the policy interest rate and the earthquake. Today, contractionary monetary and fiscal policies are implemented in order to ensure price stability. In this context, determining the effect of supply of money on inflation rate and increase in real GDP is important in predicting the macroeconomic consequences of the policies to be implemented.

In the research, the impact of supply of money on inflation and increase in real GDP is analyzed by ARDL bounds test method for monthly data for the period 2006:01-2024:03. In the long-term results, it has been determined that although the impact of money supply on economic growth and inflation is positive, the impact of supply of money on increase in real GDP is higher than inflation. While the determination that the supply of money positively impacts the increase in real GDP compatible with the studies carried out by Aslam (2016), Dingela and Khobai (2017), Gnawali (2019) and Okşak and Özen (2020), the determination that the money supply positively affects inflation is compatible with the studies conducted by Qayyum (2006), Özcan (2014), Cesur (2006), and Erkişi (2019). An advanced financial structure is needed to ensure sustainable economic growth. As a matter of fact, the significant and positive determination of the effect of M3 supply of money on increase in real GDP also reveals the effect of financial development on the real economy. It should not be ignored that the tight monetary policy implemented to control inflation may reduce economic growth performance, although it will reduce inflation. In addition to tight policies that narrow demand to reduce inflation, economic growth performance can be maintained as well as price stability by implementing measures to encourage economic growth that support supply.

References

Adak, M. (2014). Para Arzı ve Ekonomik Büyüme: Türkiye Örneği. *Bu Sayının Hakemleri, Yalova Sosyal Bilimler Dergisi, 7*, 9: 22.

Aslam, A. L. M. (2016). Impact of money supply on Sri Lankan economy: An econometric analysis. *International Letters of Social and Humanistic Sciences, 67*, 11–17.

Bulut, E., & Ulusoy M. (2019). *Makro İktisat*. Ankara: Monopol Yayınları.

Cesur, F. (2006). Para Arzı ile Enflasyon İlişkisi ve Türkiye Uygulaması (1994–2004). *Trakya Üniversitesi Sosyal Bilimler Dergisi, 8*(2), 85–104.

Chaker, M. N., Inguva, S., & Cetin, R. (2012). The impact of money supply on economic growth: a case study of the United Arab Emirates. *Journal for Global Business Advancement, 5*(1), 7–20.

Dickey, D. A., & Fuller, W. A. (1981). Likelihood ratio statistics for autoregressive time series with a unit root. *Econometrica: journal of the Econometric Society*, 1057–1072.

Dingela, S., & Khobai, H. (2017). Dynamic impact of money supply on economic growth in South Africa. An ARDL approach.

Engle, R. F., & Granger, C. W. (1987). Co-integration and error correction: Representation, estimation, and testing. *Econometrica: Journal of the Econometric Society*, 251–276.

Erkişi, K. (2019). Budget deficits, money supply and inflation: The case of fragile five countries. *Euroasia Journal of Socıal Sciences & Humanities, 6*(9), 49–60.

Friedman, M. (1988). Kapitalizm ve Özgürlük, çev. Erberk, D-Himmetoğlu, N., Altın Kitaplar Matbaası, İstanbul, 26.

George, C., Suoyai, E., Tema, L., & Boloekeye, M. (2018). Impact of money supply on some macroeconomic variables on the Nigerian economy. *Journal of Business Management and Economic Research*, 2(5), 32–46.

Gnawali, L. (2019). Money supply and economic growth of Nepal. *International Journal of Business and Management Research, 3*(9), 133–149.

Kiganda, E. O. (2014). Relationship between inflation and money supply in Kenya. *Journal of Social Economics, 2*(2), 63–83.

Kılavuz, E., & Altınöz, B. (2020). Türkiye'de Para Arzı ile Enflasyon Arasındaki İlişki: ARDL Sınır Testi Yaklaşımı. *Ekonomi Politika ve Finans Araştırmaları Dergisi*, 5(2), 242–260.

Masnan, F., Shaari, M. S., & Hussain, N. E. (2013). Relationship among money supply, economic growth and inflation: Empirical evidence from three Southeast Asian countries. *International Journal of Information, Business and Management, 5*(3), 83.

Pesaran, M. H., Shin, Y., & Smith, R. J. (2001). Bounds testing approaches to the analysis of level relationships. *Journal of applied econometrics*, *16*(3), 289–326.

Phillips, P. C., & Perron, P. (1988). Testing for a unit root in time series regression. *biometrika*, *75*(2), 335–346.

Su, C. W., Fan, J. J., Chang, H. L., & Li, X. L. (2016). Is there causal relationship between money supply growth and inflation in China? Evidence from quantity theory of money. *Review of Development Economics*, *20*(3), 702–719.

Şahin, İ., & Karanfil, M. (2015). Türkiye ekonomisinde 1980–2013 dönemi para arzının enflasyon üzerindeki etkisi. *Business and Economics Research Journal*, *6*(4), 97–113.

Okşak, Y., & Özen, E. (2020). Türkiye'de Parasal Tabanın Büyüme Üzerindeki Uzun Dönem Etkisi. *Dumlupınar Üniversitesi Sosyal Bilimler Dergisi*, *65*, 232–244.

Özcan, B. (2014). Gelişmekte olan ekonomilerde enflasyonun belirleyenleri: Dinamik panel veri analizi. *Cumhuriyet Üniversitesi İktisadi ve İdari Bilimler Dergisi*, *15*(1), 33–53.

Razia, A., & Omarya, M. (2022). The impact of the broad money supply (M2) on economic growth per capita in Palestine. *International Journal of Business Ethics and Governance*, *5*(2), 1–10.

Uğur, B., & Atılgan, D. (2021). Para Arzının Enflasyon Üzerindeki Etkisi: 5 Gelişmekte Olan Ülke Örneği. *19 Mayıs Sosyal Bilimler Dergisi*, *2*(4), 774–789.

Ünsal, E. (2013). Makro İktisat, 10. Baskı, İmaj Yayınevi, Ankara.

Qayyum, A. (2006). Money, inflation, and growth in Pakistan. The Pakistan development Review, 203–212.

Alperen Ağca[1]

A Macroeconomic Analysis: Unemployment, Health Inflation and Economic Growth

1 Introduction

For those who are interested in economic growth, inflation of health expenditure and unemployment rate, the linkage between them are indirect with respect to general output model. The mainstream view is that unemployment and economic growth has a negative relationship due to Okun's Law (1963). Inflation, on the hand, has a positive linkage through Philip's Curve (1958) and Okun's Law. In this particular paper, health inflation has been handled instead of inflation itself. Health consumer price index is a perfect proxy for health inflation. One can expect to research unemployment in healthcare system for getting a more solid example. However, those are not concern of this study due to the lack of reliability of both data and healthcare system in Türkiye. For the first time for Türkiye, this paper considers the distribution structure of the mentioned variables are not normal but asymptotic. Therefore, a residual augmented least square unit root tests have been established and analysed for cointegration.

2 Literature Review

There are many studies for explaining the relationship between the economic growth and health expenditures or unemployment rate. However, there are limited papers considering all the variables at the same time. Due to the limited data availability or lack of methods, researchers have been vicious to handle this topic especially the latter.

Brenner and Mooney (1983) have stated that the stress of unemployment in effective on health conditions and this effect is linked to economic growth by this way.

Sterret et al. (2014), have claimed that private healthcare system which has burden on employers makes them face layoffs and it would be better for

1 Ph.D., Assist. Prof., Osmaniye Korkut Ata University, Faculty of Economics and Administrative Sciences Department of Economics, alperenagca@osmaniye.edu.tr, ORCID: 0009-0007-8139-7976.

macroeconomic activity to subsidize the healthcare services. It would increase the economic growth with respect to returns to scale of business entrepreneurships.

Maruthappu et al. (2015) have reported that there is a significant link between cerebrovascular mortality, health expenditures and unemployment. The more the mortality rate is higher the less employment occurs. 1 % increase in health expenditures causes less cerebrovascular mortality.

Afzal et al. (2013) in their study which includes ARDL model for Pakistan between 1971 and 2011 have stated that higher healthy labour force leads a higher economic growth. Education, food and health expenditures causes a decrease in economic growth in Pakistan.

Barro (2013) has found that there is a significant relationship between economic growth and health expenditure through then life expectancy at birth.

3 Data and Methodology

In this study, composite leading indicator based on GDP (cli), healthcare consumer price index (hcpi) and the unemployment rate (unem) are used to analyze the long-term relationship between the variables for the time between 2005 and 2023, monthly in Türkiye. All the variables are gathered from OECD. All the variables have been transformed into logarithmic form. First, it has been established a residual augmented least squares augmented Dickey Fuller (RALS-ADF) unit root test proposed by Im, Lee and Tieslau (2014) (ILT here after) and Jarque-Bera test to determine the features of the distribution of the variables whether they are asymptomatically normal or not. Second, RALS cointegration test proposed by Lee, Lee and Im (2015) (LLI here after) has been applied to test the long-term relationship. While testing for cointegration dependent variable has been determined to be not only composite leading indicator based on GDP but also the other two variables, respectively.

It has been utilized from Hepsağ's (2022) econometric time series book for the interpretation of the equations below.

ILT (2014) RALS-ADF test follows a 2-step method to estimate.

$$\Delta y_t = \mu + \delta y_{t-1} + \sum_{i=1}^{k} \alpha_i \Delta y_{t-i} + \varepsilon_t \tag{1}$$

$$\Delta y_t = \mu + \beta t + \delta y_{t-1} + \sum_{i=1}^{k} \alpha_i \Delta y_{t-i} + \varepsilon_t \tag{2}$$

First, they have calculated the second and third moments of the residuals in traditional ADF equations (intercept only Eq1 and intercept and trend Eq2) as it can been in Eq3.

$$m_2 = \frac{\sum_{t=1}^{T} \hat{\varepsilon}_t^2}{T}, m_3 = \frac{\sum_{t=1}^{T} \hat{\varepsilon}_t^3}{T} \tag{3}$$

Afterwards, the moments have been augmented in Eq4 and Eq5.

$$\hat{w}_{2t} = \hat{\varepsilon}_t^2 - m_2 \tag{4}$$

$$\hat{w}_{3t} = \hat{\varepsilon}_t^3 - m_3 - 3m_2\hat{\varepsilon}_t \tag{5}$$

Finally, augmented variables have been included in the initial equations for RALS-ADF unit root test in Eq6 and Eq7, respectively.

$$\Delta y_t = \mu + \delta y_{t-1} + \sum_{i=1}^{k} \alpha_i \Delta y_{t-i} + \gamma_2 \hat{w}_{2t} + \gamma_3 \hat{w}_{3t} + u \tag{6}$$

$$\Delta y_t = \mu + \beta t + \delta y_{t-1} + \sum_{i=1}^{k} \alpha_i \Delta y_{t-i} + \gamma_2 \hat{w}_{2t} + \gamma_3 \hat{w}_{3t} + u_t \tag{7}$$

The statistics for testing the existing of unit root is in Eq8.

$$\tau_{RALS-ADF} = \rho \tau_{ADF} + \sqrt{1-\rho^2} Z \tag{8}$$

LLI (2015) have suggested a residual augmented cointegration technique to test the long-term relationship between the variables. They have used 4 models to test, the cointegration: error correction model, auto distributed lag, Engle-Granger and augmented Engle-Granger models.

$$\text{ECM}: \Delta y_t = d_t + \delta \hat{z}_{t-1} + \phi \Delta x_t + \sum_{i=1}^{k} \alpha_i \Delta y_{t-i} + \varepsilon_t \tag{9}$$

$$\text{ADL}: \Delta y_t = d_t + \delta y_{t-1} + \phi \Delta x_t + \sum_{i=1}^{k} \alpha_i \Delta y_{t-i} + \varepsilon_t \tag{10}$$

$$\text{EG}: \Delta \hat{u}_t = d_t + \delta \hat{u}_{t-1} + \sum_{i=1}^{k} \alpha_i \Delta \hat{u}_{t-i} + \varepsilon_t \tag{11}$$

$$\text{EG2}: \Delta \hat{u}_t = d_t + \delta \hat{u}_{t-1} + \phi \Delta x_t + \sum_{i=1}^{k} \alpha_i \Delta \hat{u}_{t-i} + \varepsilon_t \tag{12}$$

After using a similar method with RALS-ADF test by including second and third moments of the residuals to the equations. They have established RALS-ADL and RALS-EG2 cointegration tests as it is indicated in Eq13 and Eq14, respectively.

$$\Delta y_t = d_t + \delta y_{t-1} + \phi \Delta x_t + \sum_{i=1}^{k} \alpha_i \Delta y_{t-i} + \gamma_2 \hat{w}_{2t} + \gamma_3 \hat{w}_{3t} + u_t \tag{13}$$

$$\Delta \hat{u}_t = d_t + \delta \hat{u}_{t-1} + \phi \Delta x_t + \sum_{i=1}^{k} \alpha_i \Delta \hat{u}_{t-1} + \gamma_2 \hat{w}_{2t} + \gamma_3 \hat{w}_{3t} + u_t \tag{14}$$

The statistics for testing the existing of unit root are in Eq15 and Eq16.

$$\tau_{RALS-ADL} = \rho \tau_{ADL} + \sqrt{1-\rho^2} Z \tag{15}$$

$$\tau_{RALS-EG2} = \rho \tau_{EG2} + \sqrt{1-\rho^2} Z \tag{16}$$

4 Empirical Findings

Table 1 shows that first two of the variables have no unit root and $\tau_{RALS-ADF}$ values of those variables are statistically significant with respect to critical values which can be seen in ILT's (2014) paper. The last variable on the table on the other hand has unit root for both drift and trended model. One can see that the more ρ^2 is closer to *1* the probability of being stationary of a variable is higher.

It is due to the power of the unit root test which is equal to *1* whereas the test is considered to be traditional ADF test.

Table 1: Im, Lee and Tieslau RALS-ADF

	Intercept		Intercept and Trend	
	ρ^2	$\tau_{RALS-ADF}$	ρ^2	$\tau_{RALS-ADF}$
lncli	0.5	2.611**	0.2	-27.718*
lnhcpi	0.5	6.809*	0.5	4.522*
lnunem	0.8	-1.372	0.8	-1.200

* and ** represents 1 % and 5 % significancy levels respectively.

When analysing the Jarque-Bera (1987) test statistics in Table 2, none of the variables indicates a normal distribution. Nevertheless, the asymptotic distribution of the variables does not change. Probabilities of all the variables are equal to zero for both constant and intercept and trend.

Table 2: Jarque-Bera Test Statistics

	JB		p	
	Intercept	**Intercept and Trend**	**Intercept**	**Intercept and Trend**
lncli	14.371	17.110	0.001	0.000
lnhcpi	6792.074	6226.800	0.000	0.000
lnunem	469.518	456.699	0.000	0.000

As it can be seen at Table 3, there have been three models in which the variables are dependent variables respectively. For both lncli and lnhcpi the null hypothesizes suggesting that there is no cointegration between the variables have been rejected. It indicates that there is no long-term relationship between the variables when lncli and lnhcpi are the dependent variables. However, while the unemployment rate presented by lnunem is dependent variable, it can be stated that there is a long-term relationship between them since the null hypothesis cannot be rejected. Critical values of RALS cointegration for $n=3$ can be observed in LLI (2015). For rejecting the null hypothesis one can simply compare whether the absolute value of the calculated value is higher than the critical value or not.

Table 3: Lee, Lee and Im (2015) Cointegration Tests

	Intercept				Intercept and Trend			
	ρ^2_{ADL}	ρ^2_{EG2}	$\tau_{RALS-ADL}$	$\tau_{RALS-EG2}$	ρ^2_{ADL}	ρ^2_{EG2}	$\tau_{RALS-ADL}$	$\tau_{RALS-EG2}$
lncli	0.6	0.4	3.689**	3.407**	0.8	1	-4.123**	-4.522**
lnhcpi	0.4	0.4	6.144*	2.999	0.4	0.3	6.244*	4.640*
lnunem	0.8	0.8	-3.129	-3.164	0.8	0.8	-2.587	-2.556

* and ** represents 1 % and 5 % significancy levels respectively.

5 Conclusion

After worldwide pandemic, the importance of health expenditures and unemployment related to decreasing aggregate supply has come to the forefront. As the other developing countries, Türkiye's inflation become unsustainable although there were other endogenous causes of it. Besides, the healthcare expenditures have been supposed to increase after pandemic. To analyse the linkage between economic growth, unemployment and health expenditures inflation, there have been applied the RALS ADF and RALS cointegration tests in this study. As a result of unit root tests and normality tests, the data structure has indicated that the distribution of the series is not normal. Looking at the conclusion of cointegration tests, there is a long-term linkage between unemployment rate when composite leading indicator based on GDP and healthcare consumer price index have been the explanatory variables. On the other hand, putting healthcare consumer price index or composite leading indicator based on GDP on the shoes of dependent variable there is no such relationship. Policy makers should take into account that economic policies targeting employment rate would be affected by healthcare inflation and GDP in long run economic activity.

References

Afzal, M., Arshed, M. G., & Sarwar, K. (2013). Education, health, food inflation and economic growth in Pakistan. *Pakistan Economic and Social Review*, 109–138.

Barro, R. J. (2013). Health and economic growth. *Annals of Economics and Finance*, *14*(2), 329–366.

Brenner, M. H., & Mooney, A. (1983). Unemployment and health in the context of economic change. *Social Science & Medicine*, *17*(16), 1125–1138.

Hepsağ, A. (2022). Ekonometrik zaman serileri analizlerinde güncel yöntemler (WinRats Uygulamalı).

Im, K. S., Lee, J., & Tieslau, M. A. (2014). *More powerful unit root tests with non-normal errors* (pp. 315–342). Springer New York.

Jarque, C. M., & Bera, A. K. (1987). A test for normality of observations and regression residuals. *International Statistical Review/Revue Internationale de Statistique*, 163–172.

Lee, H., Lee, J., & Im, K. (2015). More powerful cointegration tests with non-normal errors. *Studies in Nonlinear Dynamics & Econometrics*, *19*(4), 397–413.

Maruthappu, M., Shalhoub, J., Tariq, Z., Williams, C., Atun, R., Davies, A. H., & Zeltner, T. (2015). Unemployment, government healthcare spending, and cerebrovascular mortality, worldwide 1981–2009: An ecological study. *International Journal of Stroke*, 10(3), 364–371.

Okun, A. M. (1963). *Potential GNP: Its measurement and significance.* Cowles Foundation for Research in Economics at Yale University.

Phillips, A. W. (1958). The relation between unemployment and the rate of change of money wage rates in the United Kingdom, 1861–1957. economica, *25*(100), 283–299.

Sterret, D., Bender, A., & Palmer, D. (2014). A Business Case for Universal Healthcare: Improving Economic Growth and Reducing Unemployment by Providing Access for All. *Health Law & Pol'y Brief*, *8*, 41.

Gülferah Ertürkmen[1] & Ayşe Eryer[2]

The Relationship Between Sectoral Employment and Economic Growth: Panel Data Analysis for E7 Countries

1 Introduction

Economic growth and employment are the most important issues for countries around the world and these issues are closely related to macroeconomics. The macroeconomic roles of countries reflect the importance of economic growth in the process of globalization. Economic growth determines not only the economic order, but also living standards, socio-cultural situation and its effects on the world (Akcan, 2018: 172). The factors of production used effectively and efficiently increase the total volume of production and the country's income, leading to economic growth (Küçükaksoy, 2020: 144).

While economic growth is an effort to maintain a certain level for developed countries, economic development is at the forefront for developing countries. Development, which gained importance especially after World War II, is a multifaceted process that includes social, cultural and political situations (Şaşmaz, 2018: 250). In countries where development is inadequate, living standards are low and this situation brings about various solutions such as change of government, new economic policies and technological investments. Unemployment is the biggest obstacle to achieving the growth targets of developed countries and the development targets of developing countries.

During the Great Depression of 1929, the inadequacy of classical economic views revealed that full employment in the economy was not possible under all circumstances. With the introduction of Keynesian policies in 1936, the importance of unemployment and wage flexibility was understood (Uslu, 2020: 103). Arthur Okun was the first person to make suggestions for the solution of the unemployment problem and to econometrically analyze the effect of unemployment on economic growth. In his 1962 study, Okun showed how

1 Ph.D., Asst. Prof., Kahramanmaraş Sütçü İmam University, Göksun School of Applied Sciences, Department of Finance and Banking, gbozkaya@ksu.edu.tr, ORCID: 0000-0003-2239-0241.

2 Ph.D., ayse_zabun46@hotmail.com, ORCID: 0000-0002-6556-1605.

important unemployment rates are for a country. Okun examined the importance of changes in the labor force participation rate, the pattern of working hours and productivity. As a result of his analysis, he determined that there was a negative relationship between unemployment and economic growth in the United States (Akcan, 2018: 173). In this context, the fact that unemployment does not decrease despite the achievement of growth targets in the economy reveals the fact that employment has not increased sufficiently and thus a growth that does not create employment (Küçükaksoy, 2020: 145).

In this study, growth and employment in E7 countries are analyzed by sectors with data from 1991 to 2022. The concepts of economic growth and employment are analyzed in terms of services, industry and agriculture sectors. After the literature review on the subject; the empirical analysis of the study is completed with the conclusion and recommendations section.

2 Previous Studies on the Subject

Barışık, Çevik and Çevik (2010) analyzed Turkey's economic growth and unemployment data for the period 1988–2008 using Markov regime chain modeling. The study revealed an asymmetric relationship between growth and unemployment rates and showed that Okun's Law was not valid in Turkey during this period.

In Şahbaz's (2014) study, data for the period 1991–2011 are analyzed and the effect of fixed capital on economic growth in Turkey and 27 European Union countries is evaluated using panel data analysis. In the study, it is determined that there is no causality relationship between economic growth and fixed capital investments in the short run, while there is a unidirectional causality relationship between fixed capital investments and employment towards economic growth in the long run.

Uçan and Çebe (2018) analyzed the relationship between economic growth, unemployment and inflation for Turkey. ARDL bounds test and Granger causality test were used in the study. According to the findings, while economic growth increases unemployment, unemployment does not have a significant effect on economic growth. These results support the interpretation that growth that does not create employment is invalid for the period between 2000 and 2016.

In his study, Pata (2018) analyzed the relationship between women's employment and economic growth for Turkey in the 1988–2015 period using symmetric and asymmetric causality tests. According to the results of Toda-Yamamoto (1995) causality test, a positive, unidirectional and statistically significant causality relationship was found from female employment to gross domestic product generated in total and services sectors. The Hatemi-J (2012) asymmetric causality test, on the other hand, indicates a unidirectional causality from negative shocks to female employment to negative shocks to total and services sector GDP. According to the results of both symmetric and asymmetric tests, there is no causality relationship between female employment and the industrial and agricultural sectors. As a result, it is revealed that women's employment supports economic growth by contributing to the gross domestic product generated in the service sector, but has no effect on the agricultural and industrial sectors.

In Ayhan's (2019) study, data for the period between January 2005 and November 2018 were analyzed using the ARDL model. The study analyzed the macroeconomic determinants of unemployment. According to the results of the analysis, CPI and interest rates have a positive effect on unemployment in Turkey, while economic growth and the number of unemployed have a negative effect.

Altun and İşleyen (2019) aimed to examine the relationship between employment in the industrial sector and economic growth in Turkey with econometric methods and to determine the direction of this relationship. The research covers data between 1991 and 2017. ARDL (Auto Regressive Distributed Lag Models) cointegration method and Granger Causality Test were used in econometric analyses. ARDL cointegration analysis revealed that there is a long-term relationship between employment in the industrial sector and economic growth. Granger Causality Test results show that employment in the industrial sector causes economic growth unidirectionally.

Azazi (2020) analyzed the period between the first quarter of 2005 and the third quarter of 2020. The study analyzed the employment rates and economic growth rates on the basis of basic sectors using Granger test. The aim of the study is to evaluate the employment creation capacity of economic growth. The results of the analysis reveal that economic growth in Turkey does not create employment in general, i.e., non-employment generating growth is valid. This finding indicates that the labor market in the services, industry and agriculture sectors has not expanded in line with economic growth. The research emphasizes that economic growth in this period has been insufficient to increase employment.

Telli Üçler (2022) analyzed economic growth and employment data in Turkey between 1992 and 2020. The relationship between the employment rates of service, industry and agriculture sectors and economic growth rates is investigated. According to the results of the Augmented Dickey Fuller Unit Root Test, employment variables were found to be stationary at the first order, while the economic growth variable was found to be stationary at the level value. According to the Granger Causality Test results, economic growth affects employment in the industrial and service sectors, but there is no causality relationship with the agricultural sector. These findings indicate that growth without employment in the agricultural sector is valid.

Yıldırım and Engeloğlu (2023) examine the effects of post-1980 economic policies on labor force by sector in Turkey. Hidden cointegration and asymmetric causality tests with data for the period 2000Q1-2022Q3 assessed the relationship between employment by sector and positive and negative shocks of economic growth. The results show that there is cointegration between increases in agricultural employment and economic growth, cointegration between decreases in construction employment and economic growth, and causality between increases in service sector employment and decreases in economic growth.

Artekin (2023) examined the relationship between the construction sector, employment and economic growth in Turkey with quarterly data between 2005 and 2023. The stationarity of the series was determined using Zivot-Andrews (ZA) and Phillips-Perron (PP) unit root tests, and it was demonstrated that employment affects economic growth using the Generalized Method of Moments (GMM). According to the Granger causality test results, bidirectional causality relationships were determined between the construction sector and economic growth, employment and economic growth, and the construction sector and employment.

3 Assessment of Empirical Findings

3.1 Data Set and Model

In this study, the relationship between sectoral employment and economic growth for E-7 countries for the period 1991–2022 is analyzed by panel data analysis method. The characteristics of the variables used in the model are summarized in Table 1.

Table 1: Variables and Explanations

Variables	Explanations	Source-Period
LGSYH	GDP per capita (US$)	WDI- 1991–2022
LH	Services Sector Employment (Share in Total Employment)	WDI- 1991–2022
LS	Industrial Sector Employment (Share in Total Employment)	WDI- 1991–2022
LT	Agriculture Employment (Share in Total Employment)	WDI- 1991–2022

The model to be used in the study;

$$LGSYH_{it} = \alpha_0 + \alpha_1 LH_{it} + \alpha_2 LS_{it} + \alpha_3 LT_{it} + \varepsilon_{it} \quad (1)$$

is in the form. In the model, LGSDP stands for per capita income level, LH stands for services sector employment, LS stands for industrial sector employment and LT stands for agricultural sector employment. Logarithmic transformations of variables are used in the model. In the model, i denotes horizontal cross-sectional units and t denotes time-sectional dimension.

3.2 Methodology and Findings

3.2.1 Methodology

Econometric analyses are evaluated in three categories based on the status of the data set. If there is more than one time interval for a unit, time series analyses are performed. If there is one time and more than one unit, horizontal cross-section econometric analyses are preferred. But when both time and unit are more than one, panel data analysis methods are chosen. Recently, the literature on panel data analysis has been expanding. This is because it provides more information than time series and horizontal cross-section methods, as well as easy access to data for units (İnci, 2014: 189).

A basic model in panel data analysis is shown as follows:

$$Y_{it} = a_i + \beta X_{it} + \varepsilon_{it} \quad (2)$$

. Burada Y, bağımlı değişkeni, X_{it},

can be written in the following form. Where Y is the dependent variable and X_{it} is the independent variable. a_i, is the fixed parameter, β is the slope parameter and ε_{it} is the error term. i=(1,2,....N) is the horizontal cross-section and t= (1,2,....N) is the time interval in this equation (Wooldrige, 2002: 10).

The models used in panel data analysis are classified as pooled least squares, fixed effects, and random effects (Nargeleçekenler, 2011: 168). Fixed effects model takes into account the individual characteristics of each horizontal section. In this model, it is assumed that the fixed coefficients for each unit are different and the corresponding slope coefficients are the same. While the expressed fixed effect differs for each cross-section, the constant of each cross-section does not change over time The functional form of the fixed effects model is as follows (Akman, 2012: 111):

$$Y_{it} = \beta_{1i} + \beta_2 X_{2it} + \beta_3 X_{3it} + \ldots \beta_k X_{it} + \varepsilon_{it} \cdot \beta_{ij} \neq \beta_{ij} \quad (3)$$

The random effects model states that the changes that occur according to unit and time effects are added as a component of the error term in this model. The advantage of the random effects model is that the degrees of freedom are not lost in the model (Özer and Çiftçi, 2009: 42–43). In this model, since the unit effect is not fixed, it is not included in the fixed parameter, but it is included in the margin of error since it is considered as random:

$$v_{it} = \mu_{it} + \mu_i$$

Here, errors represent unit errors, that is, unit differences and the change between units over time. The random effects model is characterized as an error components model and the functional representation of the model is shown as follows (Yerdelen Tatoğlu, 2020: 104):

$$Y_{it} = \beta_0 + \sum_{k-1}^{K} \beta_k X_{kit} + \mu_i + \mu_{it} \quad (4)$$

The classical model (Pooled Least Squares Method (LSM)) is stated that this method is used as a result of the fact that all coefficients between time and horizontal sections are constant (Aytun, 2012: 110).

When determining the panel data regression model, there are tests used to decide between the pooled ECM model, fixed effects model and random effects model estimators. These are F test, Breusch Pagan LM test, LR test and Hausman Test. These tests are briefly explained as follows:

- F Test: It is known as a test that compares the validity of the fixed effects model against the classical model. The null hypothesis of the F test is "unit effects are equal to zero" (Belke and Özturgut, 2020: 19; Yerdelen Tatoğlu, 2020: 213).
- The Breush-Pagan LM test is a test that compares the random effects model against the classical model. The null hypothesis of this test is "the variance of the unit effect is equal to zero" (Breush-Pagan, 1980: 239).
- LR Test: In panel data analyses, one of the tests used to determine whether the classical model is valid or not, in other words, whether there are unit and/or time effects or not, is the "Likelihood Ratio Test" (LR). In this test, the null hypothesis is set as "the classical model is true". If the null hypothesis is rejected, it is decided that the classical model is not appropriate (Şahin, 2016: 67).

- Hausman Test: The Hausman (1978) test allows for a choice between fixed effects and random effects models. The main hypothesis of the Hausman test is that there is no correlation between the error term and the independent variables, which means that the random effects method is appropriate (Yerdelen Tatoğlu, 2020: 196).

In panel data analyses, after determining the estimator models that should be preferred, it is necessary to test whether the estimator model in question meets the assumptions. In panel data analyses, the basic assumptions are changing variance, inter-unit correlation, and autocorrelation problems. The heteroskedasticity assumption can be examined by using the Levene (1960), Brown and Forsythe (1974) Test for the random effects model and the Modified Wald Test for the fixed effects model. The autocorrelation assumption is examined with the Durbin Watson (DW-d) Test of Bhargava, Franzini and Narendranathan (1982) and the Locally Best Invariant Test (LBI) Test developed by Baltagi Wu (1999). The presence of inter-unit correlation problem can be examined through Breusch Pagan Lagrange Multiplier Test (1980), Pesaran CD (2004) Test, Friedman (1937) and Frees (2004) test (Karagöz and Şener, 2023: 320).

In panel data analyses, the presence of at least one of these three assumptions causes deviations from the assumptions and distortions in the results. Therefore, it is recommended to use robust estimators to correct these deviations. In this process, the estimator to be selected is decided according to the results of panel data analysis assumptions. If there is only heteroskedasticity problem in the results, Huber, Eicker and White estimator is used; if there is autocorrelation and heteroskedasticity problem, Arellano, Froot and Rogers estimator, Wooldridge estimator and Newey-West estimator are used. If the model has both heteroskedasticity, inter-unit correlation and autocorrelation problems, the model should be re-estimated using the Driscoll-Kraay estimator, which is a robust estimator against deviations from these three assumptions (Yerdelen-Tatoğlu, 2020: 304–340). Therefore, Driscoll-Kraay estimator is used in this study.

3.2.2 Findings

In this section of the study, the model choice, testing deviations from the assumption and the results of the robust estimator that is robust to the assumptions in the panel data analysis are shown. In panel data analyses, F, LR and Hausman tests are used to test whether there is a rirm and/or time effect between variables in the model to be preferred, in other words, which of the classical model, random model and fixed effects model will be used. The results for the model created within the scope of the study are shown in Table 2.

Table 2: *F, LM ve Hausman Testi Sonuçları*

Tests	Statistic Value	Probability Value
F Test	57.13	0.0000*
LR Test	159.18	0.0000*
Hausman Test	46.95	0.0000*

Note: * 1 % significance level

According to the test results obtained in Table 2, it is understood that the classical model is not valid according to both F test and LR test results. Because the probability value of the F test is less than 1 % significance level. In other words, the classical model is not appropriate. Likewise, when we look at the LR test in the table, the probability value is less than 1 % significance level. The null hypothesis is rejected. According to this test result, it is determined that the most appropriate estimator to test the established model is random effects. According to the results of the F and LR tests in Table 2, the model in the study should not be tested with the classical model (Pooled ECT). Hausman test was used to decide between random effects and fixed effects estimator. According to the Hausman test in Table 2, the null hypothesis is rejected. It is determined that the most appropriate estimator to test the model is Fixed Effects.

As discussed in the methodology section, deviations from the assumptions of panel data studies such as heteroskedasticity, inter-unit correlation and auto-correlation problem prevent the parameters from being estimated effectively (Yerdelen Tatoğlu, 2020: 236). Therefore, deviations from the assumption should be tested at this stage of the study. Here, since the model is tested with the fixed effects estimator, appropriate deviation from assumption tests are used. The results are shown in Table 3.

Table 3: Heteroskedasticity, Autocorrelation and Inter-unit Correlation Test Results

Tests	Statistic Value	Probability Value
Modified Wald Testi	472,65	0,0000
Bhargava vd. Durbin Watson Testi	0,2750<2	
Baltagi –Wu Testi	0.3576<2	
Friedman Testi	78,589	0,0000
Pesaran Testi	6,296	0,0000

According to Table 3, heteroskedasticity, inter-unit correlation and autocorrelation problems are analyzed for the fixed effects model. The presence of heteroskedasticity problem is determined by Wald test and the probability value is less than 1 % significance level. The null hypothesis of "constant variance across units" is rejected. In other words, it is determined that there is a heteroskedasticity problem in the model. For the autocorrelation problem in the model, Bhargava, et al. and Durbin Watson Test and Baltagi-Wu's (1999) LBI test test test the null hypothesis that the autocorrelation coefficient is equal to zero. Autocorrelation test results greater than 2 indicate that there is no autocorrelation problem in the models considered, while test results less than 2 indicate that there is an autocorrelation problem (Şahbaz and Savaşan, 2020: 229). As can be seen in Table 3, both test results are less than 2. The null hypothesis of "no autocorrelation" is rejected and it is found that there is an autocorrelation problem in the model. For the inter-unit correlation problem, Friedman and Pesaran CD test was utilized. In Table 3, the probability values of both tests are less than 1 % significance level. The null hypothesis "there is no correlation between units" is rejected and it is determined that there is correlation between units in the model.

According to the basic assumption test results of panel data analysis, heteroskedasticity, autocorrelation and inter-unit correlation are detected for the model in question. In order to eliminate these problems and obtain more efficient results, tests should be conducted with robust estimators. Therefore, the model was estimated with the Driscoll-Kraay (1998) robust estimator that takes into account the heteroskedasticity, inter-unit correlation and autocorrelation problems. The results of the Driscoll-Kraay robust estimator are shown in Table 4.

Table 4: *Driscoll Kraay Test Results*

Dependent Variable:LGSDP			
	Coefficient	**Driscoll Kraay Standard Error**	**Probability Value**
LH	1.3654	0.5979	0.0300**
LS	1.3596	0.3199	0.0000*
LT	-1.4978	0.3334	0.0000*
Fixed	3.4176	2.5646	0.194***
R^2:0.7204 F_{ist}:0.000			

Note: *, ** and *** indicate 1 %, 5 % and 10 % significance levels.

According to the results of Driscoll-Kraay Standard Errors (1998) Test, the variables are found to be significant in the model obtained by using the data for the period 1995–2022 for E-7 countries. As a result of the empirical tests, the effect of services sector employment and industrial sector employment on economic growth is positive. The effect of agricultural employment, which is another independent variable, on economic growth is negative. In addition, according to the F statistic probability value showing the results of the model, the model was found significant at 1 % level.

4 Conclusion

Countries around the world attach great importance to economic policies to realize sustainable economic growth. Ensuring full employment, keeping inflation under control and monitoring the functioning of key sectors such as services, industry and agriculture make it easier for countries to achieve their growth targets. It is believed that an increase in production will bring along economic growth. In an economy where factors of production work effectively and efficiently, economic growth is expected to be realized and employment problems are expected to be avoided. However, if employment does not increase and unemployment rates do not decrease despite economic growth, this indicates the existence of a growth that does not create employment. While some studies in the literature reveal the existence of jobless growth, other studies show that economic growth increases employment.

This study analyzes the relationship between sectoral employment and economic growth in E7 countries (China, India, Brazil, Mexico, Russia, Indonesia, Russia, Indonesia and Turkey) using panel data analysis. The findings provide important information in terms of understanding the effects of sectoral employment on economic growth and making comparative analyses across countries.

The sample of the study consists of E-7 countries. In the study, the effect of employment in the service sector, industry and agriculture sector on economic growth was tested using the panel data analysis method within the scope of the 1991–2022 period data set. The dependent variable is GDP per capita, and the independent variable is service, industry and agriculture sector employment rates (share in total employment). In the study, F, LR and Hausman tests were used to determine the appropriate estimator method in the panel data analysis used to determine the effects of independent variables on the dependent variable. According to these test results, it was determined that the most appropriate estimator was the fixed effects model. After this test, it was determined whether there were deviations from the assumption in the model and it was determined that there were heteroscedasticity, inter-unit correlation and autocorrelation problems in the model. To eliminate these problems and obtain more effective results, the model was re-estimated with the Driscoll-Kraay estimator, which is a robust estimator. According to the Driscoll-Kraay estimator results, the established model is statistically significant. When the effect of individual variables on economic growth is evaluated, the effect of service and industrial sector employment level on economic growth is found to be positive and significant. These analysis results are supported by the study results of Aghion and Howitt (1998), Girma and Görg (2004), Branstetter et al. (2006), Yasar and Morrison Paul (2007). In addition, the effect of the agricultural sector employment level on economic growth is statistically significant and negative. It was determined as . These results obtained from the study are also supported by studies in the literature such as Jayaraman and Yoong (2008), Kumar and Sharma (2011), Sharma and Rumay (2015).

As a result, it is of great importance to develop policies that take into account the effects of sectoral employment on economic growth in E7 countries. Considering the contributions of the service and industrial sectors to economic growth, strategies should be created to increase employment and support sustainable growth in these sectors. In addition, considering the employment creation potential of the agriculture and construction sectors, policies that will increase the efficiency of these sectors should be adopted. In this way, E7 countries can achieve a more stable economic structure by both increasing employment rates and making their economic growth sustainable.

References

Aghion, P., & Howitt, P. (1998). *Endogenous Growth Theory*. Cambridge, MA: MIT Press.

Akcan, F. (2018). "Ekonomik Büyüme ve Sosyo-Kültürel Durumlar". *Ekonomik ve Sosyal Araştırmalar Dergisi*, 172–173.

Akcan, M. B., & Azazi, H. (2022). "İstihdam Oluşturmayan Büyümenin Sektörel Analizi: Türkiye Örneği". *Journal of Management and Economics Research*, *20*(1), 227–246. https://doi.org/10.11611/yead.1037008

Akman, E. (2012). "Sermaye Yapısını Belirleyen İşletmeye Özgü Faktörler: İMKB'de İşlem Gören Sanayi Firmaları Üzerine Bir Panel Veri Uygulaması". Doktora Tezi, Bülent Ecevit Üniversitesi, Sosyal Bilimler Üniversitesi, Zonguldak.

Altun, Y., & İşleyen, Ş. (2019). "Türkiye'de İktisadi Büyüme ile Sanayi Sektöründe İstihdam Arasındaki İlişkinin ARDL Sınır Testi Yaklaşımı ile Ekonometrik Analizi", 1991–2017. *Iğdır Üniversitesi Sosyal Bilimler Dergisi*, 17, 657–676.

Artekin, A. Ö. (2023). Türkiye'de İnşaat Sektörü, İstihdam ve Ekonomik Büyüme Arasındaki İlişkinin İncelenmesi: GMM Yaklaşımı. *Selçuk Üniversitesi Sosyal Bilimler Meslek Yüksekokulu Dergisi, 26*(2), 491–503. https://doi.org/10.29249/selcuksbmyd.1382436

Ayhan, F. (2019). "Türkiye Ekonomisinde İşsizliğin Belirleyicisi Olan Temel Makroekonomik Değişkenlerin Tespitine İlişkin Bir Uygulama". UİİİD-IJEAS, 2019 (25): 235–252 ISSN 13079832

Aytun, C. (2012). *Enformasyon ve telekomünikasyon teknolojilerinin ekonomik büyümeye etkisi: Gelişmiş ve gelişmekte olan ülkeler üzerine bir uygulama* (Yüksek lisans tezi). Çukurova Üniversitesi Sosyal Bilimler Enstitüsü, Adana.

Baltagi, B. H., ve Wu P, X. (1999). "Unequally Spaced Panel Data Reggession with AR(1) Disturbances". *Econometric Theory*, *15*, 814–823.

Barışık, S., Çevik, E. İ., & Çevik, N. K. (2010). "Türkiye'de Okun Yasası, Asimetri İlişkisi Ve İstihdam Yaratmayan Büyüme: Markov-Switching Yaklaşımı". *Maliye Dergisi*, *159*(2), 88–102.

Belke, M., ve Özturgut, R. (2020). "Doğrudan Yabancı Yatırımların Ülkeye Özgü Belirleyicileri: Yükselen Piyasa Ekonomileri İçin Panel Veri Analizi". *Akademik Araştırmalar ve Çalışmalar Dergisi*, *12*(22), 12–26.

Bhargava, A., Franzini, L., & Narendranathan, W. (1982). Serial correlation and the fixed effects model. *The Review of Economic Studies, 49*(4), 533–549. https://doi.org/10.2307/2297285.

Branstetter, L., Fisman, R., & Foley, C. F. (2006). "Do Stronger Intellectual Property Rights Increase International Technology Transfer? Empirical Evidence from U.S. Firm-Level Data". *Quarterly Journal of Economics*, *121*(1), 321–349.

Bresuch, T. S., & Pagan, A. R. (1980). "The Lagrange Multipler Test and Its Applications to Model Specification Test in Economterics". *Review of Economic Studies*, *47*, 239–253.

Brown, M. B., & Forsythe, A. B. (1974). "The Small Sample Behavior of Some Statistics Which Test the Equality of Several Means". *Technometrics, 16*, 129–132.

Driscoll, J. C., & Kraay, A. C. (1998). "Consistent Covariance Matrix Estimation with Spatially Dependant Panel Data". *Review of Economics and Statistic*, *80*(4), 549–560.

Frees, E. (2004) Longitudinal and Panel Data: Analysis and Applications in the Social Sciences. Cambridge University Press, New York. https://doi.org/10.1017/CBO9780511790928.

Friedman, M. (1937). "The Use of Ranks to Avoid the Assumption of Normality Implicit in the Analysis of Variance". *Jorunal of the Amerikan Statistical Association*, *32*, 675–701.

Girma, S., & Görg, H. (2004). Foreign Direct Investment, Spillovers and Absorptive Capacity: Evidence from Quantile Regressions. Discussion Paper No. 4794, Centre for Economic Policy Research.

Hausman, J. A. (1978). "Specification Test in Econometrics". *Econometrica, 46*(6), 1251–1271.

İnci, C. (2014). "Finansal Yönetim Kararlarının Firmanın Karlılığı ve Piyasa Değeri Üzerindeki Etkileri: BIST'deki Sanayi Şirketleri Üzerine Bir Panel Veri Uygulaması", Bülent Ecevit Üniversitesi Sosyal Bilimler Enstitüsü, Zonguldak

Jayaraman, T. K., & Yoong, J. (2008). "Is Agriculture İmportant For Economic Development? Empirical Evidence On The Relationship Between Agriculture And Economic Growth". *Australian Economic Papers, 47*(2), 119–138.

Karagöz, H. ve Şenel, S.(2023). "Sanayi Sektörü Önemini Yitirdi Mi? İhracat Verekabet Gücü Ekseninde Bir İnceleme". *İstanbul İktisat Dergisi, 73*, 307–331.

Kumar, P., & Sharma, S. (2011). "Agriculture And Economic Growth İn India: An Empirical Contribution". Economic Change and *Restructuring, 44*(1), 69–85.

Küçükaksoy, A. (2020). "Üretim Faktörlerinin Ekonomik Büyümeye Etkisi". *Makroekonomi Çalışmaları*, 144 s.

Levene, H. (1960). Robust Tests for Equality of Variances. In I. Olkin (Ed.), *Contributions to Probability and Statistics*, Stanford University Press, Palo Alto, 278–292.

Pata, U. K. (2018). "Türkiye'de Kadın İstihdamı Ve Ekonomik Büyüme İlişkisi: Simetrik Ve Asimetrik Nedensellik Testleri İle Sektörel Bir Analiz". *Uluslararası İktisadi Ve İdari İncelemeler Dergisi*, *21*, 135–150. https://doi.org/10.18092/ulikidince.391777

Nargeleçekenler, M. (2011), Hisse Senedi Fiyatları ve Fiyat/Kazanç Oranı İlişkisi: Panel Verilerle Sektörel Bir Analiz, Business and Economics Research Journal,2(2), 165–184.

Özer, M. ve Çiftçi, N. (2009). Ar-Ge Harcamaları ve İhracat İlişkisi: OECD Ülkeleri Panel Veri Analizi. Dumlupınar Üniversitesi Sosyal Bilimler Dergisi, (23), 39–49

Pesaran, M. H. (2004). "General Diagnostic Tests for Cross Section Dependence in Panels". Cambridge Working Papers in Economics, No: 435

Şahin, D. (2016). Farklı gelir gruplarındaki ülkelerde finansal gelişme ve ekonomik büyüme ilişkisinin analizi. *İktisadi ve İdari Bilimler Dergisi*, 4(4), 61–74. https://doi.org/10.18825/iremjournal.282999

Sharma, A., & Rumay, E. M. (2015). "The İmpact of Agricultural Employment on Economic Growth: A Case Study of Developing Countries". *Journal of Economic Studies, 42*(1), 58–75.

Şahbaz, A. (2014). "Sabit Sermaye Yatırımları ve Ekonomik Büyüme İlişkisi: Panel Nedensellik Analizi". *Niğde Üniversitesi İktisadi ve İdari Bilimler Fakültesi Dergisi, 7*(1), 1–12. Retrieved from https://dergipark.org.tr/tr/pub/niguiibfd/issue/19754/211452

Şahpaz, K. İ., & Savaşan, F. (2020). "Gelir Eşitsizliği ile Mücadelede Mikrokredi Uygulamalarının Etkisi, Panel Veri Analizi". *Yönetmi ve Ekonomi Araştırmaları Dergisi, 18*(2), ss. 218–235.

Şaşmaz, H. (2018). "II. Dünya Savaşı Sonrası Ekonomik Kalkınma". Kalkınma ve Büyüme Araştırmaları, s.250.

Tatoğlu, Y. F. (2020). Panel Veri Ekonometrisi, Beta Basım Yayım, 464s.

Telli Üçler, Y. (2022). "Türkiye'de Sektörler İtibari İle İstihdam Büyüme İlişkisi". *Pearson Journal, 7*(21), 148–160. https://doi.org/10.46872/pj.590

Uçan, O., ve Çebe, G. N. (2018), "2008 Krizi Öncesi ve Sonrası Türkiye'de Ekonomik Büyüme, İşsizlik ve Enflasyon İlişkisi". *Ömer Halisdemir Üniversitesi İktisadi ve İdari Bilimler Fakültesi Dergisi, 11*(3), 6–17.

Uslu, B. (2020). "Keynesyen Politikalar ve Tam İstihdam". *Ekonomi Teorileri Dergisi*, s.103.

Wooldrige, M. J. (2002). *Econometric Analysis of Cross section and Panel Data.* Cambridge Massachusetts, London England, 741s.

Yasar, M., & Morrison Paul, C. J. (2007). "Firm-Level Productivity and Technical Efficiency in MENA Manufacturing: A Comparative Analysis with OECD Countries". *Review of Middle East Economics and Finance, 3*(3), 23–47.

Yıldırım, U., & Engeloğlu, Ö. (2023). "Türkiye'de Sektörel İstihdam ve Ekonomik Büyüme İlişkisi: Saklı Eşbütünleşme ve Asimetrik Nedensellik Testi". *Fiscaoeconomia, 7*(2), 1445–1473. Doi: 10.25295/fsecon.1255703

Zekiye Örtlek[1] & Munise Ilıkkan Özgür[2]

Tourism, Foreign Direct Investment, Exports and the Impact of Trade Globalization on Economic Growth: Turkey and OECD Countries

1 Introduction

In a globalizing world, countries need a high level of external resources to make investments. To obtain these resources, countries can use various instruments such as increasing tourism revenues, encouraging exports, promoting foreign direct investments (FDI) and foreign borrowing. The tourism sector plays an important role among these instruments (Tang & Tan, 2015: 160). Therefore, since tourism is a labor-intensive sector, it provides a large number of job opportunities and reduces unemployment. In addition, visitor spending has a direct impact on the income of local people. The flow of tourists is an alternative form of export, which positively affects the country's balance of payments and increases foreign exchange reserves. The fact that the success of the tourism sector is followed by the whole world creates an incentive for other investments in the sector by bringing about efficient and rapid utilization of investments (Balaguer & Cantavella-Jorda, 2002: 878; Proença & Soukiasiz, 2008: 791). This situation significantly contributes to the economic development of the nation. FDI is defined as the capital created by individuals or institutions that invest capital from outside the borders of a country for the purpose of doing business in that country. In this context, since FDI is the main source of technology and capital transfer in the economy, it also plays an important role in promoting the economic growth of the country (Wiredu et al., 2020: 18). Exports of goods and services constitute one of the items used in the calculation of GDP through expenditures, and the increase in exports creates an increase in the volume of production at the rate of the foreign trade multiplier. Exports of goods and services are also an important determinant of a country's employment and production growth, as well as an

1 Ph.D., Lecturer, Aksaray University Eskil Vocational School, Department of Management and Organization, zekiyeortlek@aksaray.edu.tr, ORCID:0000-0003-0547-3782.

2 Ph.D., Professor, Aksaray University, Department of Economics, mozgur@aksaray.edu.tr, ORCID: 0000-0002-8711-3264.

engine of economic growth (Ramos, 2001). With the increase in globalization, countries have the opportunity to produce and consume more and reach high growth rates (Polat & Peker, 2020: 4011).

This circumstance plays a substantial role in enhancing the economic development of the nation. In this framework, the 2015–2020 period is considered and System Generalized Method of Moments (GMM), a dynamic panel estimation method, is applied for Turkey and OECD countries (USA, Australia, Belgium, Denmark, Estonia, Finland, France, Korea, Netherlands, Italy, Japan, Colombia, Luxembourg, Hungary, Mexico, Norway, Portugal).

The subsequent section, following the introductory segment, examines the scholarly literature on tourism, foreign direct investment (FDI), exports of goods and services, and the interplay between trade globalization and economic growth. The third section outlines the preparation of the dataset for the criteria spanning from 2015 to 2020. In the fourth section, econometric analysis and findings are presented and the study is completed with a conclusion.

2 Literature

Since the early 2000s, the nexus between tourism sector development and economic growth has been a prevalent subject of empirical investigation. In this regard, Table 1 provides a comprehensive summary of the extant literature on the interconnections among tourism, FDI, exports of goods and services, trade globalization and economic growth.

Table 1: Interrelations among Tourism, FDI, Exports, Trade Globalization, and Economic Growth

Author(s)	Country(s)	Period	Method	Result
Sakyi et al. (2015)	Gana	1970–2011	ARDL	Exports and FDI are an important tool in promoting economic growth.
Khamphengvong et al. (2017)	Lao PDR	1990–2015	ARDL, VECM	In the long run, FDI and trade openness positively affect economic growth.
Terzi & Akbulut Bekar (2019)	Türkiye	1974–2014	ARDL and Dolado-Lütkepohl (DL) Granger causality test	The number of tourists, openness and foreign direct investments affect each other.

Table 1: Continued

Author(s)	Country(s)	Period	Method	Result
Asamoah et al. (2019)	Sahraaltı Afrika	1996–2016	SEM	FDI does not affect economic growth while human capital and financial development positively effects economic growth.
Pata (2020)	Türkiye	1965–2017	ARDL	Sustainable economic growth requires the effective use of policy instruments such as capital stock, financial development, and trade openness along with tourism.
Banday et al. (2020)	BRICS countries	1990–2018	Dumitrescu and Hurlin granger causality test	In the long run, the impact of trade openness and FDI on economic growth is positive.
Gulistan et al. (2020)	112 country	1995–2017	Panel data	Tourism, economic growth and energy use harm the environment.
Durmuş et al. (2021)	14 Developing country	1995–2019	GMM	Trade openness, tourism revenues, and foreign direct investment (FDI) exhibit statistically significant positive impacts on economic growth.
Karaca et al. (2022)	BRICS-T ülkeleri	1992–2019	Panel data	Trade openness has an indirect effect on foreign direct investments. growth through growth.
Şahin & Temelli (2022)	Türkiye and BRICS countries	1995–2019	Driscoll and Kraay estimator, Regression model	The effect of trade openness, credit and tourism revenues on economic growth is positive and statistically significant.
Ünkaracalar (2022)	88 Selected countries	2000–2018	Dumitrescu and Hurlin panel causality test, Multiple regression model	There is bidirectional causality between political, social and economic globalization indices and their subcomponents and economic growth.
Ülker (2023)	Türkiye	1995–2020	ARDL	In the long run, the effect of trade openness and tourism on economic growth is significant and positive.

Note: ARDL: Autoregressive distributed lag bounds test, GMM: Generalized Method of Moments, SEM: Structural Equation Modeling

The table assesses the impact of openness, trade globalization, FDI, tourism and other factors on economic growth. In line with this objective, different methods such as ARDL, VECM, GMM, SEM and panel data techniques have been used. Across these studies, indicators such as FDI, trade openness and tourism revenues have a positive impact on economic growth. However, environmental damages have been observed in the relationship between energy use, tourism and economic growth. There is a bidirectional causal relationship between political, social, global and economic criteria. These studies emphasize the need for effective implementation of other policy instruments such as financial development, trade openness and capital stock along with tourism for sustainable economic growth.

This study investigates the impact of tourism, FDI, exports and trade globalization on economic growth for OECD countries and Turkey by using GDP per capita growth rate, number of tourists, FDI, exports of goods and services and trade globalization criteria. This study aims to empirically analyze the relationship between tourism, FDI, trade openness and trade globalization with economic growth. For this purpose, a panel data set was created by compiling data from OECD countries and Turkey. A data set was obtained for the period 2015–2020. The selection of these periods as the data set was determined by the fact that the most recent date of the indicator for the number of tourists started in 2015 and the last date of its compilation was 2020.

3 Data

This study aims to empirically examine the relationship between tourism, FDI, trade openness and trade globalization with economic growth. For this purpose, a panel data set was created by compiling data from OECD countries and Turkey. A data set was obtained for the period 2015–2020. The selection of these periods as the data set was determined by the fact that the most recent date of the indicator for the number of tourists started in 2015 and the last date of its compilation was 2020.

Table 2: Basic Information on the Variables Used in the Analysis

Variable	Descriptions	Reduction	Period	Source
Gross domestic product growth rate per capita	Annual (%)	pergdpgr	2015–2020	WB
Number of international tourists	Number of arrivals	ti	2015–2020	WB

Table 2: Continued

Variable	Descriptions	Reduction	Period	Source
Foreign direct investments	Net inflows (%)	fdı	2015–2020	WB
Trade in goods and services	GDP (%)	tgs	2015–2020	WB
Trade globalization	Index	tk	2015–2020	KOF

Note: Series are analyzed by taking their natural logarithms. WB: World Bank

Table 2 reports basic information on the data set used in the analysis. Variables related to GDP per capita growth rate, tourism, FDI, trade openness and trade globalization are compiled from the World Bank's database. Graph 1 shows the GDP per capita growth rates of Turkey and OECD countries between 2015 and 2020.

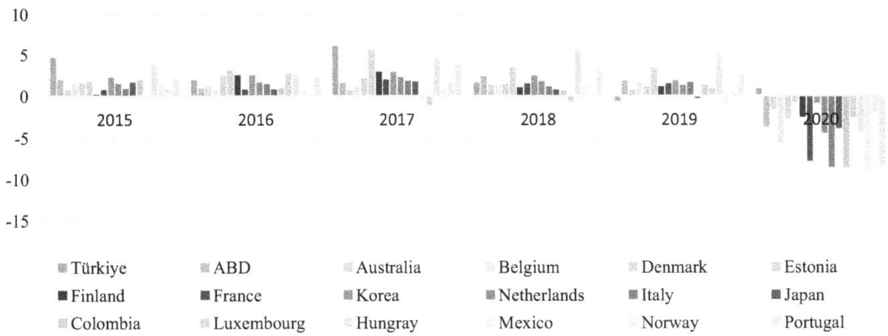

Graph 1: *GDP Per Capita Growth Rate*
Source: World Bank

As can be seen from the graph, while the GDP per capita growth rate was in the 0–6 % band range between 2015 and 2019, this band range is in the 0-(-10) percent band range in 2020. The distribution of GDP per capita growth rate in Turkey and OECD countries differs in each period. Graph 2 shows the number of tourist arrivals to Turkey and OECD countries between 2015 and 2020.

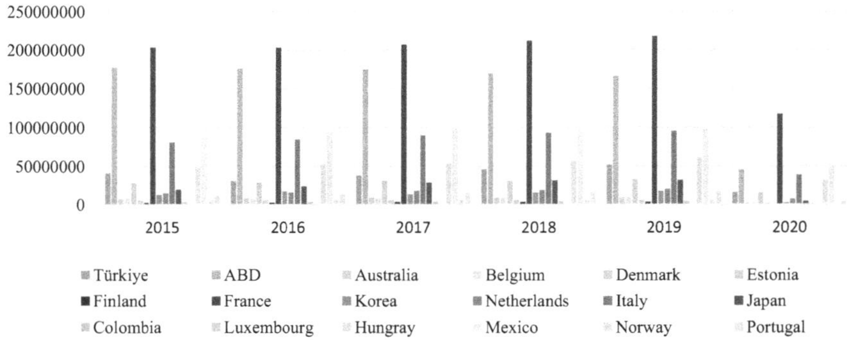

Graph 2: *Number of Incoming Tourists*

Source: World Bank

As can be seen from the graph, France is the country that attracts the highest number of tourists among OECD countries between 2015 and 2020. After that, the countries attracting the highest number of tourists are the USA, Mexico and Italy. The number of tourists coming to other OECD countries is approaching each other in these years. Graph 3 shows FDI inflows to Turkey and OECD countries between 2015 and 2020.

Graph 3: *Foreign Direct Investment (% of GDP)*

Source: World Bank

Looking at the distribution of FDI between 2015 and 2020 in the graph, it is seen that while the lowest FDI inflows among OECD countries were made to Luxembourg as of 2018, the highest FDI inflows were made to Luxembourg again in 2019. As of 2020, the OECD country with the highest FDI inflows is Hungary. Graph 4 shows exports of goods and services to Turkey and OECD countries between 2015 and 2020.

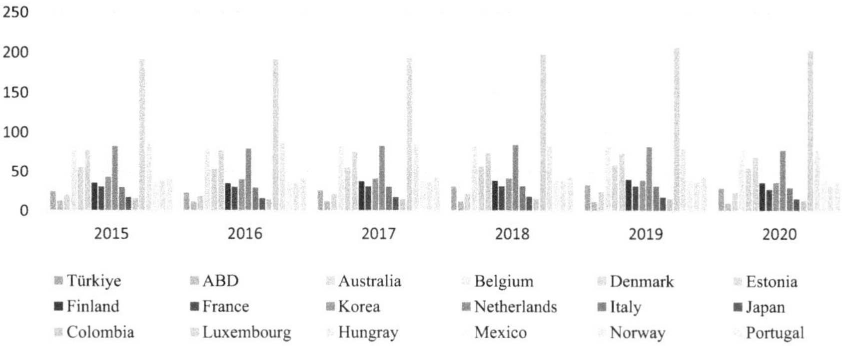

Graph 4: *Exports of Goods and Services*

Source: World Bank

Luxembourg has the highest exports of goods and services compared to other OECD countries in 2015–2020. It is followed by Hungary, the Netherlands, Belgium and Estonia. Exports of goods and services of other OECD countries are converging. Graph 5'te shows trade globalization figures for Turkey and OECD countries between 2015 and 2020.

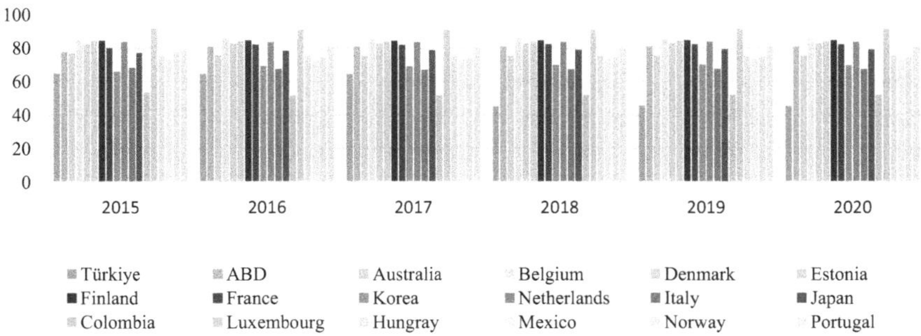

Graph 5: *Trade Globalization*

Source: KOF, https://kof.ethz.ch/en/forecasts-and-indicators/indicators/kof-globalisation-index.html (01.02.2024).

The chart shows that Luxembourg has the highest trade globalization figure for 2015–2020 compared to other OECD countries, followed by Belgium.

For this study, a model is constructed to determine the relationship between tourism, FDI, trade openness and trade globalization and economic growth in OEDC countries and Turkey. Therefore, tourism, FDI, trade openness and trade globalization variables are included in the model and their impact on economic growth is examined. GDP per capita growth rate, tourism, trade openness, foreign direct investment, and trade globalization indicators are estimated using the two-stage System GMM estimator using Stata-17 package programs with data obtained from the WB. In terms of measuring economic growth, GDP per capita growth rate is taken into account. Table 3 reports basic descriptive statistics for the data introduced.

Table 3: Basic Descriptive Statistics and Correlation Table for the Data Set

	pergdpgr	**ti**	**fdı**	**tgs**	**tk**
Ortalama	0.721357	40386038	5.205507	51.15823	76.30698
Medyan	1.322802	15711500	1.749325	37.53182	78.93480
Maksimum	6.116959	2.18E+08	234.2487	206.4116	91.59771
Minimum	-9.313503	525000.0	-117.3747	10.20923	45.16638
Standart Hata	2.850630	55886624	30.62470	42.34263	10.29010
Çarpıklık	-1.639299	1.877325	3.804970	2.236539	-1.298140
Basıklık	6.398386	5.565852	33.66007	8.252017	4.530899
Gözlem Sayısı	108	108	108	108	108
Korelasyon	**l pergdpgr**	**lti**	**lfdı**	**ltgs**	**ltk**
lpergdpgr	1.0000				
lti	0.0195	1.0000			
lfdı	0.0293	-03488	1.0000		
ltgs	0.2733	-0.3577	0.5813	1.0000	
ltk	0.0938	-0.0055	0.2680	0.4111	1.0000

The studied data are analyzed on an annual basis. This table also shows that the panel is a balanced panel. As seen in the table, the GDP per capita growth rate consists of 108 observations, while the other independent variables consist of the same number of observations. The data in the panel are not logarithmized and descriptive statistics of the raw data are shown.

The bottom panel of Table 4 reports the pairwise correlations between variables. Binary correlation values are important in terms of showing whether there is a multicollinearity problem. In Table 4, the dependent variable is the GDP per capita growth rate and the VIF values of the independent variables used in the study are tabulated.

Table 4: Multicollinearity

Değişken	**VIF**	**1/VIF**
lti	1.23	0.815841
lfdı	1.57	0.635964
ltgs	1.80	0.554958
ltk	1.24	0.803822
Mean VIF	1.46	

Since the number of international tourists 1.23, foreign direct investments 1.57, trade openness 1.80, trade globalization 1.24 and VIF values of all variables are below 5, there is no multicollinearity problem. Since the average VIF value (1.46) is not above 5, there is no multicollinearity problem.

The basic regression equation with the introduced variables is shown in Equation 1;

$$kbgdpbuy_{it=}\beta_0 + \beta_1\, kbgdpbuy_{it-1} + \beta_2\, ti_{it} + \beta_3\, fd1_{it} + \beta_4\, tgs_{it} + \beta_5\, tk_{it} + \varepsilon_{it} \quad (1)$$

In Equation 1, the dependent variable is the GDP per capita growth rate, i is the country, t is the time, and the independent variables are pergdpgr, GDP per capita growth rate; ti, number of tourist arrivals; fdı, foreign direct investment; tgs, exports of goods and services; tk, trade globalization; and ε is the random error term. In addition, to test the relationship between tourism, FDI, exports and trade globalization and economic growth, the dependent and independent variables are presented in logarithmic form. In this context, our basic model is shown in Equation 2 as follows;

$$lkbgdpbuy_{it=}\beta_0 + \beta_1 lkbgdpbuy_{it-1} + \beta_2 lti_{it} + \beta_3 lfd1_{it} + \beta_4\, ltgs_{it} + \beta_5 ltk_{it} + \varepsilon_{it} \quad (2)$$

In the empirical analysis of this study, when we consider the data set compiled because the cross-section unit size is above the time series size (T = 6 and N = 18), it is concluded that System GMM is a suitable method for analysis.

The Arellano and Bond estimator is weak when the variance of the unit effect is considerably higher than the variance of the error. At the same time, when T is small or when working with unbalanced panel data, the first difference transformation is again weak. For this reason, forward orthogonal deviations or orthogonal deviations are preferred as a different transformation method instead of the first difference transformation (Yerdelen Tatoğlu, 2020, p. 138). Arellano and Bover (1995) propose the orthogonal deviations method as an efficient instrument estimator for dynamic panel data models. This method is obtained by taking the difference of the average of all possible future values of a variable.

4 Analysis and Empirical Findings

The data set compiled for the analysis of the relationship between tourism, FDI, trade openness and trade globalization and economic growth covers 18 countries and 6 years. For this reason, panel data analysis techniques will be applied. In this panel data analysis, dynamic panel data method, two-stage System GMM method is used since the analysis is performed with small T and large N type data set. In Table 5, the dependent variable is the GDP per capita growth rate and the System GMM estimation results are given.

Table 5: System GMM Estimation Results

Dependent Variable: GDP Per Capita Growth Rate (log)				
	Coefficient	**Standard Error**	**z**	**P>z**
GDP Per Capita Growth Rate t-1 (log)	0.356	0.157	2.27	0.023***
Number of International Tourists (log)	-0.088	0.052	-1.69	0.090*
Foreign Direct Investment (log)	-0.073	0.072	-1.00	0.315
Exports of Goods and Services (log)	0.276	0.164	1.68	0.093*
Trade Globalization (log)	0.231	0.274	0.84	0.400
System GMM Estimation Results				
Number of observations	52			
Number of groups	17			
Number of vehicle variables	8			
Arellano–Bond test for AR(1) Process in First Differences	-2.43 [0.015]			
Arellano–Bond test for AR(2) Process in First Differences	0.52 [0.600]			
Sargan Test for Overdetermination Constraints	7.62 [0.055]			

Note: *, ** and *** denote statistical significance at 10 %, 5 % and 1 % levels, respectively. Values in square brackets denote probability values. Robust standard errors for the system GMM estimation are reported. Stata software package and xtabond2 command were used for estimation.

The lagged dependent variable, GDP per capita growth rate (lpergdpgr) (0.023), is significant (at 1 % significance level) and has a positive sign. However, trade openness (ltgs) (0.093) is significant and positive in explaining the dependent variable, while the number of international tourists (lti) (0.090) is negative and significant.

System GMM estimators are used to test the validity of the instrumental variables used in generalized moments estimation. According to the Sargan test statistics, the over-identification restrictions are valid as the prob value exceeds 0.05.

The lagged dependent variable GDP per capita growth rate is significant in explaining the dependent variable and its sign is positive. The autocorrelation test results confirm that there is first order autocorrelation (0.015) and no second order autocorrelation (0.600). The Sargan test (0.055) shows that the overidentification restrictions are valid.

OLS and FE estimation results of the regression equation for the impact of tourism, FDI, exports and trade globalization on economic growth are shown in Table 6.

Table 6: FE[3] ve OLS[4] Results

Dependent Variable: Δpergdpgr	**OLS**	**FE**	**GMM**
$\Delta lpergdpgr_{t-1}$	0.359	-0.085	0.356
lti	-0.055	0.868	-0.088
lfdı	-0.075	-0.263	-0.073
ltgs	0.302	-0.584	0.276
ltk	0.449	2.062	0.231

The coefficient values of Δ pergdpgr t-1parameter are -0.085 for FE and 0.359 for OLS. In this case, compared to the GMM (0.356) results (-0.085<GMM<0.359), these results indicate that the GMM estimator provides reliable and robust

3 FE: Fixed Effects Model.
4 OLS: Least Squares Method.

results in terms of the model established for the relationship between tourism, FDI, exports and trade globalization with economic growth.

5 Conclusion

In this study, the relationship between tourism, FDI, trade in goods and services and trade globalization with economic growth in Turkey and OECD countries (18 countries) between 2015–2020 is examined. The relationship between tourism, FDI, trade in goods and services and trade globalization with economic growth is analyzed with System GMM. In this context, among the criteria used in the study, GDP per capita growth rate is used as the dependent variable, while the number of tourists, FDI, trade in goods and services and trade globalization are used as independent variables. Therefore, an increase in the number of tourists (0.090) has a positive and significant effect on the GDP per capita growth rate, and an increase in exports of goods and services (0.093) has a positive and significant effect on the GDP per capita growth rate (0.023). In this case, it can be said that the number of tourists and exports of goods and services accelerate economic growth. Moreover, the number of tourists and exports of goods and services are among the determinants of economic growth. However, an increase in FDI (0.315) has a positive and insignificant effect on GDP per capita growth rate, while an increase in trade globalization (0.400) has a positive and insignificant effect on GDP per capita growth rate. An examination of the literature on the relationship between these indicators and economic growth reveals that generally, different results have been obtained.

In conclusion, in this study, the number of tourists and goods and service exports are of great importance for financial growth. For the tourism sector, an increase in exports of goods and services leads to an increase in the number of foreign tourists. In the context of Turkey and OECD countries, policy makers should develop policies that will lead to an increase in FDI by attracting more tourists to these countries. In other words, policies should be developed to direct FDI towards the tourism sector, as an important component of sustainable economic progress. As a result of the inflow of FDI into the country, the transfer inflow of high-tech products to the country will accelerate and the country's capital accumulation will increase.

References

Arellano, M., & Bover, O. (1995). Another Look at the İnstrumental Variable Estimation of Error-Components Models. *Journal of econometrics*, *68*(1), 29–51.

Asamoah, L. A., Mensah, E. K., & Bondzie, E. A. (2019). Trade Openness, FDI and Economic Growth in Sub-Saharan Africa: Do Institutions Matter?. *Transnational Corporations Review, 11*(1), 65–79.

Balaguer, J., & Cantavella-Jorda, M. (2002). Tourism as A Long-Run Economic Growth Factor: The Spanish Case. *Applied Economics, 34*(7), 877–884.

Banday, U. J., Murugan, S., & Maryam, J. (2021). Foreign Direct İnvestment, Trade Openness and Economic Growth in BRICS Countries: Evidences from Panel Data. *Transnational Corporations Review, 13*(2), 211–221.

Durmuş, S., Şahin, D., & Beşcanlar, S. (2021). Gelişmekte Olan Ülkelerde Turizm Gelirleri, Doğrudan Yabancı Yatırımlar ve Ticari Açıklığın Ekonomik Büyüme Üzerine Etkisi. *Karamanoğlu Mehmetbey Üniversitesi Sosyal ve Ekonomik Araştırmalar Dergisi, 23*(41), 517–529.

Gulistan, A., Tariq, Y. B., & Bashir, M. F. (2020). Dynamic Relationship among Economic Growth, Energy, Trade Openness, Tourism, and Environmental Degradation: Fresh Global Evidence. *Environmental Science and Pollution Research, 27*(12), 13477–13487.

Karaca, T., Güney, N., & Hopoğlu, S. (2022). Ticari Dışa Açıklık-Doğrudan Yabancı Yatırımlar-Ekonomik Büyüme İlişkisi: BRICS-T Ülkeleri Örneği. *Bingöl Üniversitesi İktisadi ve İdari Bilimler Fakültesi Dergisi, 6*(1), 109–146.

Khamphengvong, V., Enjun, X. I. A., & Srithilat, K. (2017). The Relationship among FDI, Trade Openness and Economic Growth: Empirical Evidence from Lao PDR. In *2017 4th International Conference on Industrial Economics System and Industrial Security Engineering (IEIS)* (pp. 1–6). IEEE.

Pata, U. K. (2020). Turizm, Finansal Gelişme, Ticari Açıklık ve Sermaye Stokunun Ekonomik Büyüme Üzerindeki Etkileri: Türkiye Örneği. *Çukurova Üniversitesi Sosyal Bilimler Enstitüsü Dergisi, 29*(4), 151–167.

Polat, Ç., & Peker, K. Ö. (2020). Küreselleşme ve ekonomik büyüme ilişkisi: Türkiye örneğinde bir değerlendirme. *İnsan ve Toplum Bilimleri Araştırmaları Dergisi, 9*(5), 3988–4015.

Proenca, S., & Soukiazis, E. (2008). Tourism as an Economic Growth Factor: A Case Study for SouthernEuropean Countries. *Tourism Economics,* 14(4), 791–806.

Ramos, F. F. R. (2001). Exports, Imports, and Economic Growth in Portugal: Evidence from Causality and Cointegration Analysis. *Economic Modelling,* 18, 613–623.

Sakyi, D., Commodore, R., & Opoku, E. E. O. (2015). Foreign Direct İnvestment, Trade Openness and Economic Growth in Ghana: An Empirical İnvestigation. *Journal of African Business, 16*(1–2), 1–15.

Şahin, D., & Temelli, F. (2022). Türkiye ve BRICS Ülkelerinde Turizm, Finansal Gelişme, Ticari Açıklık ve Ekonomik Büyüme Arasındaki İlişki: Panel Veri Analizi (1995–2019). *Bilecik Şeyh Edebali Üniversitesi Sosyal Bilimler Dergisi*, *7*(2), 179–191.

Tang, C. F., & Tan, E. C. (2015). Does Tourism Effectively Stimulate Malaysia's Economic Growth?. *Tourism Management*, *46*, 158–163.

Terzi, H., & Bekar, S. (2019). Türkiye'de Doğrudan Yabancı Yatırımlar, Turizm ve Dışa Açıklık Arasındaki İlişki: 1974–2014 Dönemi. *Doğuş Üniversitesi Dergisi*, *20*(1), 15–30.

Ülker, B. (2023). Ekonomik Büyüme, Turizm ve Ticari Açıklık İlişkisi: Türkiye Örneği. *Kastamonu Üniversitesi İktisadi ve İdari Bilimler Fakültesi Dergisi*, *25*(1), 254–270.

Ünkaracalar, T. (2022). Küreselleşme ve Ekonomik Büyüme İlişkisi: Seçilmiş Ülkeler Üzerine Bir Uygulama. *Karamanoğlu Mehmetbey Üniversitesi Sosyal ve Ekonomik Araştırmalar Dergisi*, *24*(43), 762–784.

Wiredu, J., Nketiah, E., & Adjei, M. (2020). The Relationship between Trade Openness, Foreign Direct İnvestment and Economic Growth in West Africa: Static Panel Data Model. *Journal of Human Resource and Sustainability Studies*, *8*, 18–34.

Yerdelen Tatoğlu, F. (2020). *İleri Panel Veri Analizi Stata Uygulamalı* (4. Baskı), Beta Yayınları.

https://kof.ethz.ch/en/forecasts-and-indicators/indicators/kof-globalisation-index.html (27.03.2024)

https://data.worldbank.org/ (27.03.2024)

Selin Dinçer[1] & Fatih Akın[2]

Analysis of the Relationship Between Economic Growth, Human Development Index and Ecological Footprint: The Case of ASEAN-5 Countries

1 Introduction

This study aims to investigate the short and long-term relationship between economic growth, Human Development Index (HDI), and Ecological Footprint (EF) across five member countries of the Association of Southeast Asian Nations (ASEAN) from 1990 to 2022 (Indonesia, Malaysia, Philippines, Singapore, and Thailand) using Panel ARDL analysis. Within this context, the study empirically analyzes the relationship among economic growth, HDI, and EF variables in ASEAN countries. While the literature extensively addresses the relationship between economic growth and the environment, studies examining the relationship between HDI, economic growth, and Ecological Footprint are relatively scarce despite HDI being a crucial development component. Hence, this study is expected to contribute to the existing literature. The findings indicate that an increase in economic growth and HDI in ASEAN-5 countries leads to a rise in Ecological Footprint. The study outlines the theoretical framework, reviews pertinent literature, and concludes with econometric methodology and empirical analysis findings.

2 Theoretical Framework

The concept of Ecological Footprint, introduced to the literature in the 1990s by William Rees and Mathis Wackernagel (1996), measures individuals' or populations' demands on global natural resources. It is particularly significant in regions with high productivity as a prominent indicator of environmental degradation.

---

1 Ph.D., Lecturer, Kırıkkale University, Keskin Vocational School, selindincer@kku.edu.tr, ORCID:0000-0003-3233-493X.

2 Ph.D., Lecturer, Erzincan Binali Yıldırım University, Refahiye Vocational School, fatih.akin@erzincan.edu.tr, ORCID: 0000-0002-7741-4004.

The relationship between economic growth and the environment gained prominence parallel to the acceleration of industrialization and environmental pollution, especially from the 19th century onwards. In 1955, Simon Kuznets initially addressed the relationship between income distribution and economic growth with the Kuznets Curve, which posits an inverted U-shaped relationship. According to this approach, there is an initial increase in income inequality and economic development, followed by a decrease.

In the 1990s, the Kuznets Curve framework was adapted to the environmental context and became known as the "*Environmental Kuznets Curve*" (EKC). According to the EKC approach, a systematic relationship exists between environmental quality and economic growth; during the early stages of economic development characterized by industrialization, ecological pollution increases. However, in later stages, pollution decreases as societies become wealthier and can afford cleaner technologies (Grossman & Krueger, 1991).

Moreover, according to the EKC approach, the initial stages of development marked by industrialization lead to rapid depletion of natural resources and increased environmental pollution. Economic decision-makers in these stages prioritize achieving higher output levels and income rather than environmental conservation. Consequently, rapid economic growth rates are accompanied by increased natural resource use, exacerbating environmental degradation (Gürlük, 2009: 2327; Manga & Gümüş Akar, 2020: 407).

Furthermore, considering the scarcity of environmental resources and the increasing global population, population growth intensifies production and consumption, raising concerns about the future depletion of natural resources. Concurrently, efforts by countries to industrialize and industrialize further contribute to environmental degradation, heightening worries about the sustainability of the environment.

Human development can be defined as expanding people's choices, including having a healthy life, access to education, decent living standards, political freedoms, and guaranteed human rights (UNDP, 1990: 10). Human development is closely associated with individuals' well-being and the expansion of their choices. The Human Development Index (HDI) is a development indicator encompassing income and the extent to which education and health needs are met. Unlike Gross Domestic Product (GDP) per capita, which focuses solely on income per person, the HDI ranks countries differently because it captures various dimensions of human life (UNDP, 1990: 12–14).

The Association of Southeast Asian Nations (ASEAN) was established on August 8, 1967, in Bangkok with the participation of Foreign Ministers from Indonesia, the Philippines, Malaysia, Singapore, and Thailand. Subsequently,

Brunei (1984), Vietnam (1995), Laos (1997), Myanmar (1997), and Cambodia (1999) joined the organization (Ministry of Foreign Affairs of Turkiye, 2023). ASEAN functions as a regional intergovernmental organization representing a process of regionalism within Asia. It aims to facilitate integration among its members in various economic, political, and socio-cultural fields, promoting intergovernmental cooperation (Jetschke, 2012; Arisman, 2018: 114–115).

The Human Development Index has been an essential indicator in economic literature since the 1990s, measuring economic performance while considering different dimensions. Therefore, examining HDI values in the ASEAN-5 countries forms the focal point of this study, which explores the relationship between HDI and the environment. Among the founding ASEAN members in 2022, Singapore had the highest HDI value at 0.94. Malaysia followed it with an HDI value of 0.807, and then Thailand and Indonesia. The country within this group with the lowest HDI value was the Philippines at 0.71 (UNDP, 2023).

3 Literature Review

The literature reveals numerous studies exploring the relationship between economic growth and the environment, predominantly grounded in the Environmental Kuznets Curve (EKC) theoretical framework. Grossman and Krueger (1991: 1–3) examined the impact of NAFTA-induced trade liberalization on environmental pollution across different countries. Their findings indicated that increased trade flow in these countries initially raises air pollution during the early stages of economic development but subsequently decreases pollution in later stages.

Abdul and Mahmud (2009) analyzed the long-term relationship between carbon emissions, income, energy consumption, and trade in China from 1975 to 2005 using the panel ARDL method. Their results demonstrated that economic growth increases carbon emissions in China, establishing a unidirectional causality from economic growth to carbon emissions. Arı and Zeren (2011) used panel data analysis to investigate the relationship between economic growth and carbon emissions in Mediterranean countries from 2000 to 2005. Their findings suggested an inverted U relationship, indicating that carbon emissions can increase even at high economic growth levels. Ahmed and Long (2012) used the ARDL method to examine the relationship between carbon emissions, economic growth, energy consumption, and population from 1971 to 2008. Their results showed that economic and population growth contributes to increased carbon emissions in the long term. Apergis and Öztürk (2015) identified the validity of the EKC hypothesis across 14 Asian countries from 1990 to 2011. Jebli et al.

(2016) confirmed the applicability of the EKC hypothesis in OECD countries from 1980 to 2010 based on variables such as carbon emissions, GDP, renewable and non-renewable energy consumption, and international trade.

Baek (2015) examined the relationship between energy consumption, economic growth, and carbon emissions in Arctic countries from 1960 to 2010 using the panel ARDL method. The results indicated that energy consumption has a detrimental effect on the environment. Tzeremes (2018) investigated the causal relationships among energy consumption, carbon emissions, and economic growth in 50 US states from 1960 to 2010. The findings suggested bidirectional causality between energy consumption, carbon emissions, economic growth, and energy consumption. Bello et al. (2018) utilized Granger causality testing to examine the impact of hydroelectric consumption on the environment in Malaysia from 1971 to 2016. Their results highlighted a reverse-U relationship between environmental degradation and real GDP, indicating that hydroelectric consumption significantly reduces environmental degradation.

Aydın et al. (2023) analyzed the relationship between ecological footprint, environmental protection expenditures, environmental taxes, natural resource rents, and economic growth using the panel cointegration method in 20 European Union countries from 1995 to 2018. The findings indicated significant long-term relationships among the variables. Koçak and Uçan (2018) used the Pedroni cointegration test to investigate the relationship between economic growth and Turkiye's Human Development Index (HDI). They selected high HDI countries (Germany, USA, Norway, etc.) from 1990 to 2015. They found evidence of a long-term relationship between HDI and economic growth in the selected countries. Using cointegration and causality tests, Bucak (2021) analyzed the relationship between the Economic Complexity Index, the Human Development Index, and carbon emissions in EU-15 countries and Turkiye from 1995 to 2014. The results indicated cointegration among the variables, with the Economic Complexity Index causing carbon emissions and the Human Development Index affecting both carbon emissions and the Economic Complexity Index.

Moran et al. (2008) examined the relationship between Human Development Index and Ecological Footprint across 93 countries from 1975 to 2003. Their findings revealed a positive correlation between HDI and Ecological Footprint, indicating that higher HDI values are associated with increased environmental impact. Gürlük (2009) focused on selected Mediterranean countries to explore the relationship between economic growth, industrial pollution, and the Human Development Index within the EKC framework. The results showed that, except for France, the EKC hypothesis was invalid in other countries studied.

Manga and Gümüş Akar (2020) used the panel ARDL method to investigate the relationship between economic growth, carbon emissions, and the Human Development Index in Mediterranean EU member countries and Turkiye from 1998 to 2014. Their findings indicated a negative relationship between the Human Development Index and carbon emissions in the studied countries. Arisman (2018) used panel regression analysis to examine the factors determining the Human Development Index in ASEAN countries from 2010 to 2015. The results highlighted that population, inflation rate, unemployment rate, and per capita income growth rate significantly influence the Index. Elistia and Syahzuni (2018) used causality analysis to analyze the impact of Human Development Levels on economic growth in ASEAN countries from 2010 to 2016. The study found a strong correlation between Human Development levels and economic growth in the studied countries. Bieth (2021) investigated the impact of the Gross Domestic Product and Human Development Index on carbon emissions using panel data regression analysis in 6 ASEAN countries and Japan from 2007 to 2018. The study found that the Gross Domestic Product had a positive but insignificant effect on carbon emissions, whereas the Human Development Index had a positive but insignificant effect.

Upon reviewing the relevant literature, it is observed that studies have yielded different outcomes depending on the group of countries examined and the variables utilized. Particularly notable is the extensive attention in the literature to the relationship between environmental factors and economic growth. Conversely, there appears to be a scarcity of studies concurrently addressing the relationship between Environmental Sustainability Index (ESI) variables and human development indicators that collectively highlight the various dimensions of development. This limitation has significantly influenced the emergence of this study.

4 Econometric Methodology and Analysis Results

In this study, the long and short-term relationships among economic growth (LGDP), ecological footprint (LEF), and Human Development Index (LHDI) are examined for five countries within the Association of Southeast Asian Nations (ASEAN) from 1990 to 2022. The model employed in the study draws on the frameworks established by Gürlük (2009) and Manga & Gümüş Akar (2020). Logarithms of the variables were taken, and econometric software packages were utilized for the analysis. The model equation specified for use in the study is presented as follows:

$$LEF_{it} = \beta_0 + \beta_{i1} LGDP_{it} + \beta_{i2} LHDI_{it} + \varepsilon_{it} \quad (1)$$

The economic growth variable (GDP per capita constant 2015 US$) data used in the model is from the World Bank (2024) database, the Human Development Index data is from the United Nations Development Program (UNDP, 2023) database, and the ecological footprint variable (gha per capita) data is from the Global Footprint Network (2024). They are obtained from the base.

First, descriptive statistics of panel data for five countries are given. Then, since the time dimension of the panel data set (T=33) was more significant than the unit dimension (N=5) (T>N), the cross-sectional dependence (CSD) of the model was determined with the LM test developed by Breusch and Pagan (1980). In the next stage, the homogeneity of the model was tested with Delta (Δ) tests developed by Pesaran and Yamagata (2008). Since cross-sectional dependency (CSD) and slope heterogeneity were detected in the model, Pesaran's (2007) CIPS test was applied to the unit root test. Since the variables are stationary at different levels, the panel ARDL method developed by Pesaran et al. (1999) was used in the short- and long-term relationship analysis.

Descriptive statistics are given in Table 1: highest average, min., and max. The variable with values is LGDP. Average and max. While the variable with the lowest values is LHDI, Min. is LEF. Highest Std. Giant. The variable with the highest value is LGDP, and the variable with the lowest value is LEF.

Table 1: Descriptive Statistics

Variables	LEF	LGDP	LHDI
Mean	0.890483	8.691730	0.726376
Median	0.772331	8.386009	0.708000
Maximum	2.121892	11.11780	0.949000
Minimum	-0.043896	7.302252	0.526000
Std. Dev.	0.654858	1.102216	0.103611
Skewness	0.335586	0.871852	0.485709
Kurtosis	1.660433	2.569796	2.502441
Jarque-Bera	15.43376	22.17586	8.189632
Probability	0.000445	0.000015	0.016659
Sum	146.9297	1434.136	119.8520
Sum Sq. Dev.	70.32963	199.2404	1.760583
Observations	165	165	165

Cross-sectional dependency and homogeneity test results are given in Table 2. The central hypothesis of cross-sectional dependency analysis is H0: There is no cross-sectional dependency, and the alternative hypothesis is H1: There is cross-sectional dependency. The central hypothesis of the homogeneity analysis is H0, Which states that the slope coefficients are homogeneous, and the alternative hypothesis is H1, Which states that the slope coefficients are heterogeneous. According to the results obtained, the H0 hypothesis ($p=0.00<0.05$), which is the main hypothesis, is rejected in both tests (cross-sectional dependence and homogeneity). In other words, the model shows cross-sectional dependence and heterogeneity.

Table 2: CSD and Homogeneity Test Results

CSD Tests	**LEF**	**LGDP**	**LHDI**
Breusch-Pagan LM	80.83324 (0.000) *	313.9214 (0.000) *	323.3045 (0.000) *
Pesaran scaled LM	15.83879 (0.000) *	67.95890 (0.000) *	70.05702 (0.000) *
Bias-corrected scaled LM	15.76067 (0.000) *	67.88078 (0.000) *	69.97890 (0.000) *
Pesaran CD	7.641919 (0.000) *	17.71610 (0.000) *	17.98054 (0.000) *
Homogeneity Tests:	**Delta Tilde**		**Delta Tilde$_{adj}$**
	5.995 (0.000) *		6.395 (0.000) *

Note: * denotes a 1 % level of significance.

The CIPS test, one of the second-generation unit root tests, was preferred due to the CSD and homogeneity test results. According to the CIPS panel unit root test results given in Table 3, the LEF and LHDI variables are stationary at the first difference. In contrast, the LGDP variable is stationary at the level.

Table 3: CIPS Unit Root Test Results

Variables	**Panel (A): Level Value**		**Panel (B):. Difference Value**
LEF	-2.202		-4.286*
LGDP	-2.629*		-
LHDI	-1.976		-3.653*
CIPS Critical Values	%1	%5	%10
	-2.55	-2.33	-2.21

Note: * denotes a 1 % level of significance.

Since the variables in the study were stationary at different levels, the panel Autoregressive Distributed Lag (ARDL) method developed by Pesaran et al. (1999) was used. Mean Group (MG) and Pooled Mean Group (PMG) are two estimators used in the panel ARDL method. The Hausman test determines which estimator to apply (Samargandi et al., 2015). Based on equation (1) in Equation (2), the equation of the panel ARDL model was created as follows.

$$\Delta LEF_{it} = a_i + \sum_{j=1}^{p-1}\beta_{ij}\,\Delta LEF_{i,t-j} + \sum_{j=0}^{l-1}\gamma_{ij}\,\Delta LGDP_{i,t-j} + \sum_{j=0}^{q-1}\delta_{ij}\,\Delta LHDI_{i,t-j} + Q_1 LEF_{i,t-j} + Q_2 LGDP_{i,t-j} + Q_3 LHDI_{i,t-j} + \varepsilon_{1,it} \quad (2)$$

The error correction equation of the panel ARDL model is given in Equation (3).

$$\Delta LEF_{it} = a_i + \sum_{k=1}^{p}\beta_{ij}\Delta LEF_{i,t-j} + \sum_{k=0}^{l}\gamma_{ij}\Delta LGDP_{i,t-j} + \sum_{k=0}^{q}\delta_{ij}\Delta LHDI_{i,t-j} + \sigma_{ij}ECM_{t-i} + \varepsilon_{it} \quad (3)$$

Panel ARDL/PMG results will be interpreted according to the Hausman test results in Table 4. According to the findings, economic growth increases the ecological footprint in both the short and long term. The Human Development Index increases the ecological footprint in the long term. Additionally, the error correction coefficient (ECM_{t-i}) was significant. As a result, economic growth and the Human Development Index increase the ecological footprint of ASEAN-5 countries.

Table 4: Panel ARDL/PMG Analysis Results

Variables	**Short Term**		**Long Term**	
Dependent Variable: LEF	**Coefficient**	**Probability**	**Coefficient**	**Probability**
LGDP	0.6707602	0.000**	0.1407139	0.029**
LHDI	1.28411	0.424	0.6289763	0.002**
Constant	-0.0111407	0.922	ARDL Model (1,1,1)	
ECM_{t-i}	-0.5107053	0.000**	Number of Country	5
Hausman Test (Chi^2)	0.85	(0.6538)	Number of Observation	165

Note: *and **denote 5 % and 1 % significance levels, respectively.

5 Conclusion

This study focuses on the short- and long-term relationship between Human Development Index, Ecological Footprint, and economic growth in ASEAN countries, based on empirical analysis using the panel ARDL method. Many studies focus on the relationship between environment and economic growth. However, including the Human Development Index, an essential development component with its multidimensional framework, into the model and examining the relationship between these variables specifically for ASEAN-5 countries will increase our scientific knowledge on the subject.

As a result of the study, it was determined that economic growth increases the ecological footprint in the short and long term in ASEAN-5 countries. On the other hand, it has been determined that the Human Development Index only increases the ecological footprint in the long term. As a result, in the long term, the increase in economic growth and the Human Development Index in ASEAN-5 countries will increase the ecological footprint. The findings of the study that economic growth increases carbon emissions appear to be compatible with the findings of the studies conducted by Arı and Zeren (2011) and Ahmed and Long (2012). The findings that an increase in the Human Development Index increases the Ecological Footprint are based on Moran et al. (2008), which appears to be compatible with the study conducted. The research findings reveal the necessity of the countries in question to turn to environmentally friendly policies to make their growth and development processes sustainable without harming the environment.

References

Abdul, J., & Mahmud, S. F. (2009). "Environment Kuznets curve for CO_2 emissions: A cointegration analysis for China". *Energy Policy*, *37*, 5167–5172.

Ahmed, K., & Long, W. (2012). "Environmental Kuznets Curve and Pakistan: An Empirical Analysis." *Procedia Economics and Finance*, *1*, 4–13.

Apergis, N., & Öztürk, İ. (2015). "Testing Environmental Kuznets Curve Hypothesis in Asian Countries". *Ecological Indicators*, *52*, 16–22.

Arı, A., & Zeren, F. (2011). "CO_2 Emisyonu ve Ekonomik Büyüme: Panel Veri Analizi". *Yönetim ve Ekonomi Dergisi*, *18*(2), 37–47.

Arisman, A. (2018). "Determinant of Human Development Index in ASEAN Countries". *Jurnal Ilmu Ekonomi*, *7*(1), 113–122.

Aydın, M., Söğüt, Y., & Altundemir, M. E. (2023). "Moving Toward the Sustainable Environment of European Union Countries: Investigating the

Effect of Natural Resources and Green Budgeting on Environmental Quality". *Resources Policy*, *83*, 1–10.

Baek, J. (2015). "Environmental Kuznets Curve for CO_2 Emissions: The Case of Arctic Countries". *Energy Economics*, *50*, 13–17.

Bello, M. O., Sakiru A. S., & Yuen, Y. Y. (2018). "The Impact of Electricity Consumption on CO_2 Emission, Carbon Footprint, Water Footprint And Ecological Footprint: The Role of Hydropower in an Emerging Economy". *Journal of Environmental Management*, *219*, 218–230.

Bieth, R. C. E. (2021). "The Influence of Gross Domestic Product and Human Development Index on CO_2 Emissions". *Journal of Physics: Conference Series*, Conf. Ser. 1808 012034, 1–12.

Breusch, T. S., & Pagan, A. R. (1980). "The Lagrange Multiplier Test and its Applications to Model Specification in Econometrics". *The Review of Economic Studies*, *47*(1), 239–253.

Bucak, Ç. (2021). "AB15 Ülkelerinde ve Türkiye'de Ekonomik Karmaşıklık Endeksi, İnsani Gelişme Endeksi ve Karbon Emisyonu: Panel Veri Analizi". *Ege Stratejik Araştırmalar Dergisi*, *12*(1), 71–88.

Elistia, E., & Syahzuni, B. A. (2018). "The Correlation of the Human Development Index (HDI) Towards Economic Growth (GDP Per Capita) in 10 ASEAN Member Countries". *Journal of Humanities and Social Studies, 2*(2), 40–46.

Global Footprint Network. (2024). "Ecological Footprint vs Biocapacity (gha per person)", Available At: https://www.footprintnetwork.org/our-work/ecological-footprint/, (15.05.2024).

Grossman, G. M., & Krueger, A. B. (1991). "Environmental Impacts of A North American Free Trade Agreement". *National Bureau of Economic Research*, w3914.

Gürlük, S. (2009). "Economic Growth, Industrial Pollution and Human Development in the Mediterranean Region". *Ecological Economics*, *68*(8–9), 2327–2335.

Jebli, M. B., Youssef, S. B., & Özturk, I. (2016). "Testing Environmental Kuznets Curve Hypothesis: The Role of Renewable and Non-Renewable Energy Consumption and Trade in OECD Countries". *Ecological Indicators*, *60*, 824–831.

Jetschke, A. (2012). Asean. In *Routledge Handbook of Asian Regionalism* (pp. 340–350). Routledge.

Koçak, E., & Uçan, O. (2018). "İnsani Gelişme Endeksi ile Büyüme ilişkisi: Pedroni Eşbütünleşme Örneği". *Journal of Politics Economy and Management*, *1*(2), 55–61.

Manga, M., & Gümüş Akar, P. (2020). "Ekonomik Büyüme, Karbon Emisyonu ve İnsani Gelişmişlik Arasındaki İlişki: Seçilmiş Akdeniz Ülkeleri". *Kahramanmaraş Sütçü İmam Üniversitesi Sosyal Bilimler Dergisi*, *17*(1), 405–419.

Ministry of Foreign Affairs of Turkiye (2023). "Güneydoğu Asya Ülkeleri Birliği (ASEAN) Uluslararası Örgüt Künyesi", Available At: https://www.mfa.gov.tr/guneydogu-asya-ulkeleri-birligi.tr.mfa (22.06.2024).

Moran, D. D., Wackernagel, M., Kitzes, J. A., Goldfinger, S. H., & Boutaudc, A. (2008). "Measuring Sustainable DevelopmentNation By Nation", *Ecological Economics*, *64*(3), 470–474.

Pesaran, M. H., & Yamagata, T. (2008). "Testing Slope Homogeneity in Large Panels". *Journal of Econometrics*, *142*(1), 50–93.

Pesaran, M. H. (2007). "A Simple Panel Unit Root Test in the Presence of Cross-Section Dependence". *Journal of Applied Econometrics*, *22*, 265–312.

Pesaran, M. H., Shin, Y., & Smith, R. P. (1999). "Pooled Mean Group Estimation of Dynamic Heterogeneous Panels", *Journal of the American Statistical Association*, *94*(446), 621–634.

Rees, W., & Wackernagel, M. (1996). "Urban Ecological Footprints: Why Cities Cannot be Sustainable – And Why They are a Key to Sustainability". *Environmental Impact Assessment Review*, *16*(4–6), 223–248.

Samargandi, N., Fidrmuc, J., & Ghosh, S. (2015). "Is the Relationship Between Financial Development and Economic Growth Monotonic for Middle Income Countries?". *World Development*, *68*, 66–81.

Tzeremes, P. (2018). "Time-Varying Causality between Energy Consumption, CO_2 Emissions, and Economic Growth: Evidence from US States". *Environmental Science and Pollution Research*, *25*, 6044–6060.

United Nations Development Programme (UNDP). (1990). "Human Development Report", Available At: https://hdr.undp.org/content/human-development-report-1990, (05.06.2024).

United Nations Development Programme (UNDP). (2023). "Human Development Index (HDI)", Available At: https://hdr.undp.org/data-center/human-development-index#/indicies/HDI, (22.06.2024).

World Bank. (2024). "GDP per capita (constant 2015 US$)", Available At: https://data.worldbank.org/indicator/, (15.05.2024).

Gökhan Dağılgan[1]

Do Renewable Energy Consumption and Industrialization Contribute to Economic Growth in OECD Countries? Assessing the Role of Financial Development, Globalization and Urbanization

1 Introduction

Energy that exists continuously is called renewable energy. 20 % of the energy obtained in the world consists of renewable energy. Dependency on fossil fuels such as oil, natural gas and coal is decreasing thanks to renewable energy. Sources that can produce energy without running out, such as sun, water, wind and wave, are called renewable energy sources. Renewable energy sources obtained from natural resources have sustainability features.

Industrialization; It refers to mechanization in goods production and the increase in the share of industry in national income. In a broad sense, it is the implementation of new techniques in production techniques, increasing the quality of produced goods, reducing production costs and creating economic, political and social changes in the country. The main goal of development economics, which developed after the Second World War, is industrialization. Economic growth is defined as the increase in the national income level and therefore the per capita income.

Energy emerges as a concept that continues to increase in importance for countries from past to present. Especially with industrialization, the need for energy has increased considerably.

After the 1850s, with the influence of the Industrial Revolution, mechanization began and the necessary power sources for steam-powered machines began to be discovered. With the developing technology, machines powered by steam power have been replaced by machines powered by electricity and fuel oil. This has increased the need for energy even further, resulting in rapid population

1 Ph.D., .Lecturer, Kütahya Dumlupınar University, Gediz Vocational School Department of Banking and Insurance, gokhan.dagilgan@dpu.edu.tr, ORCID: 0000-0002-6916-6506.

growth, colonialism, commercial entrepreneurship, the development of capitalism and increased industrial investments (Gürsucu, 2021: 71).

The issue of energy in production took its place at the top of the economic literature after the 1973 oil crisis. A transition has been made from the production approach based on oil and its derivatives to approaches based on renewable energy types. Today, the role of energy and its use in production processes remains important. Since the 1980s, when export-oriented growth models came to the fore, the effects of factors other than production factors began to be analyzed. These factors include total factor productivity, foreign trade, financial development, development levels of money and capital markets, urbanization, globalization, industrialization, public expenditures, savings, fixed capital investments, infrastructure, human and social capital, technological developments, product innovations and energy. A number of variables can be considered such as (Gövdere & Can, 2016: 210).

First of all, the renewable and non-renewable nature of energy consumption was taken into consideration when setting up the model. It is assumed in the literature that both types of energy can have an impact on economic growth. Until recent years, the relationship between fossil-based energy use and economic growth was included in models and predictions were made. In recent years, the relationship between renewable energy and economic growth has begun to be included in energy models (Emek & Polat, 2022: 334).

Along with the industrialization of countries, progress in urbanization increases energy needs. Increasing energy use causes an increase in the total production level. The increase in the level of welfare, developments in technology in the industrial sector and the rapid increase in the population in the world increase energy consumption and demand (Göv & Yılancı, 2023: 207).

The concept of financial development means eliminating the pressures on financial markets or at least reducing them to low levels. Financial liberalization positively affects growth by increasing savings and the efficiency of resources. However, one point should not go unnoticed. High interest rates can reduce investments by increasing investment costs. Based on this, it should be underlined that interest rates should be high enough to encourage savings and low enough not to discourage investments.

Industrialization is seen as the key to power, wealth and a better life. Economic growth in agriculture-based economies remains lower than in economies with industrialization. In the development phase of industrialization, the distribution of production factors should be directed towards high value-added products instead of agriculture and livestock sectors. Efficiency in industrialization is determined by infrastructure, workforce, capital, technology and entrepreneurship. Industrialization is measured by the amount of manufacturing produced by official and unofficial companies.

2 Renewable Energy Consumption and Economic Growth

Renewable energy is an energy source that renews itself as it is consumed and whose amount does not decrease. The ability to renew itself distinguishes renewable energy from fossil energy sources. It is also a sustainable energy source.

In countries targeting high economic growth after the industrial revolution, fossil-based energy consumption has become important in increasing industrial production. Changes in the understanding of production and consumption, population growth and technological developments have increased energy consumption in the production phase. Therefore, energy consumption is the main input source of production activity in growth theories.

Anser et al. (2021) says that the damages of fossil-based energy consumption have reached great levels. It states that energy supply from fossils (oil, coal and gas) tends to decrease over time. However, he emphasizes that the energy sector will face geopolitical risks and that alternative renewable energy sources should be focused on. He also stated that non-renewable energy resources are a prerequisite for economic growth, and that industrializing countries can only achieve economic growth with these energy resources.

Four different hypotheses can be mentioned in the literature to explain the causality between energy consumption and economic growth. The first is the Growth hypothesis; Energy consumption appears as a complement to labor and capital in production and contributes to economic development. If energy consumption causes economic growth, the growth hypothesis is valid. Secondly, in the Conservation Hypothesis, energy consumption does not affect economic growth, on the contrary, economic growth affects energy consumption. Thirdly, according to the feedback hypothesis, if energy consumption increases, economic growth also increases. The fourth hypothesis is the neutrality hypothesis. Energy consumption has minimal or no impact on economic development. Those who determine policies must determine policies by taking all these situations into consideration (Gürsucu, 2021: 71).

3 Industrialization and Economic Growth

Industrialization includes the processes of producing products and creating added value. These products generate income as a result of national and international sales.

The United Nations Industrial Development Organization (2020) states that there is no single development model accepted by everyone regarding industrialization as the driving force of sustainable prosperity. Even if this relationship is positive, it is not always significant or in some cases gives negative results.

Industrialization includes different factors that can increase economic growth. Therefore, it reduces poverty and inequality. In order to have a say in international trade, it must be industrialized and equipped with technological innovations.

Kaldor (1961) was the first person to investigate the relationship between economic growth and industrialization. He is among the first economists to address the role of increasing returns in economic growth. Kaldor put forward three basic growth laws to achieve economic growth. In these laws, it is assumed that industrial production captures a process that strengthens itself by contributing to the increase in productivity, thus leading to economic growth. Kaldor believes that economic policies targeting industry can support economic growth. According to Kaldor, the agriculture and services sector will not be a source of sustainable growth as they cannot increase productivity. Only the industrial sector is responsible for creating internal and self-sufficient growth (Giovanini & Arend, 2017: 192; Sarıdoğan, 2020: 1489).

4 Data Set, Method, Model and Results

4.1 Data set

The research topic is the impact of the increasing need for renewable energy consumption and the development of industrialization on economic growth in OECD economies and all over the world. Therefore, this study analyzes the impact of renewable energy consumption and industrialization on economic growth for 37 developed and developing OECD countries in the period 2001–2020 by including globalization, financial development and urbanization in the model. The data sets used in this study were obtained from the World Bank database (World Bank, 2024) and KOF Swiss Economic Institute (KOF Swiss Economic Institute, 2024). Descriptions of the variables are presented in Table 1.

Table 1: Variable Descriptions

Variables	**Symbol**	**Measurement**	**Data**
Economic growth	GDP	Per capita GDP (2015 constant US $)	WDI
Industry value added	IVA	Industry (including construction), value added (% of GDP)	WDI

Table 1: Continued

Variables	Symbol	Measurement	Data
Renewable energy consumption	REC	Renewable energy consumption (% of total final energy consumption)	WDI
Globalization	GLO	Overall globalization index	KOF SEI
Financial development	FID	Domestic credit to private sector by banks (% of GDP)	WDI
Urbanization	URB	Urban population (% of total population)	WDI

Source: Created by the author.

4.2 Method and Model

In the study, the two-stage System Generalized Method of Moments (GMM), one of the dynamic panel methods, was applied to investigate the connection between variables. It provides solutions to heterogeneity problems and controls the effects of heterogeneity on the results. The existence of a correlation between the error term and the lagged value of the dependent variable in the dynamic model creates inconsistent results when least squares methods are used. The system prevents these inconsistent results with the GMM method (Baltagi, 2008: 138). In the System GMM method, data loss resulting from the first differences method is minimized since the difference of the averages of all possible future values of the variables is taken instead of taking the difference between period t and period t-1 as in the difference GMM method (Blundell & Bond, 2000: 334). The GMM model equation is given below:

$$Y_{it} = \delta Y_{i,\, t-1+} \beta_1 X_{1,it} + \beta_2 X_{2,it} + \beta_3 X_{3,it} + \beta_4 X_{4,it} + \beta_5 X_{5,it} + u_{it} + \varepsilon_{it}$$

Another key feature of the GMM model is to control the effects of autocorrelation and heteroscedasticity. Analysis was performed by taking the logarithm of all variables. The GMM equation using variables is given as follows:

$$lnGDP_{it} = \delta lnGDP_{i,t-1+} \beta_1 lnIVA_{1,it} + \beta_2 lnREC_{2,it} + \beta_3 lnGLO_{3,it}$$
$$+\beta_4 lnFID_{4,it} + \beta_5 lnURB_{4,it} + u_{it} + \varepsilon_{it}$$

This study was carried out with the help of Stata, an econometric analysis program, and the code written for System GMM developed by Roodman (2009) was used.

4.3 Results

Table 2 gives summary statistics for each variable in the panel. The results show that the average weight of OECD economic growth is 34,656.68 per capita, with the maximum and minimum values being 112417.9 and 4007.52 respectively. The industrialization average is 24.96, the renewable energy consumption average is 19.29. At the same time, the average values of globalization, financial development and urbanization are 79.89, 96.68 and 76.68, respectively.

Table 2: Descriptive statistics

	Obs	Mean	Std. Dev.	Min	Max
GDP	740	34656.68	21987.17	4007.52	112417.90
IVA	740	24.96	5.46	10.42	40.81
REC	740	19.29	15.88	0.69	82.79
GLO	740	79.89	7.54	48.00	91.00
FID	675	96.68	46.04	0.18	304.57
URB	740	76.68	11.12	50.78	98.07

Source: Author Calculation

Table 3 shows the system GMM analysis results. It is seen that the economic growth variable gives positive significant results in all four models in its relationship with the lagged t-1 period. A 1 % increase in the T-1 period increases economic growth in the normal period by 0.94 %. This association is statistically significant at the 1 % level. Economic growth shows that OECD economies have grown more than the previous year.

When the relationship between renewable energy consumption and economic growth is examined, a 1 % increase in energy consumption increases economic growth by 0.02 %. The results obtained support the growth hypothesis.

When the relationship between industrialization and economic growth is examined, a 1 % increase in industrialization reduces economic growth by 0.18 %. The result obtained is compatible with Kaldor (1961)'s hypothesis. According to Kaldor, industrialization will not be a source of sustainable growth as it cannot increase productivity in the agriculture and services sectors.

When we examine the effects of globalization, financial development and urbanization variables on economic growth; While the increase in globalization and urbanization increases economic growth, the increase in financial development reduces economic growth.

Finally, the results of the Arellano-Bond test checking the autocorrelation of the first-order variance in the GMM model [AR (2) p=0.289] are greater than 0.1, which rejects the hypothesis that the model has level 1 autocorrelation. Thus, GMM estimation solves the endogeneity problem. According to the Hansan test results, the results of the GMM Arellano and Bond tests overcome the limitations of the model. All models are greater than 0.1.

Table 3: System GMM Analysis Results

	Reg. 1	**Reg. 2**	**Reg. 3**	**Reg. 4**
Dependent Variables				
*ln*GDP(t-1)	**0.94***** (19.79)	**0.96***** (38.17)	**0.86***** (12.04)	**0.88***** (21.75)
*ln*IVA	**-0.18***** (-2.98)	-0.17*** (-3.20)	**-0.23***** (-2.92)	**-0.07***** (-1.15)
*ln*REC	**0.02*** (1.80)	**0.01*** (1.83)	**0.02*** (1.67)	**0.03**** (1.92)
*ln*GLO	**0.31*** (1.34)	**0.01***** (2.79)	**0.39**** (1.98)	
*ln*FID	**-0.01*** (-1.57)	**-0.01**** (-1.97)		
*ln*URB	0.01 (1.23)		**0.54*** (1.69)	
Wald R^2	2553.69	4288.22	1536.17	711.52
***F*-stat**	0.000	0.000	0.000	0.000
AR(1)	-2.70	-2.80	-2.29	-2.09
***p*-value**	**(0.007)**	**(0.005)**	**(0.022)**	**(0.039)**
AR(2)	-3.13	-2.99	-3.20	-2.15
***p*-value**	**(0.289)**	**(0.305)**	**(0.198)**	**(0.320)**
Dif. Hansen	31.26	27.41	31.47	19.47
***p*-value**	**(0.270)**	**(0.260)**	**(0.490)**	**(0.222)**

Note: ***,**,* indicate the significance of the coefficients at the 1 %, 5 % and 10 % levels, respectively. Terms in parentheses indicate z statistics. AR2 Test H0: There is no second-order autocorrelation. Hansen Test H0: Overidentification restrictions apply.

5 Conclusion

Economic developments are one of the leading events that affect every segment of society and directly affect the level of welfare. In this context, economic growth or increasing per capita income is one of the main economic issues that economists, politicians and even households are interested in.

The need for energy is increasing both in developed and developing countries. Especially the supply of energy resources from limited and certain regions increases the importance of energy resources. Considering the growing economic structure of countries, industrialization rate and population growth, the need for energy is increasing. In this study, the relationship between renewable energy consumption, industrialization and economic growth within the scope of OECD countries was examined.

When the analysis results are examined on the basis of energy-economic growth in OECD countries, it is seen that the growth hypothesis is valid. According to this hypothesis, an increase in renewable energy consumption leads to an increase in economic growth. Considering countries' access to natural resources, renewable resources should be preferred instead of non-renewable resources in providing energy used as an important input in production. Orientation towards renewable energy sources will have positive results both economically and environmentally, thus contributing greatly to countries.

On the basis of industrialization-growth, it is seen that it is compatible with Kaldor's hypothesis that if industrialization takes place in the agriculture and services sectors, it will not be a source of sustainable growth as it will not provide productivity increase. OECD countries should take the steps they will take in the field of industrialization on products that create more added value.

A new revolution took place in the world after 1980, led by technology. The economic gap between the countries that realized the industrial revolution and the countries that could not keep up with this revolution has survived to this day. In order for OECD countries to achieve sustainable growth, they must continue to develop along with technological progress in industry. Thus, it is thought that it will be more beneficial for growth in the long run. In this context, it is important to implement policies such as the use of new technologies in the industrial sector, dissemination of existing technology, implementation of incentive programs for research, development and innovation activities of the private sector, and support for technological entrepreneurship.

References

Anser, M. K., Usman, M., Godil, D. I., Shabbir, M. S., Sharif, A., Tabash, M. I., & Lopez, L. B. (2021). "Does Globalization Affect the Green Economy And Environment? The Relationship Between Energy Consumption, Carbon Dioxide Emissions, and Economic Growth". *Environmental Science and Pollution Research, 21*, 51105–51118.

Baltagi, B. H. (2008). *Econometric Analysis of Panel Data* (4th ed.). Chichester: Wiley.

Blundell, R., & Bond, S. (2000). "GMM Estimation with Persistent Panel Data: An Application to Production Functions". *Econometric Reviews, 19*(3), 331–339.

Emek, Ö. F., & Polat, M. A. (2022). "Enerji Tüketimi, Küreselleşme ve Ekonomik Büyüme Arasındaki İlişki: Seçilmiş Yükselen Piyasa Ekonomileri İçin Bir Nedensellik Analizi". *Journal of Economic Policy Researches*, *9*(2), 331–351.

Göv, A., & Yılancı, V. (2023). "Ekonomik Büyüme, Enerji Tüketimi, Dış Ticaret Dengesi ve Finansal Gelişme İlişkisine Grafiksel Nedensellik Testi Yaklaşımı". *İstanbul İktisat Dergisi*, *73*(1), 203–230.

Gövdere, B., & Can, M. (2016). "Türkiye'de Enerji Tüketimi, Dışa Açıklık, Finansal Gelişme, Sabit Sermaye Yatırımları ve Diş Ticaretin Ekonomik Büyümeye Etkisi: Sinir Testi Yaklaşımı". *Niğde Üniversitesi İktisadi ve İdari Bilimler Fakültesi Dergisi*, *9*(1), 209–228.

Gürsucu, O. (2021). "Enerji Tüketimi, Ekonomik Büyüme ve Finansal Gelişme İlişkisi: Türkiye Örneği. *Siyaset*", Ekonomi ve Yönetim Araştırmaları Dergisi, *9*(2), 69–79.

Giovanini, A., & Arend, M. (2017). "Contribution of Services to Economic Growth: Kaldor's Fifth Law?". *Revista de Administração Mackenzie*, *18*(4), 190–213.

Kaldor, N. (1961). Capital Accumulation and Economic Growth. In F. A. Lutz & D. C. Hague (Eds.), *The Theory of Capital*. New York: St. Martin's Press.

KOF Swiss Economic Institute. (2024). "KOF globalisation index". Available at: https://kof.ethz.ch/en/forecasts-and-indicators/indicators/kof-globalisation-index.html, (20.06.2024)

Roodman, D. (2009). "How to Do xtabond2: An Introduction to Difference and System GMM in Stata". *The Stata Journal, 9*(1), 86–136.

Sarıdoğan, H. Ö. (2020). "Kaldor'un Birinci Yasası Çerçevesinde Sanayileşme ve Büyüme İlişkisi". *Anemon Muş Alparslan Üniversitesi Sosyal Bilimler Dergisi*, *8*(5), 1489–1496.

United Nations Industrial Development Organization. (2020). "Industrialization as the driver of sustained prosperity. Vienna". Available at: https://www.unido.org/sites/default/files/files/2020-04/UNIDO_Industrialization_Book_web4.pdf (21.06.2024)

World Bank. (2024). "World Development Indicators". Available at: https://databank.worldbank.org, (21.06.2024)

Appendix A

Table A1. OECD Economies

Australia	Estonia	Israel	Netherlands	Sweden
Austria	Finland	Italy	New Zealand	Switzerland
Belgium	France	Japan	Norway	Türkiye
Canada	Germany	Korea Rep.	Poalnd	United Kingdom
Chile	Greece	Latvia	Portugal	United States
Colombia	Hungary	Lithuania	Slovak Rep.	
Czechia	Iceland	Luxembourg	Slovenia	
Denmark	Ireland	Mexico	Spain	

Source: Created by the author.

Gözde Arslan[1]

Evaluation of Indicators Related to Innovative Human Capital Stock

1 Introduction

The concept of human capital was first introduced as an economic concept by Schultz. In his 1961 article titled "Investment of Human Capital," Schultz defined human capital as an investment in people (Schultz, 1961: 1). Later, in 1981, he expanded this concept and explained that this investment in people is a type of investment aimed at improving the quality of the population. According to Schultz, every individual can enhance their innate abilities through acquired investments. The qualities of the acquired population mentioned here are considered human capital (Schultz, 1981: 21). The concept of human capital has been explained with different definitions from history to the present. Various proxy variables have been used to measure human capital based on these definitions. However, there is no widely accepted measure of human capital in the literature, though level of education/years of schooling and training have long been considered good proxies (McGuirk et. al., 2015: 968). There are two commonly used methods in the literature for determining the concept of human capital. The first is the use of education and health inputs, and the other is the productivity and efficiency approach. These two approaches are mostly considered separately. However, unless the average years of schooling translate into increased skills, talents, productivity, and efficiency, it may not be a sufficient approach to represent human capital on its own (Laverde et al., 2018: 2). Productivity is dependent on a society's capacity for innovation. In this case, it has become necessary to add variables representing innovation to the proxy variables used to measure human capital. With the addition of the innovation variable, the concept of human capital has transformed into the concept of "Innovative Human Capital."

2 Literature

Innovative human capital is derived from the concept of human capital. Human capital explained based on education and health data is considered

1 Ph.D., Independent Researcher, gozdeiskin@hotmail.com, ORCID: 0000-0001-9763-8585.

historical human capital (Diebolt & Hippe, 2019: 543), while the type of capital that includes productivity and efficiency is considered modern human capital (Usman & Adeyinka, 2019: 30). The process referred to as modern human capital in this study will be referred to as innovative human capital. Innovative human capital is the fundamental driving force of productivity and economic growth. Innovative human capital, as a unique resource, plays a pivotal role in national economic development in the era of innovation and technology (Xu & Li, 2020: 2). In the context of high-quality development, the significance of scientific and technological innovation is on the rise, and the contribution of innovative human capital to fostering high-quality progress is becoming increasingly evident (Jin et. al., 2023: 354). Innovative human capital can be defined as a kind of heterogeneous capital that always possesses cutting-edge knowledge and skills in a specific professional field, continuously carries out innovative activities, and obtains innovative output, so that the marginal income can continue to increase (Lin et. al., 2021: 2). Innovative human capital encapsulates willingness to change in addition to education and training. It plays a decisive role in gaining a competitive advantage (McGuirk et. al., 2015: 965). According to Xu and Li (2020), this concept encompasses strong productive characteristics, including concepts such as entrepreneurship, research and development, management, and skills. Human capital is represented by education, health, and innovation (Mabrouki, 2018: 957), or more precisely, this process is referred to as the innovative human capital process. In light of all this information, human capital encompasses experience and existing knowledge. Innovative human capital, on the other hand, involves using this knowledge to generate innovations and added value.

3 Innovative Human Capital Proxies

The measurement of human capital is still highly complex because of its intangible characteristics (Aleknavičiūtė et al., 2016: 117). Therefore, it is crucial to determine the components to be used in measuring the concept optimally.

In the literature, enrollment rates in primary education (Sala-i-Martin et al., 2003: 21), secondary education (Mankiw et al., 1992: 419), and tertiary education are used as proxy variables for the education indicator. In addition to these, Barro and Lee, in their study in 1993, calculated the educational attainment of the population over the age of 25 by using the average years of schooling data alongside primary, secondary, and higher education (Barro & Lee, 1993: 383). Barro and Lee expanded this analysis in 2001 and also examined the international test scores of students (Barro & Lee, 2001: 554). This change in the analysis is actually a result of a search for efficiency rather than using numerical values for education data. Nowadays, the PISA test calculated by the OECD is a test used to measure such efficiency. It is very important to select variables that represent the quality of education when choosing the proxy variables to obtain accurate results in analyses. For example, the use of literacy rates is not a very appropriate variable for today's world. This is because this measurement only covers the primary education level, and in many countries around the world, the number of illiterate people is quite low. Therefore, the literacy rate is not a determining proxy variable for education in today's world. The higher the education level of individuals, the more likely they are to produce innovation (Akcigit et al., 2020: 26).

In light of the aforementioned information, mean years of schooling, population with tertiary education, and the PISA Index results of the 38 OECD member countries have been examined by ranking them from largest to smallest. In the 2017 ranking of population with tertiary education, Türkiye was ranked 32nd among 38 countries. In 2021, its ranking improved to 28th among 37 countries. This indicates an increase in the population with tertiary education in Türkiye. When the data on the mean years of schooling are examined, Türkiye ranked last among the 38 OECD countries in both 2017 and 2021. Although the population with tertiary education suggests an increase in the level of schooling in Türkiye, there has been no change in its ranking regarding the mean years of schooling. In the PISA index results, Türkiye was ranked 32nd among 36 countries in 2018 and 31st among 38 countries in 2022. The data related to these rankings are provided in Appendix A. When all these variables are examined, it is necessary to consider how significant the population with tertiary education is in reflecting the efficiency of education. Moreover, it should be investigated whether the increase in the PISA index, despite no increase in the mean years of schooling, is due to an increase in Türkiye's efficiency.

Figure 1 shows the average PISA test scores for the years 2015, 2018, and 2022. The gap between the average performance of Türkiye and OECD countries has steadily decreased over these three years (67.67 → 25.67 → 16). It was observed that the performance of OECD countries significantly declined from 2018 to 2022. While the decrease in the average OECD was 10.67 points, the decrease in Türkiye was found to be 1 point. In this case, it can be said that the positive picture for Türkiye is not due to an increase in Türkiye's performance, but rather due to a significant decline in the performance of OECD countries.

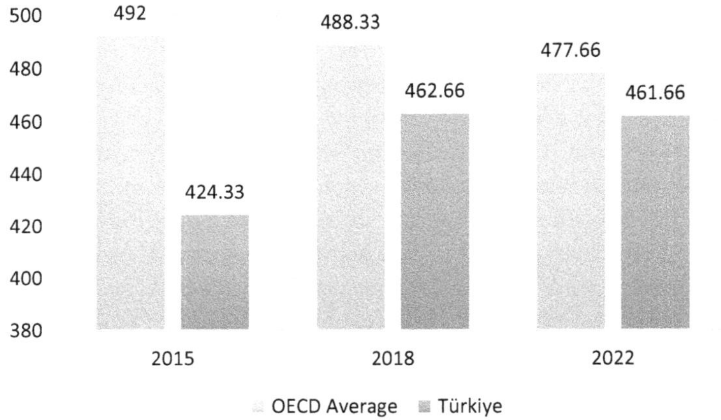

Figure 1. PISA Performance
Source: OECD (last accessed May 23, 2024)

Based on Mincer's study (Mincer, 1984: 196), which emphasizes the importance of health alongside education, the second variable considered as an indicator of human capital includes health data. The number of days spent in education is reduced due to unhealthy conditions, thereby decreasing the effectiveness of educational investments. Improving health and increasing life expectancy reduces the depreciation rate of education (Mushkin, 1962: 131). The depreciation of human capital includes deterioration in individuals' health, erosion or obsolescence of skills (Mincer, 1984: 196), or work experience being monopolized by a single person without being passed on to subsequent generations. To achieve the outcomes of investments in human capital, the individuals constituting this capital must be healthy. The indicator variables related to health include health expenditure (Mushkin, 1962: 136) and life expectancy at birth. As a result of the COVID-19 pandemic, the health variable used in human capital measurements has become more important. The cost of expenditures has increased for countries with fragile health sectors. The burden on countries has increased due to both human loss and costs arising from immediate investments rather than continuous and cumulative investments in the health sector. In this context, the average percentage of health spending of GDP in the 38 OECD countries increased from 8.69 % in 2017 to 9.57 % in 2021. The COVID-19 pandemic is among the reasons for this increase. Türkiye, included among the mentioned countries, ranked last in both years. When life expectancy data are examined, the average life expectancy of OECD member countries decreased from 80.47 years in 2017 to 79.96 years in 2021. Despite the decrease in Türkiye's average life expectancy from 77.14 years to 76.03 years, Türkiye moved from 33rd place in 2017 to 32nd place in 2021. This improvement in ranking for Türkiye, despite being last in health spending, can be considered a positive outcome. However, steps can be taken to increase life expectancy by increasing investments in health spending. Information containing data on health spending selected to represent innovative human capital is provided in Appendix B.

The education and health expenditures used to explain human capital have been mentioned. Another method used in the same calculations is the productivity and efficiency approach. The measurement of efficiency depends on innovation capacity. Innovation is not just a way to increase the wealth of nations; it also allows people to do things that have never been done before. It will determine whether the quality of life will improve or deteriorate. Producing innovation means not only producing larger quantities of the same goods but also producing a range of goods and services that did not previously exist outside our dreams (Freeman & Soete, 2003: 2). The relationship between technology, which is part of the innovation creation process, and human capital was explained by Jacob Mincer in 1984, where technology was expressed as a stock of knowledge. Human capital not only includes the transfer and embodiment of existing knowledge in people but also the production of new knowledge, which is the source of innovation and technical change. Without technological progress, capital accumulation cannot be sustained, marginal productivity begins to decline, and eventually, per capita income growth inevitably approaches zero (Freeman & Soete, 2003: 363). Innovative human capital will have a direct impact on the economy by increasing the productivity of the workforce and an indirect impact on economic growth based on increased innovation capacity. That is, sustainable growth will be achieved when the technological innovations obtained are combined with a workforce of increased quality. Therefore, it is crucial to select variables that represent a country's innovation capacity. In this context, R&D expenditures (%GDP) were selected to represent investment in skills, and the percentage of patents in force to the population was selected to represent the output, i.e., productivity, for 38 OECD countries for the years 2017 and 2021. Data on high-technology exports (HTE) were examined as an indicator of productivity.

As a result of the examination, in the data on R&D expenditures (% of GDP), Türkiye ranked 29th among 38 countries in 2017 and 28th among 34 countries in 2021 (data for Australia, Chile, Colombia, and Mexico were not available). The gap between the OECD average and Türkiye was 0.69 % in 2017 and increased to 0.85 % in 2021. This indicates that R&D expenditures relative to GDP have decreased in Türkiye. In order to observe the impact of R&D expenditures on productivity, the number of patents in force per capita was examined. However, to eliminate the effects of population differences among countries, the number of patents in force was proportioned to the populations of the countries to obtain percentages. When examining the ratio of patents in force to population, Türkiye ranked 34th among 37 countries in 2017 (data for Slovenia were missing) and remained 34th among 38 countries in 2021. In summary, while Türkiye ranks 28th and 29th in R&D expenditures, it ranks 34th in the ratio of patents in force. This indicates that Türkiye is not using its R&D expenditures efficiently. When examining the data on high-technology exports, selected as a proxy variable to represent productivity, for the same countries and years, Türkiye ranks last in both years. The numerical data related to these rankings are provided in Appendix C.

4 Conclusion

In conclusion, when analyzing the efficiency and productivity of human capital, it is necessary to approach the subject holistically. High levels of education data alone, or high health or R&D expenditures, are not sufficient indicators to measure innovative human capital. In an economy with a short life expectancy, even if the level of schooling is high, the participation rates of individuals in the workforce will decrease. Similarly, in an economy with high R&D expenditures, if patent applications or patents in force and high-technology exports are low, this indicates low productivity and efficiency. Therefore, it is crucial to select proxy variables representing innovative human capital in a way that reflects productivity and efficiency to ensure the reliability of the results.

References

Akcigit, U., Pearce, J. G., & Prato, M. (2020). "Tapping into Talent: Coupling Education and Innovation Policies for Economic Growth". *National Bureau of Economic Research*. (No. w27862).

Aleknavičiūtė, R., Skvarciany, V., & Survilaitė, S. (2016). "The Role of Human Capital for National Innovation Capability in EU Countries". *Economics and Culture*, *13*(1), 114–125.

Barro, R. J., & Lee, J. W. (1993). International comparisons of educational attainment. *Journal of Monetary Economics*, *32*(3), 363–394.

Barro, R. J., & Lee, J. W. (2001). International data on educational attainment: updates and implications. *Oxford Economic Papers*, *53*(3), 541–563.

Diebolt, C., & Hippe, R. (2019). The long-run impact of human capital on innovation and economic development in the regions of Europe. *Applied Economics*, *51*(5), 542–563.

Freeman, C., & Soete, L., (2003). *Yenilik iktisadı* (Trans. E. Türkcan). 3rd edn. TUBİTAK Yayınları.

Jin, W., Gao, S., & Pan, S. (2023). Research on the impact mechanism of environmental regulation on green total factor productivity from the perspective of innovative human capital. *Environmental Science and Pollution Research*, *30*(1), 352–370.

Laverde, H., Correa, J. C., & Jaffe, K. (2018). *"A New Index of Human Capital to Predict Economic Growth"*. arXiv preprint arXiv:1807.07051.

Lin, X., Zhao, Y., Ahmad, M., Ahmed, Z., Rjoub, H., & Adebayo, T. S. (2021). Linking innovative human capital, economic growth, and CO2 emissions: an empirical study based on Chinese provincial panel data. *International Journal of Environmental Research and Public Health*, *18*(16), 8503.

Mabrouki, M. (2018). Supporting economic growth through innovation: How does human capital influence the rate of growth. *Economics Bulletin, AccessEcon*, *38*(2), 957–972.

Mankiw, N. G., Romer, D., & Weil, D. N. (1992). A contribution to the empirics of economic growth. *The Quarterly Journal of Economics*, *107*(2), 407–437.

McGuirk, H., Lenihan, H., & Hart, M. (2015). Measuring the impact of innovative human capital on small firms' propensity to innovate. *Research Policy*, *44*(4), 965–976.

Mincer, J. (1984). "Human capital and economic growth". *Economics of Education Review*, 3(3), 195–205.

Mushkin, S. J. (1962). "Health as an Investment". NBER Chapters . In: *Investment in Human Beings* (pp. 129–157). *National Bureau of Economic Research*, Inc.

Sala-i-Martin, X., & V Artadi, E. (2003). Economic growth and investment in the Arab world.

Schultz, T. W. (1961). "Investment in Human Capital". *The American Economic Review*, *51*(1), 1–17.

Schultz, T. W. (1981). *Investing in People: The Economics of Population Quality*. University of California Press.

Usman, F. K., & Adeyinka, O. B. (2019). Effect of human capital development on economic growth of ECOWAS member states. *Advances in Sciences and Humanities*, *5*(1), 27–42.

Xu, Y., & Li, A. (2020). The relationship between innovative human capital and interprovincial economic growth based on panel data model and spatial econometrics. *Journal of Computational and Applied Mathematics*, 365, 112381

Appendix A

	Mean Years of Schooling				Population with tertiary education				PISA			
Order	Country	2017	Country	2021	Country	2017	Country	2021	Country	2018	Country	2022
1	Germany	14,1	Germany	14,1	South Korea	69,8	South Korea	69,3	Estonia	525	Japan	533
2	Canada	13,8	Switzerland	13,9	Canada	60,9	Canada	66,4	Japan	520	South Korea	523
3	Switzerland	13,7	Canada	13,8	Japan	60,4	Japan	64,8	South Korea	520	Estonia	516
4	Estonia	13,6	Iceland	13,8	Lithuania	55,6	Luxembourg	63,1	Canada	517	Canada	506
5	Iceland	13,5	United States	13,7	Ireland	53,5	Ireland	62,9	Finland	516	Ireland	504
6	United States	13,4	Estonia	13,6	Australia	52	Lithuania	57,5	Poland	513	Switzerland	498
7	Japan	13,3	Lithuania	13,5	United Kingdom	51,6	United Kingdom	57,5	Ireland	505	Australia	497
8	New Zealand	13,2	United Kingdom	13,4	Luxembourg	51,4	Netherlands	55,6	Slovenia	504	Poland	495
9	Lithuania	13,2	Japan	13,4	Switzerland	50,1	Norway	55	United Kingdom	504	Finland	495
10	United Kingdom	13,2	Israel	13,4	Norway	48,3	Australia	54,3	New Zealand	503	New Zealand	495
11	Israel	13,2	Latvia	13,3	Israel	48	Switzerland	52,3	Netherlands	502	Belgium	494
12	Latvia	13,1	Poland	13,2	Netherlands	48	United States	51,2	Sweden	502	Austria	491
13	Poland	13,1	Luxembourg	13	United States	47,8	Belgium	50,9	Denmark	501	Denmark	491
14	Denmark	13	Norway	13	Sweden	47,4	France	50,3	Germany	500	United States	489
15	Norway	13	Denmark	13	Belgium	45,7	Denmark	49,2	Belgium	500	Sweden	488
16	Czech Republic	12,8	New Zealand	13	Denmark	45,5	Sweden	49,2	Australia	499	United Kingdom	486
17	Slovakia	12,8	Slovakia	12,9	Slovenia	44,6	Spain	48,7	Switzerland	498	Czech Republic	486
18	Finland	12,7	Czech Republic	12,9	France	44,3	Slovenia	47,9	Norway	497	Slovenia	485
19	Slovenia	12,7	Finland	12,9	New Zealand	44,2	Portugal	47,5	Czech Republic	495	Latvia	484
20	Luxembourg	12,7	Slovenia	12,8	Poland	43,5	Israel	46	United States	495	Germany	482
21	Sweden	12,5	Australia	12,7	Estonia	43	Latvia	45,5	France	494	Netherlands	480
22	Australia	12,4	Sweden	12,6	Iceland	42,7	New Zealand	45,3	Portugal	492	France	478
23	South Korea	12,3	Netherlands	12,6	Spain	42,6	Greece	44,2	Austria	491	Luxembourg	478
24	Netherlands	12,3	South Korea	12,5	Greece	42,5	Estonia	43,2	Latvia	487	Portugal	478
25	Belgium	12,2	Belgium	12,4	Latvia	41,6	Austria	42,4	Iceland	481	Hungary	477
26	Austria	12,2	Austria	12,3	Finland	41,3	Iceland	41,9	Lithuania	480	Spain	477
27	Hungary	12	Hungary	12,3	Austria	40,3	Poland	40,6	Hungary	479	Lithuania	477
28	Ireland	11,4	France	11,6	Slovakia	35,1	**Türkiye**	**39,8**	Italy	477	Italy	477
29	France	11,4	Ireland	11,6	Portugal	34	Slovakia	39,5	Luxembourg	477	Norway	474
30	Chili	10,6	Greece	11,4	Czech Republic	33,8	Germany	35,9	Slovakia	469	Israel	466
31	Greece	10,5	Chili	10,9	Chili	33,7	Czech Republic	35	Israel	465	**Türkiye**	**462**
32	Italy	10,5	Italy	10,7	**Türkiye**	**31,6**	Hungary	32,9	**Türkiye**	**463**	Slovakia	458
33	Spain	10,1	Spain	10,6	Germany	31,3	Colombia	32	Greece	453	Iceland	447
34	Portugal	9,22	Portugal	9,58	Hungary	30,2	Costa Rica	30,3	Chili	438	Greece	436
35	Mexico	8,74	Mexico	9,22	Colombia	28,6	Italy	28,3	Mexico	416	Chili	435
36	Costa Rica	8,64	Colombia	8,86	Costa Rica	28	Mexico	27,1	Colombia	405	Mexico	407
37	Colombia	8,36	Costa Rica	8,8	Italy	26,8	Finland	4,06	Spain	*	Costa Rica	404
38	**Türkiye**	**8,27**	**Türkiye**	**8,63**	Mexico	22,6	Chili		Costa Rica	**	Colombia	402

Source: Mean Years of Schooling: HDI (https://globaldatalab.org/shdi/table/msch/ last accessed May 23, 2024)
Population with tertiary education: OECD (https://data.oecd.org/eduatt/population-with-tertiary-education.htm last accessed May 23, 2024)
PISA: OECD (https://www.oecd.org/pisa/data/ last accessed May 23, 2024)
* Spain's reading results were not published due to unexpected responses from students. The average score for mathematics and science is 482.
** Costa Rica was not included in the calculation as it became an OECD member on May 25, 2021.

Appendix B

	Life Expectancy				Health Spending (% GDP)			
Order	Country	2017	Country	2021	Country	2017	Country	2021
1	Japan	84,2	Japan	84,78	United States	16,768	United States	17,363
2	Switzerland	83,41	Australia	84,53	France	11,355	Germany	12,934
3	South Korea	83,29	Switzerland	83,99	Germany	11,336	United Kingdom	12,365
4	Spain	83,03	South Korea	83,7	Switzerland	11,045	Canada	12,334
5	Australia	83	Norway	83,23	Canada	10,906	France	12,308
6	Italy	82,73	Spain	83,01	Belgium	10,799	Austria	12,096
7	Israel	82,72	Sweden	82,98	Sweden	10,785	Switzerland	11,801
8	Norway	82,62	Italy	82,85	Japan	10,656	Japan	11,299
9	Iceland	82,54	Iceland	82,68	Austria	10,384	Sweden	11,247
10	France	82,47	Canada	82,66	Norway	10,23	Portugal	11,137
11	Sweden	82,44	Luxembourg	82,63	Australia	10,123	Belgium	11,043
12	New Zealand	82,22	France	82,5	Netherlands	10,108	Denmark	10,824
13	Canada	82,02	New Zealand	82,45	Denmark	10,098	Spain	10,74
14	Ireland	81,92	Israel	82,26	United Kingdom	9,596	Australia	10,59
15	Luxembourg	81,73	Finland	82,04	Portugal	9,306	Finland	10,25
16	Netherlands	81,73	Ireland	82	Finland	9,127	New Zealand	10,125
17	Austria	81,64	Belgium	81,88	Chili	9,091	Norway	9,919
18	Finland	81,5	Netherlands	81,69	New Zealand	8,97	Iceland	9,734
19	Portugal	81,5	Austria	81,58	Spain	8,945	Czech Republic	9,489
20	Belgium	81,36	Denmark	81,38	Italy	8,678	Slovenia	9,478
21	United Kingdom	81,17	Portugal	81,04	Iceland	8,261	Italy	9,384
22	Denmark	81,11	United Kingdom	80,74	Slovenia	8,185	Chili	9,343
23	Slovenia	81,07	Slovenia	80,69	Greece	8,135	South Korea	9,331
24	Germany	80,94	Germany	80,63	Colombia	7,679	Greece	9,173
25	Greece	80,85	Greece	80,11	Czech Republic	7,381	Latvia	9,045
26	Chili	80,35	Chili	78,94	Israel	7,176	Colombia	9,018
27	Costa Rica	79,38	Czech Republic	77,73	Ireland	7,125	Israel	7,905
28	Czech Republic	78,96	United States	77,2	South Korea	7,047	Lithuania	7,823
29	United States	78,82	Estonia	77,14	Costa Rica	7,045	Slovakia	7,751
30	Estonia	77,98	Costa Rica	77,02	Slovakia	6,757	Costa Rica	7,569
31	Poland	77,72	Poland	76,46	Hungary	6,744	Estonia	7,491
32	Slovakia	77,22	**Türkiye**	**76,03**	Estonia	6,597	Hungary	7,376
33	**Türkiye**	**77,14**	Slovakia	74,91	Poland	6,583	Ireland	6,717
34	Colombia	76,65	Hungary	74,53	Lithuania	6,464	Poland	6,439
35	Hungary	76,01	Lithuania	73,72	Latvia	5,965	Mexico	6,079
36	Lithuania	75,44	Latvia	73,58	Mexico	5,458	Netherlands	6,079
37	Latvia	74,81	Colombia	72,83	Luxembourg	5,135	Luxembourg	5,668
38	Mexico	74,14	Mexico	70,21	**Türkiye**	**4,18**	**Türkiye**	**4,565**
Average		80,4692		79,9561		8,69008		9,57453

Source: Life expectancy: HDI (https://globaldatalab.org/shdi/table/shdi/?levels=1&interpolation=0&extrapolation=0 last accessed May 23, 2024)-Health Spending (%GDP): OECD (https://data.oecd.org/healthres/health-spending.htm last accessed May 23, 2024).

Appendix C

	R&D (% of GDP)				Patent in Force (%Population)*				HTE			
Order	Country	2017	Country	2021	Country	2017	Country	2021	Country	2017	Country	2021
1	Israel	4,624	Israel	5,557	Luxembourg	12,0248	Luxembourg	22,495875	Korea, Rep.	32,5456	Korea, Rep.	36,1263
2	South Korea	4,292	South Korea	4,93	Ireland	3,52485	Ireland	3,9190855	Ireland	29,334	Iceland	33,4873
3	Sweden	3,363	United States	3,456	Switzerland	2,46126	Switzerland	3,0433408	Iceland	26,376	Israel	29,5924
4	Japan	3,166	Belgium	3,43	Iceland	1,92574	Iceland	2,5045635	France	25,997	Ireland	28,4778
5	Switzerland	3,077	Sweden	3,402	Korea, Rep.	1,89029	Korea, Rep.	2,2288584	Netherlands	22,9041	United Kingdom	23,8537
6	Austria	3,057	Switzerland	3,359	Austria	1,66955	Austria	1,6983971	United Kingdom	22,6067	Australia	22,3902
7	Germany	3,047	Japan	3,296	Japan	1,58593	Japan	1,6075735	Norway	21,5968	Netherlands	21,9774
8	Denmark	2,931	Austria	3,256	Belgium	1,15134	Belgium	1,3578315	Mexico	21,1661	France	21,9195
9	United States	2,904	Germany	3,129	Denmark	1,01464	Netherlands	1,2140847	Israel	21,067	Belgium	21,3797
10	Finland	2,728	Finland	2,985	Netherlands	0,96828	Denmark	1,1915175	United States	19,2588	Estonia	20,6288
11	Belgium	2,667	United Kingdom	2,915	Sweden	0,9632	Sweden	1,0756244	Costa Rica	18,4952	Norway	20,5851
12	United Kingdom	2,323	Iceland	2,805	Finland	0,92161	Germany	1,0550533	Australia	18,0894	Czechia	20,3523
13	France	2,199	Denmark	2,761	United States	0,91806	France	1,037924	Estonia	17,9345	United States	19,8939
14	Netherlands	2,179	Netherlands	2,269	France	0,84237	United Kingdom	1,0296422	Czechia	17,825	Mexico	19,7985
15	Iceland	2,084	France	2,219	Germany	0,79576	Finland	1,0259308	Latvia	17,7179	Japan	17,976
16	Norway	2,082	Slovenia	2,128	United Kingdom	0,78258	United States	1,0021758	Japan	17,5655	Latvia	16,9802
17	Slovenia	1,865	Czech Republic	1,996	New Zealand	0,75114	Norway	0,8950469	Hungary	17,1829	Costa Rica	16,967
18	Australia	1,794	Norway	1,938	Estonia	0,73707	Estonia	0,8841924	Germany	15,8473	Hungary	16,2479
19	Czech Republic	1,769	Estonia	1,752	Norway	0,6282	Slovenia	0,8682787	Sweden	15,1064	Germany	15,3814
20	Canada	1,687	Canada	1,697	Australia	0,5878	Italy	0,6928362	Canada	14,3484	Canada	15,0471
21	Italy	1,37	Portugal	1,681	Canada	0,49453	Australia	0,6124605	Switzerland	14,0234	Switzerland	14,234
22	New Zealand	1,349	Hungary	1,642	Italy	0,49172	Latvia	0,5691726	Austria	12,8819	Sweden	13,93
23	Portugal	1,319	New Zealand	1,468	Latvia	0,4535	Canada	0,5326227	Lithuania	12,5773	Denmark	13,7672
24	Hungary	1,317	Greece	1,456	Czechia	0,39272	New Zealand	0,5037563	Denmark	12,4973	Chile	12,4417
25	Estonia	1,277	Italy	1,454	Israel	0,37602	Czechia	0,4869038	Greece	12,109	Greece	12,256
26	Ireland	1,249	Poland	1,432	Portugal	0,35748	Spain	0,4760882	Slovak Republic	11,7654	Lithuania	11,5086
27	Luxembourg	1,239	Spain	1,429	Lithuania	0,35547	Lithuania	0,4540782	Belgium	11,4913	Austria	11,0064
28	Spain	1,21	**Türkiye**	**1,404**	Slovak Republic	0,32753	Portugal	0,4046293	Poland	10,6663	Finland	10,2937
29	**Türkiye**	**1,176**	Ireland	1,111	Hungary	0,26793	Slovak Republic	0,4027723	Finland	9,65604	New Zealand	9,76044
30	Greece	1,152	Lithuania	1,111	Greece	0,25046	Israel	0,3792283	New Zealand	9,2522	Poland	9,44439
31	Poland	1,038	Luxembourg	1,042	Spain	0,23336	Hungary	0,3583665	Colombia	8,96235	Spain	9,39258
32	Lithuania	0,896	Slovakia	0,915	Poland	0,20009	Greece	0,2765392	Italy	7,80431	Slovak Republic	9,0013
33	Slovakia	0,885	Latvia	0,741	Mexico	0,09168	Poland	0,2473407	Chile	7,09874	Colombia	8,20894
34	Latvia	0,511	Costa Rica	0,275	**Türkiye**	**0,08577**	**Türkiye**	**0,0979532**	Spain	6,99861	Italy	7,76932
35	Costa Rica	0,431	Australia	**	Chile	0,06745	Mexico	0,0924745	Luxembourg	6,96873	Portugal	6,25025
36	Chili	0,357	Chili	**	Costa Rica	0,01662	Chile	0,0736616	Slovenia	6,50834	Slovenia	6,20886
37	Mexico	0,328	Colombia	**	Colombia	0,01453	Costa Rica	0,0230503	Portugal	5,98808	Luxembourg	6,07486
38	Colombia	0,261	Mexico	**	Slovenia	**	Colombia	0,0164161	**Türkiye**	**3,2327**	**Türkiye**	**3,28356**
Average		1,87376		2,24826		1,09787		1,495667		15,3539		16,1551

*Patent in Force (% Population)=Patent in force/Population*100
**Missing value

Özlem Kırlı[1] & Soner Uysal[2] & Cem Mehmet Baydur[3]

Rate of Profit, Political Power and Sraffian Inflation: The Case of Türkiye

1 Introduction

The referendum held in 2017 approved the transition to a presidential government system in Türkiye, and this transition was put into practice with the 2018 general elections. The changes made to the constitution with the transition to the presidential government system have caused significant debates. The fact that the president, who is the head of the executive, is also the head of the political party in terms of legislative activities and can make important appointments to the judiciary has raised questions about the separation of powers and concerns about the economic consequences of this situation. However, these discussions were overshadowed by the Covid-19 that emerged in Wuhan Province of China in late 2019 and spread across the world in a short time. As the effects of the pandemic gradually disappeared, Türkiye's agenda turned to the 2023 general elections. In the run-up to the 2023 general elections, Türkiye faced high inflation along with a low interest rate – high exchange rate policy. In particular, there was a heated debate about the significant discrepancies between the official inflation data published by the Turkish Statistical Institute (TurkStat) and the perceived inflation data. Indeed, amid these debates, the Inflation Research Group[4] (ENAG) has started to publish its own inflation data from 2020. There are significant differences between TurkStat data and ENAG data. For example, while the inflation rate for 2020 is 14.6 % according to TurkStat, it is 36.7 %

1 Ph.D., Assoc. Prof., Muğla Sıtkı Koçman University, Department of Political Science, ozlemkirli@mu.edu.tr, ORCID: 0000-0002-2824-1725.

2 Ph.D., Muğla Sıtkı Koçman University, Department of Economics, soneruysal@mu.edu.tr, ORCID: 0000-0002-8293-5509.

3 Ph.D., Prof., Muğla Sıtkı Koçman University, Department of Economics, cbaydur@mu.edu.tr, ORCID: 0000-0002-2680-7350.

4 The Inflation Research Group (ENAG) is a community that has been calculating inflation with a different methodology than TurkStat since 2020, particularly the composition of goods in the inflation basket. Detailed information on ENAG can be found at https://enagrup.org/?hl=en.

according to ENAG. Similarly, TurkStat announced the inflation rate for 2021 as 36 %, but this rate is 83 % according to ENAG.

Many reasons are cited for the rise in inflation, including the pandemic, the Russia-Ukraine war and the government's interest rate policy. The aim of this study is to calculate inflation by taking into account the distribution and capital accumulation in Türkiye, different from the inflation rates of TurkStat and ENAG. Thus, the reliability of the inflation rates of TurkStat and ENAG will be checked. In order to do this, prices will be analyzed from a Sraffian framework in terms of distribution. In capitalist society, workers earn wages as owners of their labor and capitalists earn profits as owners of their capital. However, it should be noted that there is nothing that defines how much each of them should receive. Bhaduri (2021) notes that this is the essence of the distributional ambiguity that capital theory deals with. In the face of this uncertainty, political power steps in and constantly regulates social production and distribution. If capital is a relationship network, the political mechanism is at the center of this process. The dominant neoclassical ideology has excluded the political sphere from economic analysis. However, economic science should not be treated as a discipline independent of political science and with its own universal laws.

Neoclassical theory has recognized 'capital' as a measurable tangible asset and economic policies have been developed based on the theory of profit and interest (Bhaduri, 2021). However, if capital is not measurable due to its complex structure and relations, then we must recognize that relative scarcity as a determinant of the rate of profit is a meaningless proposition. Uncertainty about capital and its return is an insurmountable obstacle. Equilibrium requires that the rate of profit that prevails today should be expected to prevail today, and that the expectation of future profits achieved today should be expected to be achieved today when investing in any plant that exists today. The value of capital existing today is therefore equal to the supply price calculated in this way. The heavy burden that this method of capital valuation places on equilibrium assumptions emphasizes the impossibility of valuing capital in an uncertain world. In a world of unexpected events that change values, the perspectives of the businessman making investment decisions about the future and the scientist making observations about the past are irreconcilable, and all we can do is limp along blindly with a traditional method of measuring capital that satisfies neither (Robinson, 1953). Sraffa (1960) aimed to get rid of the uncertainty problem by treating the rate of profit as exogenous and independent of prices. Sraffa (1960) stated that the monetary interest rate could be taken as an indicator for the rate of profit, but went no further. Something external is necessary to complete the theory of value. This must be political power. This study investigates that the rate of profit is directly determined by governments in terms of capital accumulation and political financing.

In Türkiye in particular, political power is an important factor in the capital accumulation process. Governments concentrate on certain sectors and redistribute wealth by controlling capital accumulation. In Türkiye, the construction/housing sector is fundamental to capital accumulation. Comparing the rate of profit determined by the guarantees given to many large public investments in the construction sector with other sectors is vital for understanding inflation dynamics in Türkiye. In this study, we will try to understand inflation in Türkiye within the framework of Sraffa's (1922) work analyzing the relations between money, policy and inflation and his (1960) fundamental work on the Production of Commodities by Means of Commodities. The aim of the study is to develop a definition of inflation based on sectoral profit differentials. Also, we have a time constraint: We will try to analyze inflation in the period 2018–2023, when Türkiye transitioned to the Presidential Government System.

2 A Sraffian Model: The Case of Türkiye

In Sraffian model, there are two main determinants of prices: technology and distribution. In this study, technology will be taken as given (since it is analyzed in the short run), and thus we will focus on profits and profit expectations in order to make sense of the relationship between distribution and prices. According to Sraffa (1960), the profit rate is the exogenous variable that determines prices. In the standard model, the share of the wage depends on the profit target of capitalists, given the maximum rate of profit. To see the relationship between prices and profits, we need to go back to the standard model. The model starts with an economy with n sectors. The net national surplus is distributed between capitalists (K) and workers (L). In the model "r" denotes the rate of profit and "w" the wage.

$$\begin{aligned} &K_1(1+r_1)+L_1w=Q_1 \\ &\vdots \\ &K_n(1+r_n)+L_nw=Q_n \end{aligned} \tag{1}$$

Assuming that all surplus is taken by workers, the wage rate will be maximum $(w=w_{\max})$ and the profit rate zero $(r=0=r_1=r_2=\ldots=r_n)$.

$$
\begin{aligned}
&K_1 + L_1 w_{\max} = Q_1 \\
&\vdots \\
&K_n + L_n w_{\max} = Q_n
\end{aligned}
\tag{2}
$$

In the case where wages gradually fall $(dw < 0)$ below the maximum level, the profit rate will be positive. We can illustrate this situation as follows.

$$
\begin{aligned}
&r_1 = \frac{L_1}{K_1} dw \\
&\vdots \\
&r_n = \frac{L_n}{K_n} dw
\end{aligned}
\tag{3}
$$

Unless the $\frac{L}{K}$ ratio is equal in all sectors, the profit rate will differ across sectors. According to Sraffa (1960), inter-sectoral profit differentials are the main force driving prices. Sraffa (1960) argues that no matter how complex the pattern of price volatility arising from a change in distribution, their net result is to restore equilibrium in each sector. In this case, if equation (1) is solved for the profit rate, the following equation will be obtained.

$$
(1 + r_i) = \frac{Q_i}{K_i}\left(1 - \frac{wL_i}{Q_i}\right) \tag{4}
$$

Equation (4) above shows that it is not possible to generalize about the extent to which the rate of profit will change when wages change. To get out of the spiral of variation between wages and profits, let us choose a critical sector (e) and assume that all related sectors have equal $\frac{L}{K}$ ratios:

$$\begin{aligned} &K_e(1+r_e)+L_e w=Q_e \\ &\frac{Q_e-K_e}{K_e}=r_e+\frac{L_e}{K_e}w \\ &Q_e-K_e=Y_e \\ &\frac{Y_e}{K_e}=r_e+\frac{L_e}{K_e}w \end{aligned} \tag{5}$$

We can treat net national income as a composite good and set its value equal to 1. In this case, we can write $w=w_{\max}-dw=1-dw$ and $\frac{L_e}{K_e}=\frac{r_e}{dw}$. If these transformations are substituted in equation (5) and solved for the profit rate, a price-independent distribution relationship can be obtained:

$$r_e=\frac{Y_e}{K_e}(1-w) \tag{6}$$

In equation (6) above, the $\frac{Y_e}{K_e}$ ratio indicates the maximum profit rate $(R=r_{\max})$, in which case the distribution relationship can be rewritten as follows:

$$r=R(1-w) \tag{7}$$

Assuming that the maximum rate of profit is constant in the short run, the process of wage determination will become clearer. If equation (7) above is solved for wages, the following equation is obtained:

$$w=1-\frac{r}{R} \tag{8}$$

According to equation (8), given R, price changes will determine a wage level that is consistent with profit targets. Sraffa (1960) explains that the reason for initially choosing the wage as the independent variable is that it is assumed to consist of a basket of goods determined by physiological or social conditions independent of prices. But this assumption loses much of its force as soon as the possibility of changing the division of the product arises. The situation is reversed when the wage is taken as a given in terms of an abstract measure, which has no meaning until financial prices are determined. Sraffa (1960) argues that the rate of profit has a meaning independent of prices and can be taken as given before prices are determined. Thus, it can be said that the rate of profit is an independent variable in Sraffa. Sraffa (1960) argues that the profit rate is affected by the monetary interest rate, but goes no further. However, a different Sraffa emerges in his previous studies.

It can be said that the political and economic structure, which creates complex networks of relations, determines profits, rents and thus prices (Sraffa, 1922; Panico, 2021). While Sraffa recognizes that monetary interest can be effective in determining rents, the basic assumption of this study is that in many developing countries, the political power of the government and the opposition affect/determine rents/profits. To incorporate these conditions, we will write a "housing distribution equation" where the rate of profit is determined by politics and a "housing distribution equation" in a market without political favoritism. And these equations will be compared with each other.

Governments have many instruments such as budget, borrowing and asset sales. Governments also have monopoly power through legal regulations. Minsky (1986) argues that this monopoly power gives not only governments but also capitalists in industries the power to set prices in line with their profit targets. For example, political power can provide land and financing to the construction industry under favorable conditions. Equation (9.1) below shows distributional relations determined by political power, while equation (9.2) shows distributional relations without market intervention for construction sector.

$$K(1+r_{pol})+\phi_{pol}=P_{pol} \tag{9.1}$$

$$K(1+r_{m})+\phi_{m}=P_{m} \tag{9.2}$$

In the above equations, K is the value of capital invested in the construction sector in terms of labor and other inputs, P is the price per square meter of housing units produced and l is the value of land (rent). Considering that land value can be obtained very cheaply or free of charge due to political power, $\phi_{pol}=0$ can be taken. In this case, equation (9.1) is subtracted from equation (9.2) to obtain the price change equation "based on sectoral profit rate differences".

$$K(r_{pol} - r_m) - \phi_m = P_{pol} - P_m \tag{10}$$

Equation (10) above is written for the construction sector only. For any "i" sector in the economy, the same equation can be generalized as follows.

$$\begin{aligned} r_{pol} - r_m &= \Delta r_i \\ P_{pol} - P_m &= \Delta P_i \\ K_i \Delta r_i &= \Delta P_i + \phi_m \end{aligned} \tag{11}$$

In equation (11) above, K_i denotes the value of capital tied to inputs and labor in sector "i". Given the initial price of capital P_{0i}, we can write $P_{0i} = K_i$. This will give equation (12) below.

$$P_{0i} \Delta r_i = \Delta P_i + \phi_m \tag{12}$$

In equation (12) above, if both sides of the equation are divided by P_{0i}, the inflation rate for the relevant sector is obtained:

$$\frac{\Delta P_i}{P_{0i}} = \Delta r_i - \frac{\phi_m}{P_{0i}} \tag{13}$$

The inflation defined for sector "i" in equation (13) can be generalized for an economy with "n" sectors as follows. α_i in the equation shows the weights of sectors in the total economy.

$$\pi = \sum_{i=1}^{n} \alpha_i \frac{\Delta P_i}{P_{0i}} \tag{14}$$

π above shows the inflation rate in an economy. The change in distribution, i.e., the differentiation of profit rates between sectors, will cause prices to change. This is why the relevant inflation rate is called "Sraffian Inflation". Each sector needs to increase prices in order to preserve its capital. Otherwise it is impossible for the capitalist to maintain his capital and his share in total capital. From a Sraffian point of view, differences in profits across sectors are the main force driving prices for the survival of capital. From a political point of view, if the rate of profit reaches a level that the economy cannot sustain with the guarantees given, prices can increase indefinitely to reach the rate of profit necessary for the survival of capital. In Türkiye there is a model of capital accumulation based on the construction sector. At the center of the model are the profit rates determined by political power. Public guarantees in Build-Operate-Transfer projects guarantee a significant rate of profit. According to a report of a member of parliament, the Osman Gazi Bridge was commissioned in 2016 and will be handed over to the state in 2035. The toll for the bridge, which is guaranteed to pass 40 thousand vehicles per day, is 35 dollars plus tax[5]. The cost of the Osman Gazi Bridge is estimated at $1.3 billion. To find the present value of the bridge revenues, the 10-year treasury long-term interest rate of USA (4.01 %) is used as the discount rate. The present value of the revenue of the Osman Gazi Bridge is $6.3 billion. The profit rate of the Osman Gazi Bridge is 384 %. The present value of the 25-year revenue of the Eurasia Tunnel, with a guaranteed daily passage of 70,000 vehicles and a fare of $4 per vehicle, is $1.5 billion. The cost is $1.4 billion. The profit rate of this project is 7 %. Yavuz Sultan Selim Bridge has an annual return of $593 million and the company will receive this money whether 135 vehicles pass or not. It is under warranty for 7 years and 8 months. It can be taken approximately 8 years. The present value of the guaranteed return is $3.96 billion. The cost of the bridge is $2.5 billion. The profit rate is 58.8 %. *The average profitability of these three public projects is 149 %.* High profit rates on such public projects can cause serious deficits on the budget due to future devaluation due to dollar guarantees. With this average profit rate, two sectors will be compared: the construction sector and the manufacturing sector in Istanbul.

Since Istanbul is the heart of Türkiye's economy, we calculated the profit rate for the 2019–2022 period based on Istanbul housing data. House price/capital ratio data was obtained from a private real estate appraisal company in Istanbul. Profit ratios were calculated based on landowners' advertisements for 50 % housing supply. For example, the profit rate for the Istanbul housing sector for 2021 is calculated in equation (15) below. The 2020 Istanbul housing sector profit rate was found to be 62 %.

5 For detailed information on the cost and profit margin of the Osman Gazi Bridge and Eurasia Tunnel, please refer to https://siyasalparadigmalar.org/wp-content/uploads/2021/01/Yap-Islet-Kar-Et.pdf.

$$
\begin{aligned}
&K(1+r_m)+\phi_m=P_m \\
&\frac{K(1+r_m)}{P_m}+\frac{\phi_m}{P_m}=1 \\
&0.231(1+r_m)+0.5=1 \\
&r_m=1.164(\%116)
\end{aligned}
\qquad (15)
$$

It can be said that profit rates in the manufacturing industry increased from 34 % in 2020 to 48 % in 2023. The difference between the profit rates of these two sectors and the political profit rate is shown in Table 1. The inflation that will arise due to the sectoral profit differential, with the weight of both sectors at 50 %, is shown in the last row of the table. The Sraffian inflation should be seen as a pressure to change prices as long as the sectoral profit differential persists. In light of this information, in 2020 Sraffian inflation was 73 %, in 2021 it was 92 %, in 2022 it was 67.5 % and in 2023 it was 87.5 %. In 2020, TurkStat announced annual inflation at 14.60 %, while ENAG calculated it at 36.72 %. TurkStat itself declared that the inflation felt by the citizens was 1.5 times higher than the inflation measured by itself[6]. In 2020, the perceived inflation according to TurkStat is 22 %. In 2020, the Sraffian inflation calculated by us was 73 %. For 2021, TurkStat's official inflation rate of 36.08 %, while ENAG calculations put it at 82.81 %. Our Sraffian inflation calculation is 92.7 % and TurkStat perceived inflation is 55 %. In 2022 and 2023, TurkStat inflation is 64 %, while ENAG calculates inflation at 137 % and 127 %. While TurkStat's perceived inflation is 96 %, our Sraffian inflation is 67.5 % and 87.5 % respectively. Looking at this data, we can say the following. At this point, we can say that Sraffian inflation is close to the TurkStat's perceived inflation rate and the ENAG inflation rate.

Table 1: Official, Perceived and Sraffian Inflation

Years	2020	2021	2022	2023
$P(TL/m^2)$	6481	10094	27886	44000
$K(TL/m^2)$	1500	3100	6500	11150

(*continued on next page*)

6 TurkStat states that inflation for 2023 is calculated at 64.77 %, while consumers' expectations are 96 %. For details of this statement, please visit https://www.tuik.gov.tr/media/announcements/t%C3%BCfe%20duyuru%2024%20ocak.pdf.

Table 1: Continued

Years	2020	2021	2022	2023
$\frac{K}{P}$	0,231	0,31	0,233	0,253
Inflation by TurkStat (Official Inflation)	14,6	36	64,2	64
Inflation by ENAG	36,7	82,81	137,5	127
Perceived Inflation by TurkStat (1,5x Official)	22	54	96,1	96
Profit rate for construction in İstanbul (%)*	116	62,5	114,5	75
Profit rate of manufacturing in İstanbul (%)	34	54	48	48**
Some public projects average profit margin (%)	149	149	149	149
Profit difference for the construction industry	33	86,5	34,5	74
Profit difference for the manufacturing sector	113	99	101	101
Sraffian Inflation % $(\alpha_i = 0.5)$	73	92,7	67,5	87,5

*** The profit rate for all years is calculated as in equation (15).**
****It is assumed that the profit rate of 2022 will be maintained in 2023 and the data is taken from the Istanbul Chamber of Commerce report.**

In fact, according to the data obtained, it can be said that inflation in Türkiye is a result of the capital accumulation process. It is important to develop an anti-inflation program in line with this accumulation process. There is a significant gap between TurkStat inflation and the medium-term inflation target of 36 %. As long as this gap is maintained, inflation will not fall in Türkiye. Unless there is a change in fiscal law to equalize the profit rates in build-operate-transfer projects to market profit rates, the full burden of a contractionary fiscal policy could fall on workers, taxpayers and competitive market firms, leading to a serious loss of production and unemployment.

3 Conclusion

Based on the idea that the rate of profit is determined by political power in terms of capital accumulation and financing of politics, especially for some sectors, it seems necessary to reduce the influence of political decisions in economic processes in order to stabilize inflation in the Turkish economy. One of the claims of this study is that the main cause of inflation in Türkiye is the necessity to equalize the difference between the politically determined profit rates and the market profit rates through inflation. The inflation rate based on the difference in profit rates, which is necessary to protect the existence of capital and its distribution in different sectors, can also be seen as a limit to inflation. Sectors and capitals that fail to meet this rate lose power and position and eventually cease to exist. The persistence of high Sraffian inflation can also be interpreted as an indicator of income and wealth transfer between individuals, firms and sectors in Türkiye. As long as the politically determined profit gap between sectors persists, prices will rise and the economy will continue to destabilize. The more political interference, the greater the likelihood of a deeper crisis. The economy will eventually correct itself through a crisis and prices will come back to where capital wants them to be. In other words, prices will continue to rise until capital is able to continue its accumulation process and earn a rate of profit that will protect its structure.

References

Bhaduri, A. (2021). A Simple Course in Capital Theory: A Misunderstood Debate. *Sage Journals*, *69*(2).

Minsky, H. (1986). *Stabilizing an Unstable Economy*. New Haven, Conn.: Yale University Press.

Panico, C. (2021). Sraffa's Monetary Writings, Objectivism and the Cambridge Tradition, *Palgrave Studies in the History of Economic Thought.* In: Ajit Sinha (ed.), *A Reflection on Sraffa's Revolution in Economic Theory* (pp. 419–453). Palgrave Macmillan.

Robinson, J. (1953). The Production Function and the Theory of Capital, *Review of Economic Studies*, *21*(2), 81–106.

Sraffa, P. (1922). The Bank Crisis in Italy, *Economic Journal*, *32*, 178–197.

Sraffa, P. (1960). *Production of Commodities by Means of Commodities*. Cambridge, UK: Cambridge University Press.

Sevdagül Dengiz[1]

The Conflict Inflation Approach: A Review of Literature

1 Introduction

The conflict theory of inflation focuses on the relationship between the income expectations of people in different social classes and the current income needed to meet these expectations. This theory presents a perspective that inflation arises not only from the supply and demand of money but also from social and political conflicts. According to the perspective of this theory, inflation is associated also with historical, demographic, social, and political dynamics, alongside with economical factors (Erceg et al., 2000).

Since the 1960s, economists have proposed various views on the acceleration of inflation in the US economy. While some economists focus on the money supply, others emphasize the tightness in product and labor force markets, external shocks affecting price expectations, and the power of labor and the labor force market. Marxist political economists differ among themselves, but many of them use a "conflict theory" framework to explain inflation in modern capitalist economies. Since the conflict between workers and capitalists holds a central importance in the Marxist analysis of capitalism, it is considered natural for Marxists to model inflation within a framework that emphasizes class conflict over the production and distribution of income (Rosenberg & Weisskopf, 1981).

In this study, the conflict theory of inflation and the evolving approaches to this theory such as contributions by Rowthorn, Martin, Michl, and Lorenzoni & Werning will be discussed. Within the consideration of economic and social impacts of conflict inflation, the concept will be comprehensively analyzed within a theoretical framework.

2 What is Inflation Conflict Theory?

The term "conflict theory" is often used to refer to a theory that explains inflation as a result of the conflict between labor and capital. It is also a variant of

1 Ph.D., Independent Researcher, sevdaguldengiz@gmail.com, ORCID: 0000-0003-0506-0275.

institutional economics. Conflict theory has many features. Conflicting parties interact within a certain market framework while pursuing their goals. This framework consists of incremental and uncoordinated markets where various actors have different competencies over price (wage) determination. The existence of these markets encourages the formation of wage-wage, wage-price, or price-price spirals. In these spirals, the profits obtained by some intermediaries in the market are reversed in another market and partially or completely ceased. This causes conflicting goals to emerge as inflation (Vera, 2010).

The purpose of income policies, which were popular in the 1970s, was to achieve the targeted income distribution and coordinate wage and price decisions, aiming to reach to the result without inflation. Conflict theory is associated with class conflict, same as in the wage-price spiral. The strife between oil-producing countries and oil-consuming countries is an example. If oil-consuming countries raise their own prices in response to high oil prices, oil-producing countries retaliate by raising oil prices again, and it goes on.

Martin, Mich, and Rowthorn define the relevant actors as workers and firms. Lorenzoni and Werning (L&W) ignore the identity of the intermediaries and focus only on the market in which the intermediaries interact. Due to this, their contributions should be classified as a general basis rather than a conflict theory (Dutt, 1992).

Rowthorn examined conflict theory in the context of recent inflation and the cost of living encountered in the UK. He studied demands on national output in terms of foreign trade, taxation, profits, and wages. The war in Ukraine and Covid triggered inflation, resulting in a sharp increase in world commodity prices due to supply chain disruptions. In 2022, when the cost of living became a political issue, import prices in the UK increased by 12.2 % and export prices by 11.7 %. The net cost of this increase to the UK is seen to be equivalent to 3.4 % of the national wage. High world prices were reflected in high consumer prices. Since money wages could not keep up with prices, real wages fell by 3.4 % throughout 2022. Hence, the loss incurred by the UK due to high world prices corresponds to the entire decrease in real wages throughout 2022. When official data for 2021 and 2022 were evaluated, it was emphasized that the share of private sector profits in GDP remained almost constant during this period but increased towards the end. As Kalecki pointed out, a typical firm passes high costs on to the consumer to keep a normal profit margin. In summary, it is stated that the burden of the increase in world prices was largely burdened by wage and salary workers, and private firms maintained their profit margins by passing high costs on to consumers. These state of affairs occurred during a period of economic stagnation. In the post-Covid period, GDP growth slowed to an average of 0.05 %.

Given stagnant productivity, rising import prices, and the flexibility of profits and taxes, a decline in real wages was inevitable. Money wages increased but were not sufficient to compensate for the cost of living (Rowthorn, 2024)

3 Approaches to the Conflict Theory of Inflation

Following the first oil shock, in the light of discussions with Robinson and other Keynesians, Rowthorn (1977) developed a simple model known as the conflict theory of inflation. Following Rowthorn, Dutt (1992) used conflict theory to analyze accumulation and crises. Carlin and Soskice examined the conflict between firms and workers using a three-equation model to develop the conflict theory of inflation. In the New Keynesian framework, Erceg et al. (2000) used conflict theory to examine monetary policy. A few of these theories are explained below (Rowthorn, 2024).

Rowthorn: Rowthorn states that there is a single type of final output in the economy. The total output is constant and equal to 1. He classifies the competitors for output as workers (wages), firms (profits), foreigners (net import costs), and the government (taxes). The shares of import costs and taxes in the final output are exogenous and equal to F and T, respectively. The establishment of real wages is a two-stage process. In the first stage, workers in the labor market negotiate for a wage that takes future inflation into account. If prices rise as expected, workers receive a share of the final output equal to W^{lab}. In the second stage, firms set prices in the product market that give them a share of the final output equal to Π^{prod}. This means that after firms set their prices, the share received by workers is as follows (Rowthorn, 1977).

$$W^{prod} = 1 - \Pi^{prod} - F - T \quad (1)$$

The difference between these two quantities is defined as the 'aspiration gap' in the following equation.

$$gap = W^{lab} - W^{prod} = W^{lab} + \pi^{prod} + F + T - 1 \quad (2)$$

If prices increase more than anticipated during wage bargaining, the aspiration gap becomes positive. With the default pricing behavior in Rowthorn's mode, unexpected inflation is proportional to the aspiration gap.

$$p^{un} = \lambda x\, gap \tag{3}$$

If agents respond to unexpected inflation by upwardly revising their price expectations, the result can potentially be an explosive wage-price spiral. The model does not determine the actual rate of inflation, but only its unanticipated component. If all wages and prices increase by an additional %x per year, the above equation remains valid.

To examine the impact of a shock on import prices and taxes, let the unemployment rate (u) be an indicator of economic stagnation, and assume that both W^{lab} and Π^{prod} are negatively related to (U). When there is economic stagnation, workers receive lower wages and firms increase prices less. Combining these elements leads to the emergence of an aspiration gap.

$$gap(U,F,T) = W^{lab}(U) + \pi^{prod}(U) + F + T - 1 \tag{4}$$

$$p^{un}(U,F,T) = \lambda x\ gap(U,F,T)$$

By keeping (F) and (T) constant in the above equation, a downward sloping Phillips curve that links unexpected inflation to unemployment is obtained. These equations have implications in terms of public policy. One of the common reactions in the 1970s was for the government to implement price and wage controls, thereby reducing claims on national output and closing the aspiration gap. A popular alternative today is for the central bank to use monetary policy to increase economic stagnation and reduce the inflation rate. The goal is to create a period of stagnation or near-stagnation, causing workers and firms to adjust their targets and accept lower wage and price increases. This type of policy uses the pressure of economic stagnation to close the aspiration gap and eliminate undesired inflation (Rowthorn, 2024).

Martin: He examines a model where expectations play a significant role. He uses this model, which he calls primitive, to highlight the fundamental elements of conflict theory that are often hidden in the details. He begins by addressing two fundamental regimes: price-focused and wage-focused. In both regimes, workers and firms have targets equal to X* and Z*, respectively. The target gap is X* Z*. If X* Z* > 1, there is inflation and conflict. In the wage-focused regime, all workers receive a money wage that allows them to reach their target based on the new average price. However, due to contractual reasons, only a small portion of firms can set a new price to reach their target. The remaining firms sign contracts to sell at the previous price. If X* - Z* > 0, there is inflation and conflict. Martin has been criticized for neglecting the idea that pricing and wage-setting could have an error rectification dimension (Martin, 2023).

Michl: In this case, a dynamic form of the expectation gap based on growth rates is appropriate. Firstly money wages, secondly prices are assessed. The growth rate of money wages is shown by the following equation:

$$\widehat{W} = R^{w} + \alpha_{w}\left(e - e^{*}\right) \tag{5}$$

Here, (e) is the employment rate (e*) and it is constant, R^w is the assumed inflation rate during wage negotiations, and it is considered the expected inflation rate. The growth rate of prices is as follows:

$$\widehat{P} = R^{p} + \alpha_{p}\left(u - 1\right) \tag{6}$$

In the equation above, R^p represents the inflation rate firms expect when setting their prices, and (u) represents utilization of capacity. The third equation is as follows:

$$R^{p} = \delta R^{w} + \left(1 - \delta\right)\widehat{W} \tag{7}$$

This equation shows how firms' expectations of inflation are affected by the scale of wage agreements. If prices increase at the rate projected in the labor market, real wages rise in line with expectations (Michl, 2023).

Lorenzoni and Werning (L&W): They have used a continuous-time model to analyze conflict inflation in an economy. They consider an N-sector network where sectors are interconnected through an input-output matrix $M=[m_{ij}]$. Some sectors provide labor services, hence their sales prices are tied to wages. The matrix is irreducible and rows sum up to 1. There is a strictly positive y vector for each j, which weights various sectors (Lorenzoni and Werning, 2023).

$$y_{j} = \sum_{i} y_{i} m_{ij} \; and \sum_{i} y_{i} = 1 \tag{8}$$

To understand the concept of staggered decision-making, L&W assume Calvo pricing; where individual agents in sector (i) can reset their sales prices with a probability of λ_i per unit time. This probability varies across sectors but remains constant over time. Prices are measured logarithmically.

At time T, an agent in sector (i) chooses the following reset price:

$$P_{it}^{*} = \alpha_{it} + \sum_{j} m_{ij} P_{jt} \tag{9}$$

Here, α_{it} represents the target of agents in the sector. For the economy, the aspiration gap is defined as follows:

$$gap_t = \sum_{i} y_i \alpha_{it} \tag{10}$$

If $gap_t > 0$,there is a conflict in expectations. The inflation rate in sector i is given in the following equation:

$$\pi_{it} = \lambda_i (p_{it}^{*} - p_{it})$$

$$\text{Average inflation rate} = \bar{\pi}_t = \lambda \; x \; gap_t \qquad \lambda = \left(\sum_{i} \frac{y_i}{\lambda_i} \right)^{-1} \tag{11}$$

Since the average inflation is proportional to the aspiration gap, L&W characterize average inflation as conflict inflation.

In Rowthorn's study unexpected inflation is proportional to the aspiration gap. The same holds true for Michl and under specific restrictive conditions for L&W. Martin is concerned with the compensatory process where firms or workers attempt to make up for past losses. Therefore, he does not take into account expectations. In Martin's model, actual inflation is proportional to the aspiration gap (Rowthorn, 2024).

4 The Economic and Social Effects of Conflict Inflation

Conflict inflation is a complex phenomenon with a wide range of economic effects. The most fundamental impact of it is the increase in prices of goods and services. This situation reduces consumers' purchasing power and erodes their real incomes. For fixed-income or low-income groups, this effect is particularly more apparent. High inflation generally prompts central banks to raise interest rates which causes an increase in borrowing costs and this affects investors' loan and investment decisions. Investors tend to be more cautious in high inflation environments to minimize risks. Conflict inflation depreciates the national currency and thus encourages exports while making imports more expensive. This affects the country's balance of foreign trade and reduces its competitive capacity (Morlin, 2021).

Increases in production costs lead employers to pass costs on to consumer prices and cause pressures on wages in the labor market. This situation also affects unemployment rates because employers may limit labor or restrict wage increases to balance cost increases. High inflation adversely affects economic growth by creating uncertainty and influencing consumers' and businesses' spending decisions. Additionally, conflict inflation generally has a negative impact on income distribution, as price increases deepen income inequalities. As a result, fiscal policies play a significant role in managing inflation during periods of conflict inflation. Governments attempt to balance demand and supply pressures through these policies while striving to maintain economic stability. However, the effectiveness of these policies varies depending on the structural characteristics of the economy, the determination of policymakers, and society's response to policy implementations. In addition to its economic effects, conflict inflation is a phenomenon which has effects on social and societal dynamics significantly. Rapid price increases in high inflation environments raise living costs for those with fixed incomes. This amplifies income inequality, since lower-income individuals must allocate a portion of their income to meet basic needs, while higher-income individuals are affected by these increases less. High inflation often leads to an increase in poverty rates and social unrest. Especially, it makes difficult for fixed-income and low-income families to meet basic needs, increase their poverty rates and threats social stability. Increases in business costs alongside high inflation also affect business employment policies. Businesses tend to reduce labor costs to balance their total costs and it leads to higher unemployment rates. These social effects go beyond the economic effects of conflict inflation, causing unfathomable and wide-ranging consequences in societal fabric. Therefore, policymakers fighting against inflation must consider not only economic factors but also social and societal dynamics (Carlin & Soskice, 2014).

5 Conclusion

In a capitalist economy, inflation can increase profits by reducing the purchase power of other types of income. The key point here is expectation and foresight. Expectation refers to a state of mind, while foresight refers to actual behavior. To expect something means to trust that it will happen, while to foresee means both to expect and to act according to that expectation. Hence, inflation cannot be fully foreseen by everyone.

Conflict is endemic in an economy and relates to all aspects of economic life. Conflict inflation leads to wage bargaining which affects income distribution and thus real profit margins. Conflict theories explain inflation as a result of conflicting demands in markets. The recent cost of living crisis experienced in many developed and developing countries can be usefully examined from the perspective of conflict theory. Conflict occurs between labor class and capital owners in developed countries, while each sector conflicts with one another in developing countries.

The monetarist approach addresses the fundamental causes and solutions of inflation through money supply and central bank policies and examines conflict inflation in this context. The Keynesian approach, on the other hand, bases its economic analyses of conflict inflation on macroeconomic fundamentals. Since every economy has its own unique dynamics, multiple economic and social policy tools need to be used together to keep conflict inflation under control.

References

Carlin, W., & D. Soskice. (2014). *Macroeconomics: Institutions, Instability, and the Financial System*. Oxford: Oxford University Press.

Dutt, A. K. (1992). "Conflict, Inflation, Distribution, Cyclical Accumulation and Crises." *European Journal of Political Economy, 8*(4), 579–597.

Erceg, C., Henderson, D., & Levin, A. (2000). "Optimal Monetary Policy with Staggered Wage and Price Contracts." *Journal of Monetary Economics, 46*(2), 281–313.

Lorenzoni, G., & Werning, I. (2023). "Inflation is Conflict." https://conference.nber.org/conf_papers/f188954.pdf.

Martin, B. (2023). "Rudimentary Inflation Conflict Models: A Note." Centre for Business Research, University of Cambridge, Working Paper No. 535: 1–21.

Michl, T. (2023). "Inflation Stabilization and Normal Utilization." *Journal of Post-Keynesian Economics.*

Morlin, G. S. (2021). Inflation and Conflicting Claims in the Open Economy, https://www.researchgate.net/publication/355983948.

Rosenberg, S., & Weisskopf, T. E. (1981). A Conflict Theory Approach to Inflation in the Postwar U.S. Economy. *The American Economic Review*, *71*(2), 42–47.

Rowthorn, R. E. (1977). "Conflict, Inflation and Money." *Cambridge Journal of Economics, 1*(3), 215–239.

Rowthorn, R. E. (2024). "The Conflict Theory of Inflation Revisited". *Review of Political Economy*: 1–12.

Vera, L. (2010). Conflict Inflation: An Open Economy Approach. *Journal of Economic Studies*, *37*(6), 597–615.

Süleyman Gürbüz[1]

Effects of Housing Interest Rate, Inflation and Google Search Index on Housing Price Index: A Review for Türkiye

1 Introduction

Housing refers to structures designed to fulfill the essential need for shelter. Throughout human history, people have developed diverse methods to satisfy this necessity. With population growth, efforts have been made to optimize limited space by constructing taller buildings on smaller plots of land. This increasing demand for housing has, over time, led to significant rises in housing prices.

The House Price Index (HPI) has been developed to monitor changes in housing prices. The HPI measures the average price of houses sold within a specific period (Bianconi &Yoshino, 2013, p. 27). However, the calculation of the HPI is complicated by factors such as the location and characteristics of the properties (Karaağaç & Altınırmak, 2018, p. 225). To address these challenges, the Central Bank of the Republic of Turkey (CBRT) began publishing the Hedonic House Price Index in 2016. This approach eliminates the effects arising from differences in the characteristics, quality, and other attributes of houses (CBRT, 2024).

The House Price Index (HPI) can be influenced by various factors, including housing supply and demand, inflation, per capita disposable income, interest rates, economic crises, wars, and natural disasters. In the case of Turkey, particularly following the sharp rise in exchange rates since 2020, factors such as increased costs and the COVID-19 pandemic have contributed to a reduction in supply, leading to a rapid increase in inflation. The rise in inflation rates is associated with increased housing prices, which can lead to a decrease in housing demand (Yılmaz, 2020). Conversely, high inflation rates may drive investors towards assets such as land and housing, potentially causing an increase in demand and, consequently, a rise in the HPI. Furthermore, inflation-induced increases in interest rates can lead to higher mortgage rates, raising the costs for

1 Ph.D., Yozgat Bozok University, Faculty of Economics and Administrative Sciences Department of Economy suleyman.gurbuz@bozok.edu.tr, ORCID: 0000-0003-1651-2310.

prospective homebuyers. In such a scenario, a decline in housing demand and a subsequent decrease in housing prices may be expected (Öztürk, 2023).

In the aforementioned scenario, it may not be possible to directly ascertain whether the predominant factor driving the increase in the House Price Index (HPI) is a rise in costs or an increase in demand. This study utilizes Google search data to identify the demand-driven component of the HPI increase. The rapid development of internet services provides a valuable tool for analyzing public behavior. Google search data is considered a valid and accurate measure of public interest when appropriate search terms are selected (Mavragani et al., 2018, p. 2). For instance, Subramaniam and Chakraborty (2021) defined searches on Google for the term "COVID-19" as the "COVID-19 fear index." Similarly, Gunay et al. (2022) employed Google search data as an index representing investor behavior in their study. Drawing on these examples, the Google search index for the phrase "house prices" has been chosen as a variable to gauge public sentiment towards housing prices and to observe the state of demand.

In the following part of the study, some studies in the related literature will be discussed. Then, the results of the analyses using house price index, inflation, interest rate and Google Trend data will be presented. Finally, the results will be discussed and recommendations will be made based on the study's findings.

2 Literature Review

The literature contains numerous studies analyzing various factors influencing the House Price Index (HPI). Research conducted across different countries or country groups has yielded diverse results, reflecting the unique internal dynamics of each region. In compiling the literature review section, the variables selected for this study are classified accordingly.

The first section will review studies analyzing the effects of inflation on the House Price Index (HPI). Poterba (1992), in his study on the United States, concluded that expected inflation leads to higher real house prices. Paksoy et al. (2014) examined the situation in Turkey and its southeastern cities, finding that while the HPI does not generally exert inflationary pressure across Turkey, it does in some specific cities. Kuang and Liu (2015) explored the relationship between inflation and house prices in major Chinese cities. Their study concluded that rising inflation rates drive up housing prices by increasing construction costs and that demand for housing also increases as people seek to hedge against inflation, further elevating prices. Karaağaç and Altınırmak (2018) identified a unidirectional causality from the Consumer Price Index (CPI) to the HPI in Turkey, both nationally and in certain sub-regions. Rahman and Ridzuan

(2020) analyzed factors affecting house prices in Malaysia, finding that inflation has a negative long-run effect on house prices.

The other group of studies consists of articles analyzing the relationship between the House Price Index (HPI) and interest rates. Wheaton (1983) concluded that interest rates have a significant effect on the HPI in the United States. Badurlar (2008) found that income and interest rates negatively impact the HPI in Turkey in both the short and long run. Çankaya (2013) analyzed data from 75 countries in different regions and concluded that there is a positive relationship between the HPI and annual loan interest rates. Benito and Poveda (2019) discovered that a one-unit increase in interest rates in Spain decreases the HPI by 5 %. Çetin (2021) examined the factors affecting the HPI in Turkey and found that the Consumer Price Index (CPI) negatively affects the HPI, as do mortgage interest rates.

The final group of studies examines the relationship between Google searches and house prices. Dietzel (2016) concluded in his study on the United States that Google Trends is a reliable variable for forecasting house prices. Belej (2022), in his research on Poland, found that searches for the term "Housing" on Google influence house prices. Bulczak (2021), in his study on the United Kingdom, determined that Google Trends data can predict potential changes in the housing market.

3 Data and Methodology

This study analyzes the variables 'House Price Index,' 'Inflation,' 'Mortgage Interest Rate,' and 'Mortgage Loan Interest Rate' using monthly data for Turkey from January 2012 to March 2024. Additionally, the variable 'Google Trends,' which represents the frequency of searches for the term 'House Prices' on Google, is included. The natural logarithm is applied to all variables except the interest rates. "Google Trends data are sourced from Google (Google, 2024), while the other variables are obtained from the Central Bank of the Republic of Turkey (CBRT), Electronic Data Distribution System (EDDS, 2024).

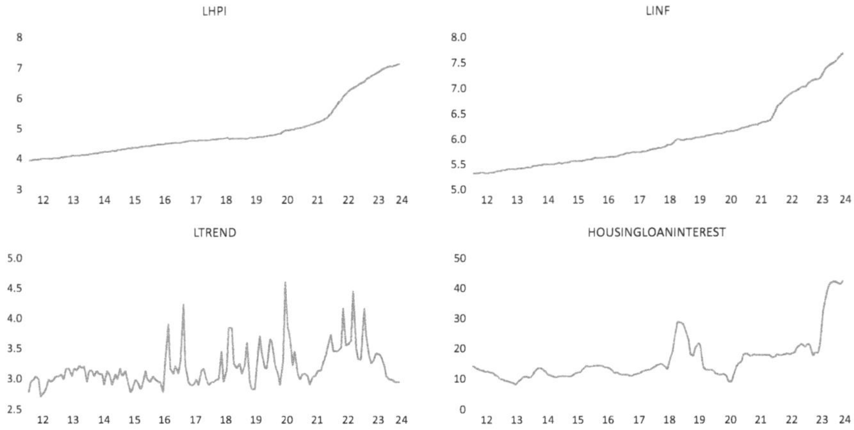

Figure 1: Graphs of the variables

When analyzing the graphs presented in Figure 1, it is observed that the 'House Price Index' and the inflation variable exhibit a nearly synchronous movement. Price increases notably accelerated during the COVID-19 pandemic, with this momentum intensifying further after 2021. In early 2020, the 'housing interest rate' reached its lowest point, remaining around 20 % until 2023, before entering a significant upward trend thereafter. The Central Bank of the Republic of Turkey (CBRT) maintained a low policy rate from the onset of the pandemic until 2023, which was subsequently replaced by a rising policy rate. Additionally, it is noted that the frequency of Google searches for the term 'house prices' peaked during the period when the housing loan interest rate was at its lowest. This suggests that the increase in the House Price Index may be attributed to both rising inflation and heightened demand. The fundamental regression equation to be established for the aforementioned variables is outlined below.

$$LHPI_t = a_0 + LInf_t + LTrend_t + HLI_t + \varepsilon_t \tag{1}$$

In Equation 1, LHPI is the dependent variable, LInf, Trend and HLI are the independent variables and εt is the error term in period t.

4 Analysis Results

This study examines factors that may influence the House Price Index (HPI) in Turkey, including inflation, mortgage interest rates, and the Google search index for the term 'house price.' Prior to conducting the primary analysis to explore the relationships between these variables, it is crucial to test the stationarity of the time series. Analyzing non-stationary series can lead to spurious regression results. Additionally, determining whether the series is stationary at level or after differencing is essential for selecting the appropriate analytical method.

Various unit root tests are available to determine the stationarity of a time series. A review of the literature reveals that the Augmented Dickey-Fuller (ADF) and Phillips-Perron (PP) unit root tests are among the most commonly employed methods. These tests assess the series in its entirety. However, certain variables in some countries may be influenced by economic crises, abrupt policy changes, or significant events, leading to sudden shifts in the series' trajectory (Yavuz, 2006, p. 165). Unit root tests that do not account for these structural breaks may yield inaccurate conclusions about the stationarity of the series. Consequently, several unit root tests that incorporate structural breaks have been developed. One such method is the Zivot-Andrews (1992) test, which allows for the endogenous detection of a single structural break. In this study, the Zivot-Andrews test, which accounts for structural breaks, will be utilized. The results of the unit root tests for the level values and first differences of the series are presented in Tables 1 and 2.

Table 1: Zivot-Andrews Structural Break Unit Root Test for Level Values of the Series

Variables	LHPI	LInf	LTrend	HLI	%1 c.v (*)	%5 c.v (**)	%10 c.v. (***)
Intercept	-6,264 (4)* (10/2021)	-3,490 (4) (12/2021)	-3,936 (3) (07/2018)	-2,941 (1) (05/2022)	-5,34	-4,93	-4,58
Trend	-4,795 (4)** (10/2020)	-3,873 (4) (01/2021)	-4,054 (3) (05/2022)	-4,521 (1)** (05/2022)	-4,80	-4,42	-4,11
Both	-5,220 (4)** (06/2020)	-3,550 (4) (12/2020)	-5,151 (3)*** (05/2022)	-4,706 (1) (08/2021)	-5,57	-5,08	-4,82

According to the results of the Zivot-Andrews unit root test for the level values of the variables, the House Price Index (HPI) is stationary at the 1 % significance level in the constant model, and at the 5 % significance level in both the trend model and the model with both trend and constant. The inflation variable is non-stationary in all three models. The Google Trends variable is stationary at the 10 % significance level in both models, while the housing interest rate is stationary at the 10 % significance level in the model with trend. In all other models, the variables are found to contain unit roots. To ensure more robust results, this study will focus on the 5 % and 1 % significance levels. Therefore, all variables, except for the HPI, exhibit unit roots at their level values.

Table 2: Zivot-Andrews Structural Break Unit Root Test for First Differences of the Series

Variables	ΔLHPI	ΔLInf	ΔLTrend	ΔHLI	%1 c.v (*)	%5 c.v (**)	%10 c.v. (***)
Intercept	-3,743 (3) (05/2021)	-6,396 (4)* (11/2021)	-8,798 (4)* (09/2017)	-7,165 (3)* (11/2018)	-5,34	-4,93	-4,58
Trend	-2,777 (3) (07/2017)	-4,747 (4)** (03/2020)	-8,955 (4)* (05/2022)	-7,068 (3)* (05/2022)	-4,80	-4,42	-4,11
Both	-4,931 (4)*** (10/2021)	-6,910 (4)* (12/2021)	-9,275 (4)* (09/2021)	-7,640 (3)* (02/2019)	-5,57	-5,08	-4,82

The results of the Zivot-Andrews unit root test applied to the first differences of the series indicate that the inflation, trend, and housing loan interest rate (HLI) variables become stationary after differencing. Specifically, the inflation variable is stationary at the 5 % significance level only in the trend model, and at the 1 % significance level in the other models. The House Price Index (HPI) variable is stationary at the 10 % significance level only in the model with both trend and constant, while it remains non-stationary in the other models. The trend and HLI variables are stationary at the 1 % significance level in all three models.

Since the series are stationary at different levels, it is not feasible to apply Johansen or Engle-Granger cointegration tests to these variables. These cointegration tests require that the series be non-stationary at level or stationary at first differences. Consequently, the Autoregressive Distributed Lag (ARDL) model, as developed by Pesaran, Shin, and Smith (2001), will be utilized. The ARDL model allows for the detection of cointegration relationships irrespective of the level at which the series are stationary. Both the ARDL bounds test and the restricted error correction model are employed in this analysis. According to the Akaike Information Criterion (AIC), the optimal lags for the ARDL model are identified as (4,3,2,0). The results of the bounds test are presented in Table 3.

Table 3: ARDL Bounds Test Result

F- Statistic	**I(0) %1**	**I(1) %1**
6,1056	4,29	5,61
k=3		

Upon analyzing the results presented in Table 3, it is observed that the F-statistic value from the analysis exceeds the upper bounds of both the lower I(0) and upper I(1) threshold values at the 1 % significance level. This finding suggests the presence of a potential cointegration relationship between the series. To further investigate, the model was estimated to determine the long-run cointegration relationship among the variables. The results of this estimation are presented in Table 4.

Table 4: Long and Short Run Cointegration Coefficients

Long Run Coefficients				
Variable	Coefficient	Std. Error	t-Statistic	Prob.
LINF(-1)	1,6809	0,2132	7,8822	0,0000*
LTREND(-1)	-0,4991	0,2338	-2,1344	0,0345**
HLI	-0,0453	0,0205	-2,2029	0,0292**
Short Run Coefficients				
ECM Coeffi.	-0,0189	0,0038	-4,9986	**0,0000***
ΔLHPI(-1)	0,5635	0,0849	6,6326	**0,0000***
ΔLHPI(-2)	-0,1234	0,0899	-1,3718	0,1724
ΔLHPI(-3)	0,3530	0,0727	4,8562	**0,0000***
ΔInf	0,1980	0,0561	3,5298	**0,0006***
ΔInf(-1)	0,3297	0,0576	5,7224	**0,0000***
ΔInf(-2)	-0,1186	0,0611	-1,9419	0,0543***
ΔTrend	-0,0014	0,0026	-0,5594	0,5768
ΔTrend(-1)	0,0043	0,0026	1,6411	0,1031
C	-0,0564	0,0114	-4,9406	**0,0000***
R^2	0,8950	**Adj. R^2**		0,8879
ARCH LM Test	0,9686	**Breusch-Godfrey Serial Correlation Test**		0,7690

Analysis of the long-run coefficients reveals that the inflation variable exerts a positive and statistically significant effect at the 1 % significance level. In contrast, both the trend and housing loan interest rates have negative effects, significant at the 5 % level. In examining the unrestricted error correction model, which captures short-term effects, it is found that the first and third lagged values of the House Price Index (HPI), inflation, one lagged value of inflation, and the constant term are significant at the 1 % level. The trend variable, however, does not have a statistically significant effect in the short run. The coefficient of the Error Correction Term (ECM) is statistically significant and negative, indicating that approximately 1.8 % of the deviations from equilibrium in the short run are corrected within one period. The results of the autocorrelation and heteroskedasticity tests further confirm the reliability of the model.

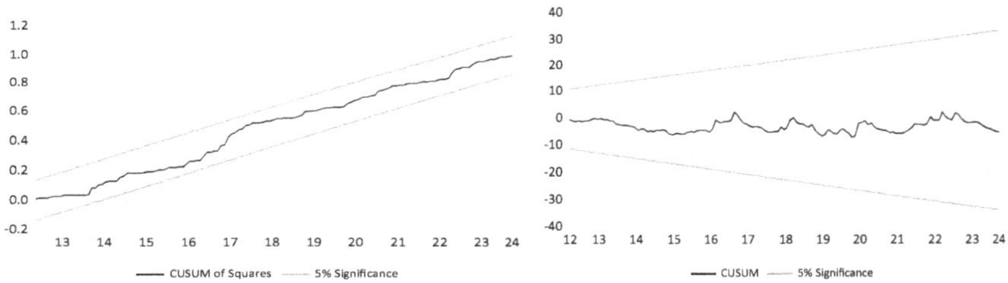

Figure 2: CUSUM and CUSUM of Squares Graphs

CUSUM and CUSUMQ graphs were constructed to assess the stability of the ARDL model. The resulting statistics indicate that the lines remain within the critical bounds at the 5 % significance level. This suggests that the coefficients in the ARDL model are stable.

5 Results

The global economy experienced a significant crisis in 2008, centered around the U.S. mortgage crisis. The sharp decline in housing loans led to the formation of a substantial bubble in the real estate sector. This unforeseen development precipitated a severe financial crisis. Such circumstances necessitate a thorough analysis of the real estate sector.

In this study, the effects of inflation, the frequency of Google searches for the term 'house prices,' and mortgage interest rates on the House Price Index (HPI) in Turkey for the period from January 2012 to March 2024 are analyzed. The results of the unit root test indicate that the HPI is stationary at level, while the other variables are stationary at their first differences. These findings suggest that the Autoregressive Distributed Lag (ARDL) method is appropriate for examining the relationships between the variables.

The ARDL bounds test reveals a cointegration relationship among the variables. Further analysis to determine the long-run relationship shows that inflation has a positive and statistically significant effect on the HPI, whereas Google Trends and housing loan interest rates have statistically significant negative effects. In the short run, the Google Trends data does not exhibit a statistically significant impact.

The results of the study indicate that inflation increases housing costs, which in turn raises house prices. Contractionary monetary policies aimed at reducing inflation, coupled with higher interest rates, may lead to a decrease in the House Price Index (HPI). However, it is important to consider that middle- and low-income individuals may be disproportionately affected by adverse economic conditions. High interest rates could reduce housing demand, potentially leading to a decrease in housing supply. This reduction in production may result in job losses for those employed in housing construction and related sectors. To mitigate these negative effects, subsidies designed to boost housing supply could be a more viable policy option.

The study also suggests that Google Trends data, used to gauge housing demand, is a statistically and econometrically valid variable. Incorporating this variable can facilitate predictions regarding housing demand, providing valuable insights for policymakers, investors, and prospective homeowners. This can help these stakeholders make informed decisions about future house prices.

References

Badurlar, M. (2008). Short- and long-run effects of income and interest rates on house prices. *Journal of Economic Research, 33*(127), 133–152.

Bełej, M. (2022). Does Google Trends show the power of social interest as a determinant of house price dynamics? *Sustainability, 14*(9), 5601.

Benito, A., & Poveda, E. (2019). The impact of interest rates on house prices: Evidence from Spain. *Economic Systems, 58*, 100661.

Bianconi, M., & Yoshino, J. A. (2013). House price indices and cyclical behaviour. *International Journal of Housing Markets and Analysis, 6*(1), 26–44.

Bulczak, G. M. (2021). Use of Google Trends to Predict the Real Estate Market: Evidence from the United Kingdom. *International Real Estate Review, 24*(4).

Çankaya, S. (2013). A Global Perspective on the Relationship between House Prices and Macroeconomic Factors. *Finance and Finance Writings*, *1*(100), 143–154.

CBRT (2024), https://www.tcmb.gov.tr/wps/wcm/connect/e7fe7b68-74a3-4162-ae3b-bbf40d0b26fd/KFE-Uygulama-Degisiklikleri.pdf?MOD=AJPERES&-CACHEID=ROOTWORKSPACE-e7fe7b68-74a3-4162-ae3b-bbf40d0b26fd-orGbPzj, (Date of Access: 13/ 07/ 2024).

Çetin, A. C. (2021). Analysing the factors affecting house prices in Turkey. *Mehmet Akif Ersoy University Journal of Applied Sciences*, *5*(1), 1–30.

Dietzel, M. A. (2016). Sentiment-based predictions of housing market turning points with Google trends. *International Journal of Housing Markets and Analysis*, *9*(1), 108–136.

EDDS (2024), https://evds2.tcmb.gov.tr/ (Date of Access: 19/ 07/ 2024).

Google (2024), https://trends.google.com/ (Date of Access: 21/ 07/ 2024).

Gunay, S., Muhammed, S. & Kata, V. (2022). Identifying the Role of Investor Sentiment Proxies in NFT Market: Comparison of Google Trend, Fear-Greed Index and VIX, Annual Event of Finance Research Letters, CEMLA Conference: New Advances in International Finance, Mexico City.

Karaağaç, G. A., & Altınırmak, S. (2018). Causality Relationship between Turkey House Price Index and Level-Based House Price Indices and Selected Variables. *Black Sea International Scientific Journal*, *39*(39), 222–240.

Kuang, W., & Liu, P. (2015). Inflation and House Prices: Theory and Evidence from 35 Major Cities in China. *International Real Estate Review*, *18*(2).

Mavragani, A., Ochoa, G., & Tsagarakis, K. P. (2018). Assessing the methods, tools, and statistical approaches in Google Trends research: Systematic review. *Journal of Medical Internet Research*, *20*(11), e270.

Öztürk, S. (2023). Determinants of house price index in Turkey: An econometric modelling study. *Journal of Economic Research*, *58*(2), 457–472.

Paksoy, S., Yöntem, T., & Büyükçelebi, B. (2014). The Relationship Between House Price Index and Inflation (An Empirical Study on Trc1, Trc2 and Trc3 Level Regions). Assam International Refereed Journal, 1(2), 54–69.

Pesaran, M. H., Shin, Y., & Smith, R. J. (2001). Bounds testing approaches to the analysis of level relationships. *Journal of applied econometrics*, 16(3), 289–326.

Poterba, J. M. (1992). Vergilendirme ve konut piyasaları. Kanada-ABD Vergi Karşılaştırmaları içinde (s. 275–294). Üniversite

Rahman, M. F. A., & Ridzuan, A. R. (2020). Malezya'da konut fiyat endeksini belirleyen faktörler. *International Journal of Academic Research in Accounting, Finance and Management Sciences*, *10*(1), 351–359.

Subramaniam, S., & Chakraborty, M. (2021). COVID-19 korku endeksi: borsa getirileri için önemli mi? *Review of Behavioral Finance, 13*(1), 40–50.

Wheaton, C. (1983). Dayanıklı tüketim malları, konut ve kalıcı gelir hipotezi. *Journal of Monetary Economics, 12*(2), 279–311.

Yavuz, N. Ç. (2006). Testing the effect of tourism revenues on economic growth in Turkey: Structural break and causality analysis. *Journal of Dogus University, 7*(2), 162–171.

Yılmaz, A. (2020). Enflasyon ve konut kredi faiz oranlarının konut fiyatları üzerindeki etkisi: Bir meta analiz çalışması. *İstanbul Ticaret Üniversitesi Ekonomi ve İşletme Fakültesi Dergisi, 20*(2), 345–362.

Zivot, E., & Andrews, D. W. K. (1992). Further evidence on the great crash, the oil- price shock, and the unit- root hypothesis. *Journal of Business & Economic Statistics*, 20(1), 25–44.

Ipek Tekin Gölpınar[1]

Understanding the Causal Link Between Inflation and Distribution: An Empirical Analysis for BRIC-T Countries

1 Introduction

The primary objective of the modern central banks is price stability which is generally related to the fact that high inflation harms social welfare and creates negative socioeconomic consequences by reducing purchasing power of individuals. On the other hand, since the 1970s and 1980s, world economies on average have witnessed growth in income inequality (UTIP-EHII, 2024) that policymakers need considerable attention regarding its causes and the social, economic, and political consequences it creates.

In recent decades, the economic performance of the BRICS nations has been crucial for understanding the future path of the global economy. In the same vein, the term BRICS neatly captures the growing importance of emerging economies globally (Council on Foreign Relations, 2017). These countries have been the main drivers of global economic growth (Acheampong et al., 2023: 620). Once experiencing high economic growth, these countries also have stood as a group of countries that have challenged instability in inflation and deterioration in income distribution. As Ivins (2013) reports, income inequality within all BRICS nations significantly surpasses the average observed in OECD countries. This disparity has been particularly pronounced since the early 1990s. As highlighted in OECD (2010: 37), while income inequality in China and India remains lower than the peak levels, it still surpasses the average for OECD nations. This contrasts with Brazil, where the Gini coefficient actually decreased by 9 % between 1993 and 2008. In contrast, China, India, and South Africa all saw increases in inequality during this period, with China experiencing the most significant rise (24 %), followed by India (16 %).

Aside from the union expanding with the participation of several more countries at the 2023 BRICS summit, Turkiye is similar to the former countries in terms of inflation and distribution dynamics. For that reason, it is also included

---

1 Ph.D., Assoc. Prof., Cukurova University, Faculty of Economics and Administrative Sciences Department of Economics, itekin@cu.edu.tr, ORCID: 0000-0001-8547-9185.

in our investigation. This chapter aims to contribute to empirical literature by investigating the two-way causality between inflation and distribution for BRIC-T countries rather than just replicating extensive literature on the effects of inflation on income inequality. Within this scope, the causal relations between inflation rate and three different distribution variables are analyzed using a heterogeneous panel causality test for BRIC-T countries. The content of this chapter organized as follows: Section 2 is devoted to a summary of theoretical and empirical discussions. Section 3 presents the empirical analysis consisting of data, methodology and results. Section 4 concludes the chapter.

2 A Brief Literature Review

The impact of inflationary processes on income distribution has been a prominent research topic in the economic literature for many years. Ribeiro et al. (2022) summarize the existing redistributive mechanisms of inflation in three titles: Firstly, when prices rise, wages typically take time to catch up. This "lag" means workers' real wages decrease, essentially transferring wealth from workers to capitalists. Secondly, inflation can erode the disposable income and savings of middle-class households. This reduces their purchasing power and can lead to higher poverty rates and a wider gap between rich and poor. Third, price increases hurt borrowers and benefit lenders. Since wealthier individuals have superior access to information and financial expertise, they can potentially benefit from such economic shifts. Hu et al. (2024) mention inflation complexity arising from two interacting mechanisms: Higher domestic inflation can contribute to increased global interest rates. This, in turn, can dampen economic growth, potentially leading to a rise in income inequality. Second, inflation can erode the value of financial assets, such as bonds. This disproportionately affects individuals who rely on these assets, often retirees. However, it can also increase the value of certain real assets, potentially benefiting higher-income individuals.

Most of the empirical works agreed on the negative distributional consequences of inflation (Li and Zou, 2002; Albanesi, 2007; Thalassinos, Uğurlu & Muratoğlu, 2012; Berisha et al., 2020; Ünal & Doğan, 2021: Sintos, 2023). Among them, Berisha et al. (2020) investigate the effects of inflation, interest rate and income on income inequality in BRICS countries and find evidence on the distorting effect of inflation on income distribution. Ünal & Doğan (2021) concluded that inflation in the 2005–2018 period in Turkiye had a significant impact on the deterioration of income distribution and prevented the income gap between high and low income groups from closing. Other strand of empirical literature supports the nonlinear Kuznets type relationship between inflation and inequality (e.g., Bulíř (2001); Monnin, 2014; Siami-Namini & Hudson, 2019. Bulir (2001) examined how inflation affects income inequality in 75 countries over two decades (1970–1991). The research suggests that lower inflation tends to lead to less income inequality. This effect is especially strong in situations of hyperinflation. Siami-Namini & Hudson (2019) also finds a nonlinear relationship which differs for developed (inverse U-shape) and developing countries (U shape). Toda Yamamoto causality test results indicate that inflation does not affect income inequality in neither developed nor developing countries. Interestingly, in developing countries, income inequality seems to predict future inflation.

However, similar evidences on the inverse direction has been scarce so far. The political economy of inflation underlines the importance of government policies on inflation. As Desai et al. (2003) state, there are two opposing ideas about why inflation happens. One view, *populist approach*, says inflation is caused by people wanting government handouts funded by taxes that come from inflation itself. It also suggests that competitive elections make inflation worse. The other view, *the state capture approach*, says inflation is caused by rich and powerful people who benefit when more money is created. This view argues that competitive elections can actually help control inflation. Albanesi (2007) from another point of view, even if similarly states the cross-country evidences on positive inflation-income inequality nexus, contributes to the literature by presenting a political economy model that shows this relationship as an outcome of a distributional conflict affected by government policies. The long-term level of inflation (equilibrium inflation) is positively associated with the extent of income disparity within a society, as low-income households are more vulnerable to inflation.

Contrary to mainstream economics, which considers inflation to be demand-pull and focuses on the consequences of inflation rather than its sources, heterodox approaches examine the distributional dynamics of inflation. The role of distributive conflict in price and wage determination is emphasized by Amitrano & Vasconcelos (2019). In that sense, the key distinction of inflation lies in its close association with the specific institutional framework of each society. Ribeiro et al. (2022) also try to determine the inflation-distribution nexus simultaneously in a Post-Keynesian model. By using Panel VAR model for a sample of 133 countries for 1990–2014 period, they concluded that increases in labor share is associated with decreasing future inflation. This mechanism is achieved primarily through the positive effect of an increased wage share on productivity growth which subsequently strengthens the competitive position of firms, thereby enabling them to reduce prices.

Heterodox views also possess the literature on the distributional consequences of low inflation policies of the monetary authorities as novel studies empirically analyzed this view. For example, Altunbaş & Thornton (2022) assesses the impact of inflation targeting on income inequality for 121 developed and developing countries. They concluded that inflation targeting strategy is associated with worsening income distribution measured by Gini and labor income share. Similarly, the results obtained by Clavijo-Cortes (2024) suggest that setting interest rates through monetary policy ultimately benefits owners of capital (wealth) in the fight against inflation, potentially exacerbating existing economic inequalities.

3 Empirical Analysis

3.1 Data

The empirical analysis relies on an annual balanced panel data of five BRIC[2]-T (Brazil, Russian Federation, India, China and Turkey) countries over the period of 1995–2021. Annual data for inflation rate is based on consumer price index and gathered from World Bank World Development Indicators (WB-WDI) database. A variety of income distribution data is used in the analysis. Firstly, Gini coefficient as the primary inequality data is from Standardised World Income Inequality Database (SWIID) provided by Solt (2019). Gini coefficient has been criticized as being more sensitive to transfers in the middle of the distribution, but less sensitive to transfers between very rich or very poor groups (Alvaredo, 2011: 274). Today another debate on inequality is focusing on the increasing gap between the segments receiving the lowest share of income and the segments receiving the highest share. For this purpose, pre-tax national income share of the highest %1, namely Top 1 % (P99/P100) distribution data from World Inequality Database (WID) is additionally used in the empirical analysis. Since a change in the functional distribution of income against labor income would indicate inequality in the income distribution between labor and capital, labor income share (labsh) is also used as another indicator.[3]

2 The fact that South Africa differs widely from the other five countries in terms of both inflation and income distribution indicators is the reason why the country is not included in the analysis.

3 Labor share is extracted from Penn World Table (PWT 10.01) database. The data is available until 2019. Therefore, the period is limited to 1995–2019 whenever this data is used.

Figure 1 presents three illustrations showing the relationship/correlation between inflation rate and distribution indicators for BRIC-T countries. At a glance, while there is a positive correlation between the inflation rate and inequality indicators of Top1 share or the labor share of income (a decrease in labor share means an increase in inequality) it turns negative once the distribution indicator becomes the Gini coefficient. Therefore, the need for using different indicators in measuring income distribution becomes obvious.

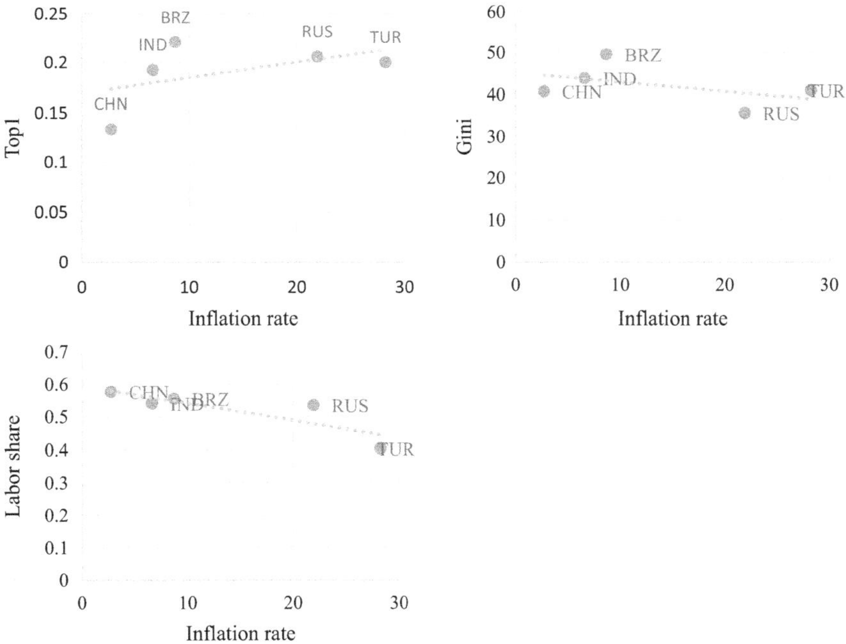

Figure 1: The relationship between inflation rate and income distribution indicators for BRIC-T countries

Source: WB-WDI (2024), WID (2024), SWIID version 9 & PWT version 10.01. (1995–2021 average; the last curve is produced for 1995–2019 average)

Table 1 provides the correlation matrix summarizing the correlation coefficients among the pairs of variables. Notably, these coefficients in Table 1 show either negative or positive correlations, but with a weak pattern. This table also suggests a rationale for using different inequality metrics.

Table 1: Correlation matrix (1995–2019)

	inf	gini	top1	labsh
inf	1.0000			
gini	-0.0794	1.0000		
top1	0.0929	0.2874	1.0000	
labsh	-0.3737	0.0817	-0.3824	1.0000

3.2 Methodology

In the first stage of the methodological process, the cross-sectional dependence (CD) and the slope homogeneity are investigated. Then, unit root tests and estimation methods are selected as a result of determining whether there is a correlation between cross-sections and whether the slope parameters are homogeneous. Based on the outcomes of these tests, a causality test is applied to determine the direction of the relationship between the inflation rate and each representative variable for income distribution.

3.2.1 Preliminary Tests

One of the tests to investigate the existence of cross-sectional dependence is the Lagrange Multiplier test (LM) developed by Breusch & Pagan (1980). In addition, Pesaran et al. (2008) alternatively proposed the bias-corrected LM test (LM_{adj}). When T>N, the test statistic has an asymptotic standard normal distribution. According to small sample findings based on Monte Carlo simulations, LM_{adj} test can strongly control size; its power is good in panels with exogenous variables and normally distributed error terms (Pesaran et al., 2008: 120–121). For each CD test, the H_0 hypothesis refers to "no cross-sectional dependence".

Cross-sectional dependence test results in Table 2 indicate the rejection of the H_0 hypothesis according to the LM and LM_{adj} test statistics. According to the findings, there is cross-section dependence in the models at the 1 % significance level. In addition, estimation methods are chosen depending on whether the slope parameters are homogeneous or heterogeneous by cross-sections. Pesaran and Yamagata (2008) referred to Swamy (1970)'s attempt for a slope homogeneity test and developed an adjusted slope homogeneity test for large panels. This applies regardless of the number of observations (N) or the number of time periods (T). It even works well when the data doesn't necessarily follow a normal distribution. According to the test results in Table 2, at least one test rejects the H_0 hypothesis that the slope parameters are constant.

Table 2: Cross-sectional dependence and homogeneity test results

Model	LM (Breusch & Pagan, 1980)	LM_{adj} (Pesaran, Ullah & Yamagata, 2008)	Slope homogeneity test	
			Swamy S test	$\widetilde{\Delta}_{adj}$
inf = f (gini)	71.944 (0.000)	2.361 (0.009)	162.77 (0.000)	7.204 (0.000)
gini = f (inf)	264.240 (0.000)	4.246 (0.000)	577.37 (0.000)	0.982 (0.326)
inf = f (top1)	61.145 (0.000)	2.052 (0.020)	74.80 (0.000)	4.415 (0.000)
top1 = f (inf)	254.121 (0.000)	3.381 (0.000)	323.08 (0.000)	4.147 (0.000)
inf = f (labsh)	42.590 (0.000)	7.945 (0.000)	45.69 (0.000)	1.511 (0.131)
labsh = f (inf)	247.676 (0.000)	8.663 (0.000)	811.85 (0.000)	0.501 (0.616)

* The values in parentheses are the probability values of the test statistics. Hypotheses regarding cross-sectional dependence: For all t's and i≠j, H_0: cov(εit εjt) = 0, while for at least one pair i≠j, H_1: cov(εit εjt) ≠ 0. As homogeneity test suggests; for all i H_0: βi = β while H_1: βi≠ βj, i≠j. Slope homogeneity test is calculated for 1995–2019 period for all models.

Based on the results in Table 2, *second-generation* unit root tests were utilized in searching the integration level of the series. For this purpose, IPS (Im et al., 2003), and CADF (Pesaran, 2007) unit root tests were applied. These tests take heterogeneity and heterogeneity with cross-sectional dependence into consideration, respectively. The results are presented in Table 3. According to the panel CADF unit root test results, only inflation and Top1 variables are stationary for constant model at 5 % significance level. As for the IPS test, inflation is stationary both for constant and constant + trend model, while Top1 is stationary for constant model, and labor share is for constant + trend model. Gini is not stationary with respect to any of the results. The following causality test is independent of the stationarity degree of the series.

Table 3: IPS and CADF Panel Unit Root Test Results

Variables	IPS IPS (Im et.al., 2003)		CADF (Pesaran, 2007)	
	CIPS stat		Z {t-bar} stat	
	Constant	Constant + Trend	Constant	Constant + Trend
inf	-3.006 [-2.22]	-2.857 [-2.82]	-2.240 (0.013)	-0.501 (0.308)
gini	-0.453 [-2.25]	-1.214 [-2.94]	0.842 (0.800)	0.474 (0.682)
top1	-3.008 [-2.22]	-2.343 [-2.94]	-3.567 (0.000)	-1.349 (0.089)
labsh	-1.639 [-2.25]	-2.988 [-2.94]	0.629 (0.735)	-0.960 (0.169)

Notes: For IPS and CADF tests, lag-length is 1. For CADF test, probability values of Z [t-bar] statistics are in parenthesis. For IPS test, 5 % significance level critical values for CIPS test are in square parenthesis. 1995–2019 period is taken for all series.

3.2.2 A Heterogeneous Panel Causality Test

In the last stage of our analysis, the bootstrap causality test is used which is developed by Kónya (2006) for heterogeneous and cross-sectional dependent panel data models. A cross-country investigation into the potential for Granger causality between inflation rate (inf) and income distribution (Gini, Top1, labsh) could be conducted by employing a finite-order bivariate vector autoregressive (VAR) model. However, Kónya (2006) differently uses the SUR estimator proposed by Zellner (1962) and considers the following equations:

$$y_{1,t} = \alpha_{1,1} + \sum_{l=1}^{py_1} \beta_{1,1,l} y_{1,t-1} + \sum_{l=1}^{px_1} \mu_{1,1,l} x_{1,t-1} + \varepsilon_{1,1,t}$$

$$y_{2,t} = \alpha_{1,2} + \sum_{l=1}^{py_1} \beta_{1,2,l} y_{2,t-1} + \sum_{l=1}^{px_1} \mu_{1,2,l} x_{2,t-1} + \varepsilon_{1,2,t}$$

.

$$y_{N,t} = \alpha_{1,N} + \sum_{l=1}^{py_1} \beta_{1,N,l} y_{N,t-1} + \sum_{l=1}^{px_1} \mu_{1,N,l} x_{N,t-1} + \varepsilon_{1,N,t}$$

and

$$x_{1,t} = \alpha_{2,1} + \sum_{l=1}^{py_2} \beta_{2,1,l} y_{1,t-1} + \sum_{l=1}^{px_2} \mu_{2,1,l} x_{1,t-1} + \varepsilon_{2,1,t}$$

$$x_{2,t} = \alpha_{2,2} + \sum_{l=1}^{py_2} \beta_{2,2,l} y_{2,t-1} + \sum_{l=1}^{px_2} \mu_{2,2,l} x_{2,t-1} + \varepsilon_{2,2,t}$$

.

.

$$x_{N,t} = \alpha_{2,N} + \sum_{l=1}^{py_2} \beta_{2,N,l} y_{N,t-1} + \sum_{l=1}^{px_2} \mu_{2,N,l} x_{N,t-1} + \varepsilon_{2,N,t}$$

here index i refers to the cross-sections (i= 1, …, 5), t to the time period (t= 1, …, 27) and *l* to the lag length. $\varepsilon_{1,i,t}$, $\varepsilon_{2,i,t}$ are error-terms. The equations under consideration here constitute a system of SUR rather than a VAR model. This method offers two key benefits. Firstly, it avoids the assumption of panel homogeneity, enabling individual Granger causality tests on each panel member. Secondly, it eliminates the need for pre-testing unit roots and co-integration, streamlining the analysis. Due to the application of country-specific bootstrap critical values, the stationarity of the variables y_t and x_t is not a prerequisite. Table 4 shows the causality test results separately for each pair of variables.

Table 4: Panel Heterogeneous Causality Test Results

	H_0 *inflation* ↛ *Gini*		H_0 *Gini* ↛ *inflation*		**Individual Results**
Countries	**WALD**	**p-value**	**WALD**	**p-value**	
Brazil	1.116	0.136	0.055	0.799	No causality
Russian Federation	3.889	0.028	2.591	0.105	*INF* → *GINI*
India	3.768	0.181	1.006	0.573	No causality
China	7.786	0.004	3.668	0.035	*INF* ↔ *GINI*
Turkiye	0.225	0.998	2.131	0.228	No causality
	H_0 *inflation* ↛ *Top*1		H_0 *Top*1 ↛ *inflation*		**Individual Results**
Countries	**WALD**	**p-value**	**WALD**	**p-value**	
Brazil	0.043	0.823	1.330	0.303	No causality
Russian Federation	0.016	0.917	6.601	0.043	*TOP*1 → *INF*
India	6.669	0.003	0.447	0.297	*INF* → *TOP*1
China	0.316	0.383	4.738	0.022	*TOP*1 → *INF*
Turkiye	4.590	0.246	1.021	0.210	No causality
	H_0 *inflation* ↛ *labshare*		H_0 *labshare* ↛ *inflation*		**Individual Results**
Countries	**WALD**	**p-value**	**WALD**	**p-value**	
Brazil	2.092	0.054	0.635	0.405	*INF* → *LABSH*
Russian Federation	0.417	0.662	1.957	0.182	No causality
India	2.542	0.276	0.016	0.833	No causality
China	3.641	0.078	5.164	0.096	No causality
Turkiye	3.473	0.414	4.384	0.007	*LABSH* → *INF*

*The sign "→" represents causality in one direction and the sign "⟷" represents bi-directional causality. dmax = 1. Number of bootstrap replications is 1000.

According to Table 4, inflation rate significantly affects inequality; in other words, it is the cause of inequality -measured by Gini coefficient only in Brazil and China. In China, there is also an inverse causality from Gini to inflation rate which points to the inflationary effects of unequal distributions. However, inflation rate is the cause of Top1 income share only in India. On the other hand, Top1 shares in Russia and China are the causes of inflation. In measuring functional income distribution, labor share of income is affected by inflation rates in Brazil. Interestingly, only Turkiye's data support the inverse causality from labor share to inflation rate. Labor share in Turkiye has a significant effect on the inflation rate.

4 Concluding Remarks

This study intended to investigate whether there is any causality relationship between inflation and distributional indicators in BRIC-T countries. In that context, a heterogeneous causality test by Kónya (2006) was deployed for the 1995–2021 annual period.

The findings support the effect of inflation on income distribution. Besides, the fact that one of the reasons for inflation in Russia, China, and Turkiye is income distribution reveals the necessity of alternative strategies in the fight against inflation. In the case of Turkiye, when the negative relationship between the labor income share and inflation (Figure 1) is evaluated together with the causality results obtained, it is obvious that the low labor share is the cause of high inflation. Similarly, referring to Figure 1 and causality results, the high Top 1 % level is the cause of high inflation in Russia. All in all, it is important that monetary policy authorities take into account country-specific factors rather than uniform strategies, include strategies that improve income distribution depending on the distribution indicator, and prioritize redistribution on their agenda in coordination with fiscal policies. In addition, the policies that limit the increase in the shares of those at the top of the income distribution relative to those with the lowest shares, and regulations aimed at improving functional distribution in favor of the labor share can also provide effective results in the fight against inflation.

The findings obtained from this study also reveal the need to examine the way this relationship occurs in more detail and by modeling it with different control variables. However, the intention here was to provide evidence simply about the direction of this relationship and a starting point for further studies, particularly in individual countries.

References

Acheampong, A. O., Adebayo, T. S., Dzator, J., & Koomson, I. (2023). Income inequality and economic growth in BRICS: Insights from non-parametric techniques. *The Journal of Economic Inequality*, *21*(3), 619–640.

Albanesi, S. (2007). Inflation and inequality. *Journal of Monetary Economics*, *54*(4), 1088–1114.

Altunbaş, Y., & Thornton, J. (2022). Does inflation targeting increase income inequality?. *Journal of Post Keynesian Economics*, *45*(4), 558–580.

Alvaredo, F. (2011). A note on the relationship between top income shares and the Gini coefficient. *Economics Letters*, 110(3), 274–277.

Amitrano, C. R., & Vasconcelos, L. (2019). Income distribution, inflation and economic growth: A post-Keynesian approach. *Panoeconomicus*, 66(3), 277–306.

Berisha, E., Gupta, R., & Meszaros, J. (2020). The impact of macroeconomic factors on income inequality: Evidence from the BRICS. *Economic Modelling*, *91*, 559–567.

Breusch, T. S., & Pagan, A. R. (1980). The Lagrange multiplier test and its applications to model specification in econometrics. *The Review of Economic Studies*, *47*(1), 239–253.

Bulíř, A. (2001). Income inequality: does inflation matter?. *IMF Staff papers*, *48*(1), 139–159.

Clavijo-Cortes, P. (2024). Conflict inflation and the role of monetary policy. *Review of Political Economy*, 1–24.

Council on Foreign Relations. (2017). "How the BRICS Got Here" (A. Ayres), Available at: https://www.cfr.org/expert-brief/how-brics-got-here (01.07.2024).

Desai, R. M., Olofsgård, A., & Yousef, T. M. (2003). Democracy, inequality, and inflation. *American Political Science Review*, *97*(3), 391–406.

Hu, R., Wang, J., Zheng, Z., & Yang, Y. (2024). Inflation and income inequality in an open-economy growth model with liquidity constraints on R&D. *Journal of International Money and Finance*, 103127.

Im, K. S., Pesaran, M. H., & Shin, Y. (2003). Testing for unit roots in heterogeneous panels. *Journal of econometrics*, *115*(1), 53–74.

Ivins, C. (2013). Inequality matters: BRICS inequalities fact sheet: Oxfam.

Kónya, L. (2006). Exports and growth: Granger causality analysis on OECD countries with a panel data approach. *Economic modelling*, *23*(6), 978–992.

Li, H., & Zou, H. F. (2002). Inflation, growth, and income distribution: A cross-country study. *Annals of Economics and Finance*, *3*(1), 85–101.

Monnin, P. (2014). Inflation and income inequality in developed economies. *CEP Working Paper Series.*

Organisation for Economic Co-operation and Development. (2010). Tackling Inequalities in Brazil, China, India and South Africa: The Role of Labour Market and Social Policies. OECD Publishing. http://dx.doi.org/10.1787/9789264088368-en.

Pesaran, M. H. (2007). A simple panel unit root test in the presence of cross-section dependence. *Journal of Applied Econometrics*, *22*(2), 265–312.

Pesaran, M. H., & Yamagata, T. (2008). Testing slope homogeneity in large panels. *Journal of econometrics*, 142(1), 50–93.

Pesaran, M. H., Ullah, A., & Yamagata, T. (2008). A bias-adjusted LM test of error cross-section independence. *The econometrics journal*, *11*(1), 105–127.

Ribeiro, R. S., D'Amato, S., & Pereira, W. M. (2022). The inflation-distribution nexus: A theoretical and empirical approach. *Review of Political Economy*, *34*(1), 146–164.

Siami-Namini, S., & Hudson, D. (2019). Inflation and income inequality in developed and developing countries. *Journal of Economic Studies*, *46*(3), 611–632.

Sintos, A. (2023). Does inflation worsen income inequality? A meta-analysis. *Economic Systems*, *47*(4), 101146.

Solt, F. (2019). "The Standardized World Income Inequality Database Versions 8-9". Harvard Dataverse, V11.

Swamy, P. A. (1970). Efficient inference in a random coefficient regression model. *Econometrica: Journal of the Econometric Society*, 311–323.

Thalassinos, E., Ugurlu, E., & Muratoglu, Y. (2012). Income Inequality and Inflation in the EU. *European Research Studies*, 15(1), 127–140.

UTIP-EHII. (2024). University of Texas Inequality Project-Estimated Household Income Inequality Dataset. Available at: https://utip.gov.utexas.edu/datasets.html (01.07.2024)

Ünal, İ. B., & Doğan, M. K. (2021). Türkiye'de enflasyonun gelir eşitsizliğine etkisinin mikro düzeyde analizi. *Ankara Üniversitesi SBF Dergisi*, *77*(2), 309–332.

Zellner, A. (1962). An efficient method of estimating seemingly unrelated regressions and tests for aggregation bias. *Journal of the American Statistical Association*, *57*(298), 348–368.

Veysel Karagöl[1]

How Does Energy-Related Uncertainty Affect Global Food Prices? Asymmetric Evidence from the Frequency Domain

1 Introduction

Uncertainty is not desirable at any time in life. Uncertainty can have undesirable effects on people or societies by creating an unsafe environment. The economy has also had its share of uncertainties. Many macroeconomic variables such as financial markets, economic policies, foreign trade, employment, spending or investment decisions, and prices are affected by uncertainty. However, the focus of this study is food prices. Food prices are affected by factors such as war, population, climate conditions, agricultural productivity, natural disasters, trade policies, supply chain problems, macroeconomic uncertainties, and energy costs. This study examines the relationship between energy-related uncertainties and food prices.

Dang et al. (2023) calculated an index that measures energy-related uncertainties. In their study, they stated that no current energy-related uncertainty index was available and that by doing so they would contribute to the literature in three ways. First, this new index covers not just a single country but many developed and developing economies. Second, when creating the index, monthly country reports of the Economist Intelligence Unit are used, not newspaper news, as in many other studies. This ensures the comparability of the index across countries. Third, they provide new evidence on the economic impacts of energy-related uncertainty (Dang et al., 2023).

The energy-related uncertainty index captures many important energy-related peaks and troughs[2]. However, since it is still quite new, the empirical literature on this index is quite shallow. Even so, there are many studies examining the relationships between food prices and variables such as energy prices, energy price shocks, energy price volatility, and other measures of energy uncertainty.

---

1 Ph.D., Assoc. Prof., Van Yüzüncü Yıl University, Erciş Faculty of Business Administration, Department of Economics, veyselkaragol@gmail.com, ORCID: 0000-0001-9939-0173.

2 For detailed information, please see: Dang et al. (2023).

One of them, Aye (2015), states that uncertainty in oil prices has a positive and significant effect on food prices. Studies such as Alghalith (2010), Taghizadeh-Hesary et al. (2018), Zmami and Ben-Salha (2019), and Alnour et al. (2023) also express that variables such as energy prices, energy price volatility, and energy price shocks trigger an increase in food prices. Kirikkaleli and Darbaz (2021) and Raza et al. (2022) emphasize that there is a unidirectional causality between energy prices and food prices, while Baumeister and Kilian (2014) find that the relationship between the variables in question is negligibly insignificant. Unlike other studies, Olayungbo (2021) implies that in oil-exporting developing countries, the direction of causality is from food prices to oil prices. Aye (2015), Zmami and Ben-Salha (2019), Karakotsios et al. (2021), and Alnour et al. (2023) also confirm that the relationship between energy price shocks and food prices is asymmetric. Nazlioglu et al. (2013) and Saucedo et al. (2015) compare the before and after of the Global Financial Crisis. According to these studies, the spillover levels and risk transfer between the energy market and agricultural commodity markets were low before the crisis and quite high after the crisis. Similar to these studies, Darwez et al. (2023) emphasize that the first impact of the increase in oil prices is experienced by agricultural commodity markets due to increasing costs. In addition, Han et al. (2023) and Ulussever et al. (2023) prefer a machine learning algorithm to identify the determinants of food prices. Han et al. (2023) find that the most important determinants of food prices are the increase in oil prices and the US dollar supply. Ulussever et al. (2023) list the most important determinants of food prices as raw material prices, fertilizer prices, and oil prices.

This study aims to reconsider the relationship between energy uncertainties and global food prices for the post-Global Financial Crisis period, which is quite common in the empirical literature. Do energy-related uncertainties affect food prices? If so how? The motivation for this study is the increasing food prices all over the world in recent years. The study is expected to contribute to the macroeconomic literature with the use of a new energy-related uncertainty index and a different methodological basis.

The rest of the paper is organized as follows: Section 2 covers methodology. Then, Section 3 includes the data and preliminary analysis. Section 4 presents the findings. Section 5 ends with the conclusion.

2 Methodology

If X_t provides useful information in the prediction of Y_t from two randomly distributed variables in a certain period t, it can be said that X_t is a significant Granger cause of Y_t (Granger, 1969). The noteworthy point here is that the existence of Granger causality is being investigated for a certain period t. Such relationships are analyzed by time-domain causality tests. However, in such tests, producing a single test statistic for the entire period may ignore the possibility that causality may vary according to different frequencies (Bozoklu and Yilanci, 2013; Kırca et al. 2020). Moreover, empirical findings show that macroeconomic variables are not stationary and involve structural changes. For this reason, it is thought that it would be reasonable to examine the relationship between macroeconomic variables at different frequencies. The method developed by Breitung and Candelon (2006), considering the studies of Geweke (1982) and Hosoya (1991), allows this. Geweke (1982) and Hosoya (1991) propose a measure of causality at a given frequency based on a decomposition of spectral density. Thus, Breitung and Candelon (2006) have developed the causality test in the frequency domain, based on the studies of Granger (1969), Geweke (1982), and Hosoya (1991). This test is also based on VAR models (Sims, 1980). Then, Ranjbar et al. (2017) took the causality test in the frequency domain one step further through the asymmetric methodology in the study of Hatemi-J and Uddin (2012). Ranjbar et al. (2017) developed the asymmetric frequency domain causality test, which considers the relationships at different frequencies of positive and negative shocks of variables. The cumulative forms of the variables that are decomposed into positive and negative shocks can be written as follows:

$$Y_t^+ = \sum_{i=1}^{t} e_{1i}^+,\ Y_t^- = \sum_{i=1}^{t} e_{1i}^-,\ X_t^+ = \sum_{i=1}^{t} e_{2i}^+,\ X_t^- = \sum_{i=1}^{t} e_{2i}^- \tag{1}$$

In the asymmetric frequency domain causality test, a representative[3] finite-order VAR model including positive and negative shocks of the variables, where $z_t = [x_t, y_t]'$ is a two-dimensional time series vector, is as follows:

$$\Theta(L)z_t = \varepsilon_t \tag{2}$$

It is a lag polynomial of size $2x2$, with $\Theta(L) = I - \Theta_1 L - \ldots - \Theta_p L^p$ in Equation 2. ε_t is the error term with white noise. Due to this characteristic of Σ, which is

3 In order to provide a simple presentation, a single representative equation has created that includes the positive and negative shocks of the variables.

positive definite and symmetric, there is a Cholesky decomposition for $G'G = \Sigma^{-1}$. Under the assumption that the system is stationary, the representation of moving averages as $\Phi(L) = \Theta(L)^{-1}$ and $\Psi(L) = \Phi(L)G^{-1}$ is as in Equation 3:

$$z_t = \Phi(L)\varepsilon_t = \begin{bmatrix} \Phi_{11}(L) & \Phi_{12}(L) \\ \Phi_{21}(L) & \Phi_{22}(L) \end{bmatrix} \begin{bmatrix} \varepsilon_{1t} \\ \varepsilon_{2t} \end{bmatrix} = \Psi(L)\eta_t = \begin{bmatrix} \Psi_{11}(L) & \Psi_{12}(L) \\ \Psi_{21}(L) & \Psi_{22}(L) \end{bmatrix} \begin{bmatrix} \eta_{1t} \\ \eta_{2t} \end{bmatrix}$$
(3)

Here, the spectral density of Y_t can be expressed as in Equation 4:

$$f_y(\omega) = \frac{1}{2\pi}\left\{ \left\| \Psi_{11}(e^{-iw}) \right\|^2 + \left\| \Psi_{12}(e^{-iw}) \right\|^2 \right\} \tag{4}$$

Hence, the measurement of causality in frequency is defined in Equation 5:

$$M_{X \to Y}(\omega) = log\left[1 + \frac{\left\| \Psi_{12}(e^{-iw}) \right\|^2}{\left\| \Psi_{11}(e^{-iw}) \right\|^2} \right] \tag{5}$$

If $\left| \Psi_{12}(e^{-iw}) \right|^2 = 0$ in Equation 5, the result is 'X_t is not the Granger cause of Y_t at ω frequency'. Breitung and Candelon (2006) add the following linear constraints to the model:

$$\sum_{k=1}^{p} \theta_{12,k} \cos(k\omega) = O \text{ and } \sum_{k=1}^{p} \theta_{12,k} \sin(k\omega) = O \tag{6}$$

To simplify the formula, by writing $\alpha_j = \theta_{11,j}$ and $\beta_j = \theta_{12,j}$, the representative VAR equation Y_t for Y_t^+ ve Y_t^- can be expressed as:

$$Y_t = \alpha_1 Y_{t-1} + \ldots + \alpha_p Y_{t-p} + \beta_1 Y_{t-1} + \ldots + \beta_p Y_{t-p} + \varepsilon_{1t} \tag{7}$$

The existence of Granger causality can be tested at any ω frequency value with the approach of Breitung and Candelon (2006). Accordingly, the hypotheses for each frequency value are as follows:

$$H_0 : R(\omega)\beta = 0, \beta = \left[\beta_1, \ldots, \beta_p\right] \text{ve}\, R(\omega) = \begin{bmatrix} \cos(\omega) & \ldots & \cos(p\omega) \\ \sin(\omega) & \ldots & \sin(p\omega) \end{bmatrix} \quad (8)$$

Wald test statistics with χ^2 distribution are used to test the null hypothesis. Critical values are obtained by bootstrap simulations. If the probability values of the test statistics are less than 10 %, 5 % and 1 %, the null hypothesis is rejected. Accordingly, the significance of the probability value of ω at a frequency of 0.5 and 2.5 indicates the existence of long-term and short-term causality, respectively.

3 Data and Preliminary Analysis

This study aims to examine whether energy-related uncertainty has an impact on global food prices for the monthly period 2008:01-2022:10[4]. Dang et al. (2023) have created indices using text analysis of the Economist Intelligence Unit monthly reports of 28 developed and developing countries. These indices react strongly to oil shocks, the Global Financial Crisis, European debt crises, the Covid-19 pandemic, and the Russia-Ukraine War. Energy-related uncertainty indices ERU_GDP and ERU_E are GDP weighted and equally weighted, respectively. FAO food price index (FPI), which is calculated by the Food and Agriculture Organization of the United Nations (FAO) with a basket of different foods, is also preferred to represent global food prices. All variables are used in their logarithmic form $(LFPI, LERU\_GDP, LERU\_E)$. Positive and negative superscript on variables indicate positive or negative components of the relevant variable.

4 Food Price Index and energy-related uncertainties data are obtained from FAO's website (Date of Access: 12 May 2024) and Economic Policy Uncertainty website (Date of Access: 12 May 2024), respectively.

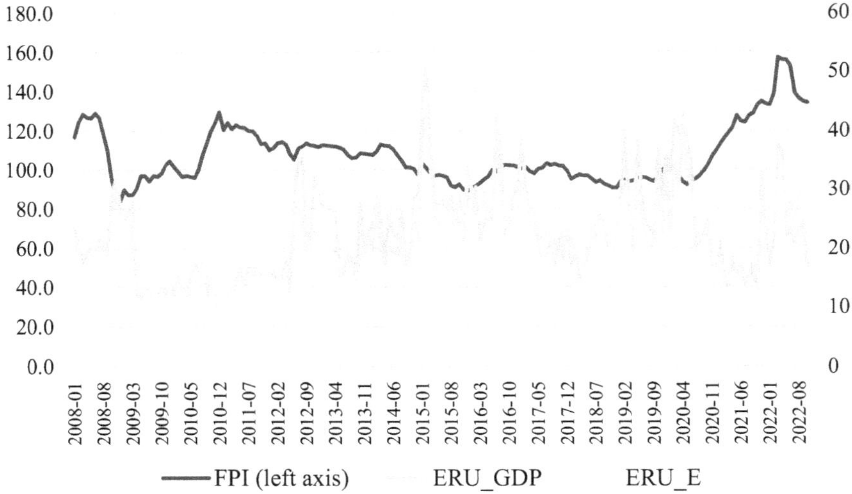

Graph 1: The Monthly Data of FPI, ERU_GDP and ERU_E Series

Graph 1 contains a monthly presentation of the raw data. It can be said that especially in recent years, energy-related uncertainties have increased and food inflation has reached an all-time peak. Time series graphs of variables hint that one of the drivers of changes in global food prices may be energy-related uncertainties.

Table 1: Summary Statistics and Normality Test

	LFPI	LERU_GDP	LERU_E
Mean	4.666087	3.061242	3.029839
Median	4.628377	3.076511	3.052590
Maximum	5.054949	3.925120	3.767157
Minimum	4.426579	2.247700	2.392669
Std. Dev.	0.128047	0.345453	0.297683
Skewness	0.801086	0.063779	-0.043907
Kurtosis	3.206810	2.373749	2.435103
Jarque-Bera	19.35544	3.029419	2.423913
Probability	0.000063	0.219872	0.297614
Observations	178	178	178

Table 1 presents summary statistics and normality test for variables in logarithmic form. The skewness of $LFPI$ and $LERU\_GDP$ points out that these series are long right-tailed, while the skewness of $LERU\_E$ indicates that it is long left-tailed. Furthermore, $LFPI$ exhibits leptokurtic property, while $LERU\_GDP$ and $LERU\_E$ have platykurtic property. $LERU\_GDP$ and $LERU\_E$ are normally distributed but $LFPI$ is not. A total of 178 months of observation has used in the study.

Table 2: Traditional and Breakpoint Unit Root Test Results

Variable	ADF Test	Breakpoint Test
$LFPI^+$	-1.3628 (0.8684)	-2.2811 (>0.99) – Break Date: 2017M08
$\Delta LFPI^+$	-9.9703*** (0.0000)	-11.028*** (<0.01) – Break Date: 2010M12
$LFPI^-$	-4.1653*** (0.0062)	-7.5454*** (<0.01) – Break Date: 2008M07
$\Delta LFPI^-$	-	-
$LERU\_GDP^+$	-3.6941** (0.0252)	-4.9685* (0.0835) – Break Date: 2012M05
$\Delta LERU\_GDP^+$	-16.354*** (0.0000)	-17.279*** (<0.01) – Break Date: 2012M06
$LERU\_GDP^-$	-2.6528 (0.2577)	-5.5974** (0.0157) – Break Date: 2015M04
$\Delta LERU\_GDP^-$	-15.951*** (0.0000)	-17.006*** (<0.01) – Break Date: 2009M04
$LERU\_E^+$	-1.6927 (0.7506)	-3.2864 (0.8992) – Break Date: 2013M09
$\Delta LERU\_E^+$	-16.042*** (0.0000)	-16.779*** (<0.01) – Break Date: 2021M11
$LERU\_E^-$	-1.5930 (0.7921)	-3.3366 (0.8824) – Break Date: 2011M11
$\Delta LERU\_E^-$	-15.753*** (0.0000)	-17.031*** (<0.01) – Break Date: 2009M04

***, **, and * denote 1 %, 5 %, and 10 % significance levels respectively. The values in parentheses indicate probability values.

An important preliminary step in time series analysis is the unit root tests for the detection of stationary. In this study, whether the series contains a unit root is important in determining the appropriate lag length. The trend and intercept structures of the variables are considered in the unit root tests results in Table 2. The findings of both traditional and structural break unit root tests indicate that all variables are difference stationary, that is, $I(1)$, except for the negative component of the $LFPI$. Therefore, the maximum cointegration degree $(d_{max} = 1)$ is added to the appropriate lag lengths in the VAR models. Then, $VAR(p + d_{max})$ models are established and asymmetric causality is tested in the frequency domain.

4 Findings

First, all variables are decomposed into positive and negative components. Table 3 shows the short- and long-term findings of asymmetric causality relationships from $LERU\_GDP$ to $LFPI$. According to these findings, the only significant relationship is from $LERU\_GDP^{-}$ to $LFPI^{-}$ in the long run. In other words, a significant causal relationship has been detected from the negative shocks of $LERU\_GDP$ to the negative shocks of $LFPI$ in the long run.

Table 3: Asymmetric Causality Test Findings in the Frequency Domain ($LERU\_GDP$ and $LFPI$)

H_0	Long term (ω=0,5)		Short term (ω=2,5)		VAR(p+d_{max})
	Test Stat.	Prob.	Test Stat.	Prob.	
$LERU\_GDP^{+} \nrightarrow LFPI^{+}$	0.0030	(0.9985)	0.8378	(0.6578)	3
$LERU\_GDP^{-} \nrightarrow LFPI^{-}$	14.252***	(0.0008)	3.4237	(0.1805)	3
$LERU\_GDP^{-} \nrightarrow LFPI^{+}$	0.0335	(0.9834)	0.6236	(0.7321)	5
$LERU\_GDP^{+} \nrightarrow LFPI^{-}$	2.3740	(0.3051)	2.2966	(0.3172)	3

***, **, and * denote 1 %, 5 %, and 10 % significance levels respectively.

The test is repeated using $LERU\_E$ instead of $LERU\_GDP$, as it is thought to make the findings robust. Table 4 presents the short- and long-term findings of asymmetric causal relationships from $LERU\_E$ to $LFPI$. Here, a weak causal relationship is observed from $LERU\_E^{-}$ to $LFPI^{-}$ and $LFPI^{+}$ in the short term. However, there is a significant and strong causal relationship from $LERU\_E^{-}$ to $LFPI^{-}$ in the long term. This makes the findings of the test applied with $LERU\_GDP$ robust.

Table 4: Asymmetric Causality Test Findings in the Frequency Domain ($LERU\_E$ and $LFPI$)

H_0	Long term (ω=0,5)		Short term (ω=2,5)		VAR(p+d_{max})
	Test Stat.	Prob.	Test Stat.	H_0	
$LERU\_E^{+} \nrightarrow LFPI^{+}$	1.3372	(0.5124)	0.1267	(0.9386)	3
$LERU\_E^{-} \nrightarrow LFPI^{-}$	13.847***	(0.0010)	5.5514*	(0.0623)	3
$LERU\_E^{-} \nrightarrow LFPI^{+}$	0.6908	(0.7079)	5.3617*	(0.0685)	4
$LERU\_E^{+} \nrightarrow LFPI^{-}$	1.4158	(0.4927)	2.7447	(0.2535)	3

***, **, and * denote 1 %, 5 %, and 10 % significance levels respectively.

All of the findings provide evidence for the existence of a long-run asymmetric causal relationship from negative shocks of energy-related uncertainties to negative shocks of global food prices. These findings can be interpreted as global food prices will decrease in the long term when energy-related uncertainties decrease.

The empirical findings of this study are in line with Aye (2015) in that energy price uncertainty has a positive and significant impact on food prices and an asymmetric relationship is found between the variables. This study is also similar to other studies such as Zmami and Ben-Salha (2019), Karakotsios et al. (2021), and Alnour et al. (2023), which emphasize a positive and asymmetric relationship between the variables.

5 Conclusion

This study aims to examine whether energy-related uncertainties affected global food prices in the post-Global Financial Crisis period. Findings from the asymmetric frequency domain causality test indicate that energy-related uncertainties affect global food prices asymmetrically. Negative shocks of energy-related uncertainty have an impact on negative shocks of global food prices in the long run. Although this gives a hint that there is a same-directional relationship between the two variables, it essentially enables the interpretation that global food prices decrease when energy-related uncertainties decrease. These findings have some notable implications. The first of these is that energy-related uncertainties are an important driving force of global food prices and it should be taken into account in new models on global food prices. Second, the relationship between energy-related uncertainties and global food prices is asymmetric and therefore it should be examined with asymmetric methods. Third, energy-related uncertainties need to be reduced in the fight against global food prices in the long term. Finally, it is thought that any policy that will increase the security of energy supply and ensure environmental sustainability will contribute to the fight against global food prices.

References

Alghalith, M. (2010). The interaction between food prices and oil prices. *Energy Economics*, *32*(6), 1520–1522.

Alnour, M., Altıntaş, H., & Rahman, M. N. (2023). Unveiling the asymmetric response of global food prices to the energy prices shocks and economic policy uncertainty. *World Development Sustainability*, *3*, 100083.

Aye, G. C. (2015). The effect of oil price uncertainty on food price in South Africa. *International Journal of Economics and Management Engineering*, *9*(5), 1380–1385.

Baumeister, C., & Kilian, L. (2014). Do oil price increases cause higher food prices?. *Economic Policy*, *29*(80), 691–747.

Bozoklu, S., & Yilanci, V. (2013). Energy consumption and economic growth for selected OECD countries: Further evidence from the granger causality test in the frequency domain. *Energy Policy*, *63*, 877–881.

Breitung, J., & Candelon, B. (2006). Testing for short-and long-run causality: A frequency-domain approach. *Journal of Econometrics*, *132*(2), 363–378.

Dang, T. H. N., Nguyen, C. P., Lee, G. S., Nguyen, B. Q., & Le, T. T. (2023). Measuring the energy-related uncertainty index. *Energy Economics*, *124*, 106817.

Darwez, F., Alharbi, F., Ifa, A., Bayomei, S., Mostfa, E., Lutfi, A., & Alrawad, M. (2023). Assessing the impact of oil price volatility on food prices in Saudi Arabia: Insights from nonlinear autoregressive distributed lags (NARDL) analysis. *ECONOMICS-Innovative and Economics Research Journal*, *11*(2), 5–23.

Economic Policy Uncertainty, www.policyuncertainty.com (Date of Access: 12 May 2024).

Food and Agriculture Organization of the United Nations (FAO), https://www.fao.org/worldfoodsituation/foodpricesindex/en/ (Date of Access: 12 May 2024).

Geweke, J. (1982). Measurement of linear dependence and feedback between multiple time series. *Journal of American Statistical Association*, *77*(378), 304.

Granger, C. W. (1969). Investigating causal relations by econometric models and cross-spectral methods. *Econometrica: Journal of the Econometric Society*, *37*(3), 424–438.

Han, X., Yuan, T., Wang, D., Zhao, Z., & Gong, B. (2023). How to understand high global food price? Using SHAP to interpret machine learning algorithm. *Plos one*, *18*(8), e0290120.

Hatemi-J., A., & Uddin, G. S. (2012). Is the causal nexus of energy utilization and economic growth asymmetric in the US?. *Economic Systems*, *36*(3), 461–469.

Hosoya, Y. (1991). The decomposition and measurement of the interdependency between second-order stationary processes. *Probability Theory and Related Fields*, *88*(4), 429–444.

Karakotsios, A., Katrakilidis, C., & Kroupis, N. (2021). The dynamic linkages between food prices and oil prices. Does asymmetry matter?. *The Journal of Economic Asymmetries*, *23*, e00203.

Kırca, M., Canbay, Ş., & Pirali, K. (2020). Is the relationship between oil-gas prices index and economic growth in Turkey permanent? *Resources Policy*, *69*, 101838.

Kirikkaleli, D., & Darbaz, I. (2021). The causal linkage between energy price and food price. *Energies*, *14*(14), 4182.

Nazlioglu, S., Erdem, C., & Soytas, U. (2013). Volatility spillover between oil and agricultural commodity markets. *Energy Economics*, *36*, 658–665.

Olayungbo, D. O. (2021). Global oil price and food prices in food importing and oil exporting developing countries: A panel ARDL analysis. *Heliyon*, *7*(3), e06357.

Ranjbar, O., Chang, T., Nel, E., & Gupta, R. (2017). Energy consumption and economic growth nexus in South Africa: Asymmetric frequency domain approach. *Energy Sources. Part B: Economics, Planning, and Policy*, *12*(1), 24–31.

Raza, S. A., Guesmi, K., Belaid, F., & Shah, N. (2022). Time-frequency causality and connectedness between oil price shocks and the world food prices. *Research in International Business and Finance*, *62*, 101730.

Saucedo, A., Brümmer, B., & Jaghdani, T. J. (2015). The dynamic pattern of volatility spillovers between oil and agricultural markets. In *Proceedings of the International Conference–Food in the Bio-based Economy; Sustainable Provision and Access, Wageninge, The Netherlands* (pp. 27–29).

Sims, C.A. (1980). Macroeconomics and reality. *Econometrica*, *48*(1), 77–97.

Taghizadeh-Hesary, F., Rasoulinezhad, E., & Yoshino, N. (2018). *Volatility linkages between energy and food prices: Case of selected Asian countries* (No. 829). ADBI Working Paper.

Ulussever, T., Ertuğrul, H. M., Kılıç Depren, S., Kartal, M. T., & Depren, Ö. (2023). Estimation of impacts of global factors on world food prices: A comparison of machine learning algorithms and time series econometric models. *Foods*, *12*(4), 873.

Zmami, M., & Ben-Salha, O. (2019). Does oil price drive world food prices? Evidence from linear and nonlinear ARDL modeling. *Economies*, *7*(1), 12.

Sefer Uçak[1] & Jiyan Kılıç[2]

Industrial Production Index and Foreign Trade Relationship in Türkiye: ARDL Model

1 Introduction

Countries can increase their welfare levels through increases in production and productivity. The development of foreign trade will also enhance countries' competitive power and contribute to their economic growth according to their comparative advantages.

Türkiye, especially from the early years of the Republic until the decisions of January 24, 1980, attempted to meet domestic demand by increasing domestic production with import-substitution policies. After 1980, with the export-oriented industrialization model it implemented, it aimed to contribute to its economic growth with increasing export revenues parallel to the increase in domestic industrial production (Şahin, 2016: 194).

In Türkiye, the distribution of imports shows that intermediate goods account for 70.6 %, capital goods for 14.8 %, and consumer goods for 14.5 %. The predominance of intermediate goods is due to energy and raw materials. The main reason for this, although it varies by sector, is that Türkiye's total manufacturing industry import dependency rate is around 24 % (Küçük Karaman, 2023: 697).

Looking at the distribution of exports, it can be observed that the share of manufacturing industry is 93.9 %, agriculture, forestry, and fishing sector is 4.1 %, and mining and quarrying sector is 1.5 %. In terms of technology intensity, while the share of high-technology product exports in manufacturing industry foreign trade is 3.6 %, it is 10.2 % in imports. Among the main reasons for this are; the low value-added nature of our exports and the low share of R&D expenditures in GDP (TURKSTAT). One of the leading indicators of economic growth is the Industrial Production Index announced by TURKSTAT. It is an economic growth

1 Ph.D., Asos.Prof., Balıkesir University, Sındırgı Vocational School, Department of Foreign Trade, 10200, Balıkesir/Türkiye. seferucak@balikesir.edu.tr, ORCID: 0000-0002-9251-9584.

2 Ph.D., Lect., Balıkesir University, Sındırgı Vocational School, Department of Foreign Trade, 10200, Balıkesir/Türkiye, jiyan.kilic@balikesir.edu.tr, ORCID: 0000-0002-9904-8420.

indicator that allows comparative monitoring of the situation in the industrial sector and changes in production activities over time (Koç et al., 2016: 43). This index shows developments in the industrial sector and their short-term effects. The Industrial Production Index covers three sub-sectors: Mining and Quarrying (B), Manufacturing (C), and Production and Distribution of Electricity, Gas, Steam, and Air Conditioning (D). According to the 2010 reference index value, the manufacturing industry sector accounts for 81.51 %, the production and distribution of electricity, gas, steam, and air conditioning sub-sector for 12.44 %, and the mining and quarrying sector for 6.05 % (Koç et al., 2016: 43).

2 Empirical Literature

The relationship between economic growth and foreign trade is an extremely popular research area. When the literature is examined, it is observed that it has been studied in every period in terms of various data sets and periods considered. In Table 1 below, the relationship between foreign trade and economic growth is summarized in the literature using different methods by creating various variables and data sets at various times.

Table 1: Summary of Literature

Author	Period	Country	Variables	Method	Result
Terzi & Oltulular (2004)	1987:Q2-2001:Q3	Türkiye	Economic Growth and Industrial Production Index	Granger Causality Test – ECM	It has been observed that the causality relationship between growth and industrial production index is positive and bidirectional relationship.
Hayaloğlu (2015)	1990–2012	India, Indonesia, Brazil, Türkiye, South Africa	GDP Per Capita, Domestic Loans to the Private Sector, Domestic Loans to the Private Sector by Banks, Population Size, Trade (Export and Import), Foreign Direct Investments, Gross Capital Formation	Dynamic Panel Data Analysis	There is a positive relationship between financial development and economic growth.

Table 1: Continued

Author	**Period**	**Country**	**Variables**	**Method**	**Result**
Korkmaz & Aydın (2015)	2002:Q1 – 2014:Q2	Türkiye	Export unit value index, Import unit value index, and Economic growth	VAR Analysis, Granger Causality Test	A bidirectional causality relationship between imports and economic growth has been obtained.
Şerefli (2016)	1975–2014	Türkiye	Foreign Trade and Economic Growth	Granger Causality Test	A causal relationship between growth and foreign trade could not be determined.
Gövdere & Can (2016)	1970–2011	Türkiye	Growth, Energy Consumption, Financial Development, Fixed Capital Investments, Openness, Exports, Imports	ARDL bounds testing approach (ARDL)	In the long run, energy consumption, openness, exports and imports affect growth positively. No findings were found regarding the impact of fixed capital investments and loans on growth.
Alancıoğlu & Bayrak-tutan (2023)	1980–2021	Türkiye	GDP, exports, imports and gross capital formation	VAR Analysis	There is a unidirectional relationship from exports to growth and a bidirectional causality between imports and growth.
Yılmaz & Albayrak (2023)	1980–2020	Türkiye	Economic growth, Imports and exports	ARDL bounds testing approach	While the effect of exports on economic growth is negative, the effect of imports is positive.

(*continued on next page*)

Table 1: Continued

Author	Period	Country	Variables	Method	Result
Kazaz & Uzunoğlu (2023)	2013:01-2023:03	Türkiye	imports, exports, real effective exchange rate, industrial production index, and imported intermediate goods	ARDL bounds test and VAR Granger causality	The findings show that there is a significant and positive relationship between intermediate goods imports and imports in the short and long term. While the relationship between intermediate goods imports and exports is negative in the short term and positive in the long term. and shows that there is a statistically significant relationship.

Source: Compiled by the author.

3 Data Set and Methodology

In this study, monthly data for Türkiye from January 2013 to March 2024 were used to test the relationship between the Industrial Production Index (IPI), the Export Unit Index, Import Unit Index and the Real Effective Exchange Rate. The dependent variable in this study is the Industrial Production Index (IPI), while the independent variables are determined as the Export Unit Index (EXP), Import Unit Index (IMP), and Real Effective Exchange Rate (REER). The Industrial Production Index, Export Unit Index, and Import Unit Index data for the period under consideration were obtained from the TURKSTAT (2024) database, while the data set for the Real Effective Exchange Rate was obtained from the CBRT (2024) database. Econometric analyses of the data set consisting of 135 observations were conducted using the EViews 10 software package. The econometric model established is shown in the equation below:

$$IPI_t = \alpha_1 + \alpha_2 EXP_t + \alpha_3 IMP_t + \alpha_4 REER_t + \varepsilon_t \tag{1}$$

In this equation, (IPI) represents the Industrial Production Index, (EXP) represents the Export Unit Index, (IMP) represents the Import Unit Index, (REER) represents the Real Effective Exchange Rate, (α_1) is the constant term, and (ε_t) is the error term.

3.1 Unit Root Tests

It is essential to investigate whether the stochastic process in a time series changes over time. If the nature of the probabilistic process does not change over time, the series is stationary; otherwise, it is non-stationary. Therefore, the non-stationarity of the series leads to spurious regression. (Kutlar, 2007: 284).

Table 2: ADF and PP Unit Root Test

			Augmented Dickey Fuller Unit Root Test		**Phillips-Perron Unit Root Test**	
			Level	**1st Difference**	**Level**	**1st Difference**
IPI I(0)	**Intercept**	**%1**	-3.479656	-3.480038	-3.479656	-3.480038
		%5	-2.883073	-2.883239	-2.883073	-2.883239
		%10	-2.578331	-2.578420	-2.578331	-2.578420
		t-statistic	-1.230925	-13.19957	-0.733099	-19.03699
		(Prob.)	(0.6599)	**(0.0000)***	(0.8337)	**(0.0000)***
	Trend and Intercept	**%1**	-4.027959	-4.028496	-4.027959	-4.028496
		%5	-3.443704	-3.443961	-3.443704	-3.443961
		%10	-3.146604	-3.146755	-3.146604	-3.146755
		t-statistic	-4.166393	-13.15408	-4.119351	-19.35483
		(Prob.)	**(0.0065)***	**(0.0000)***	**(0.0076)***	**(0.0000)***
	None	**%1**	-2.582334	-2.582465	-2.582334	-2.582465
		%5	-1.943229	-1.943247	-1.943229	-1.943247
		%10	-1.615134	-1.615122	-1.615134	-1.615122
		t-statistic	1.028563	-13.08070	3.051607	-14.10644
		(Prob.)	(0.9198)	**(0.0000)***	(0.9994)	**(0.0000)***
EXP I(1)	**Intercept**	**%1**	-3.480038	-3.480038	-3.479656	-3.480038
		%5	-2.883239	-2.883239	-2.883073	-2.883239
		%10	-2.578420	-2.578420	-2.578331	-2.578420
		t-statistic	-1.584745	-9.215246	-1.496039	-9.303319
		(Prob.)	(0.4876)	**(0.0000)***	(0.5327)	**(0.0000)***

(continued on next page)

Table 2: Continued

			Augmented Dickey Fuller Unit Root Test		Phillips-Perron Unit Root Test	
	Trend and Intercept	**%1**	-4.027959	-4.028496	-4.027959	-4.028496
		%5	-3.443704	-3.443961	-3.443704	-3.443961
		%10	-3.146604	-3.146755	-3.146604	-3.146755
		t-statistic	-1.396358	-9.537315	-1.454090	-9.530659
		(Prob.)	(0.8579)	**(0.0000)***	(0.8403)	**(0.0000)***
	None	**%1**	-2.582465	-2.582465	-2.582334	-2.582465
		%5	-1.943247	-1.943247	-1.943229	-1.943247
		%10	-1.615122	-1.615122	-1.615134	-1.615122
		t-statistic	-0.309730	-9.247520	-0.304928	-9.333993
		(Prob.)	(0.5724)	**(0.0000)***	(0.5742)	**(0.0000)***
IMP I(1)	**Intercept**	**%1**	-3.480818	-3.480818	-3.479656	-3.480038
		%5	-2.883579	-2.883579	-2.883073	-2.883239
		%10	-2.578601	-2.578601	-2.578331	-2.578420
		t-statistic	-1.574477	-4.292572	-1.345559	-11.79450
		(Prob.)	(0.4928)	**(0.0007)***	(0.6070)	**(0.0000)***
	Trend and Intercept	**%1**	-4.029595	-4.029595	-4.027959	-4.028496
		%5	-3.444487	-3.444487	-3.443704	-3.443961
		%10	-3.147063	-3.147063	-3.146604	-3.146755
		t-statistic	-2.104933	-4.403631	-1.797059	-11.86871
		(Prob.)	(0.5379)	**(0.0031)***	(0.7009)	**(0.0000)***
	None	**%1**	-2.582734	-2.582734	-2.582334	-2.582465
		%5	-1.943285	-1.943285	-1.943229	-1.943247
		%10	-1.615099	-1.615099	-1.615134	-1.615122
		t-statistic	-0.167491	-4.308770	-0.113512	-11.82679
		(Prob.)	(0.6240)	(0.**0000**)*	(0.6428)	**(0.0000)***
REER I(0)	**Intercept**	**%1**	-3.480425	-3.480425	-3.479656	-3.480038
		%5	-2.883408	-2.883408	-2.883073	-2.883239
		%10	-2.578510	-2.578510	-2.578331	-2.578420
		t-statistic	-1.248251	-9.224384	-1.207880	-8.699097
		(Prob.)	(0.6521)	**(0.0000)***	(0.6700)	**(0.0000)***
	Trend and Intercept	**%1**	-4.029041	-4.029041	-4.027959	-4.028496
		%5	-3.444222	-3.444222	-3.443704	-3.443961
		%10	-3.146908	-3.146908	-3.146604	-3.146755
		t-statistic	-2.825197	-9.220489	-2.703177	-8.696277
		(Prob.)	(0.1910)	**(0.0000)***	(0.2371)	**(0.0000)***

Table 2: Continued

		Augmented Dickey Fuller Unit Root Test			Phillips-Perron Unit Root Test	
	None	%1	-2.582599	-2.582599	-2.582334	-2.582465
		%5	-1.943266	-1.943266	-1.943229	-1.943247
		%10	-1.615111	-1.615111	-1.615134	-1.615122
		t-statistic	-2.060847	-8.967438	-2.598530	-8.067693
		(Prob.)	**(0.0382)****	**(0.0000)***	**(0.0096)***	**(0.0000)***

Note: *(*) Significant at the 1 % level and (**) significant at the 5 % level.*

Source: *Compiled by the author*

In this study, the ADF (Augmented Dickey Fuller) and PP (Phillips Perron) Unit Root Tests, which are commonly used in the literature to test the stationarity of the series, are employed. For each variable, the stationarity status at the Level level with intercept, trend and intercept, and none cases at various significance levels was initially investigated, followed by a stationarity test in the first differences. According to the results of the ADF and PP unit root tests in Table X, it was determined that the series are stationary at different orders (I(0) and I(1)). Therefore, the ARDL (Auto Regressive Distributed Lag) method was used to investigate the cointegration relationship between variables.

3.2 ARDL Long Run Form and Bounds Test

Various cointegration tests are employed to test the existence of long-term relationships between variables. Some of these tests are the Johansen and Engle-Granger Cointegration tests. According to these tests, it is conditioned on the series not being stationary at the level values and being stationary when the first differences of all series are taken. In other words, it is expected that the series for the variables are not at I(0) but at I(1). Thus, the existence of a long-term relationship can be tested.

By relaxing the above-mentioned stationarity assumption developed by Peseran et al. (2001), they demonstrated that the long-term relationship can be investigated at different degrees of integration. To overcome this limitation, they introduced the concept of Bounds Testing to the literature. Once the stationary levels of the variables are identified as I(0) and I(1), testing for long-term relationships becomes possible by establishing an Auto Regressive Distributed Lag (ARDL) model. In the Bounds Test, since the ARDL Model is established, it does not consider any endogeneity problem that the variables may create.

Table 3: ARDL Long Run Form and Bounds Test (1,4,1,1)

Conditional Error Correction Regression				
Variable	**Coefficient**	**Std. Error**	**t-Statistic**	**Prob.**
C	4.530255	8.046198	0.563031	0.5745
@TREND	0.147515	0.036394	4.053273	0.0001
IPI(-1)*	-0.338175	0.067265	-5.027514	0.0000
EXP(-1)	0.022848	0.098496	0.231969	0.8170
IMP(-1)	0.045303	0.052561	0.861925	0.3905
REER(-1)	0.088597	0.059387	1.491865	0.1384
D(EXP)	-0.328486	0.218017	-1.506695	0.1345
D(EXP(-1))	0.407356	0.190096	2.142903	0.0342
D(EXP(-2))	0.136599	0.191427	0.713585	0.4769
D(EXP(-3))	-0.378984	0.196486	-1.928811	0.0561
D(IMP)	0.386392	0.107082	3.608364	0.0005
D(REER)	-0.086001	0.105775	-0.813058	0.4178
F-Bounds Test				
Test Statistic	**Value**	**Signif.**	**I(0)**	**I(1)**
F-statistic	5.284607	**10 %**	2.97	3.74
k	3	**5 %**	3.38	4.23
		1 %	4.3	5.23

Source: *Compiled by the author*

According to the hypotheses of the Bounds Test: H_0: There is no cointegration relationship among the variables. H_1: There is a cointegration relationship among the variables.

In Table 3, it is observed that the F-statistics are above the critical values. Since the F statistic value (5.284607) above is greater than the lower I (0) and upper I (1) limit critical values given at different significance levels (1-5-10 %), H_0 was rejected and it was concluded that there was a cointegration relationship. Therefore, it is concluded that there is a long-term cointegration relationship among the series. Once the existence of cointegration relationship is determined, an ARDL Error Correction Regression Model is established to determine short-term relationships.

Table 4: ARDL Error Correction Regression (1,4,1,1)

ECM Regression Case 4: Unrestricted Constant and Restricted Trend				
Variable	**Coefficient**	**Std. Error**	**t-Statistic**	**Prob.**
C	4.677770	0.878791	5.322963	0.0000
D(EXP)	-0.328486	0.205415	-1.599129	0.1124
D(EXP(-1))	0.407356	0.185194	2.199619	0.0298
D(EXP(-2))	0.136599	0.185275	0.737278	0.4624
D(EXP(-3))	-0.378984	0.189860	-1.996126	0.0482
D(IMP)	0.386392	0.099008	3.902646	0.0002
D(REER)	-0.086001	0.101754	-0.845186	0.3997
CointEq(-1)*	-0.338175	0.064710	-5.226012	**0.0000**
		Akaike info criterion		5.090442
F-statistic	6.381686	**Schwarz criterion**		5.266026
Prob(F-statistic)	0.000002	**Hannan-Quinn criter.**		5.161790

Source: *Compiled by the author.*

As seen in Table 4, the error correction model of the ARDL model was tested using the error correction term ECR equation. The term CointEq(-1) in the table is referred to as the error correction variable (lagged by one period). This value indicates how much of the short-term deviations from equilibrium can be corrected in the long term. Looking at the test statistic results, it implies that a deviation occurring in the short term can be corrected after 1.77 months, reaching the long-term equilibrium. This coefficient, which tests the existence of short-term relationships, was found to be statistically significant and negatively signed. (Pazarlıoğlu & Gülay (2007)). In the short term, a 1 % change in the lagged value of exports results in a 0.41 % increase in the industrial production index. However, when looking at the third lagged value of exports, a 1 % change results in a 0.38 % decrease in the industrial production index. For the import variable, a 1 % change in the lagged value leads to a 0.39 % increase in the industrial production index.

Table 5: Pairwise Granger Causality Tests and Wald Test

Hypothesis	**Obs**	**F-Statistic**	**Prob.**
EXP does not Granger Cause IPI	133	1.67576	0.1912
IPI does Granger Cause EXP		7.99796	**0.0005****

(continued on next page)

Table 5: Continued

Hypothesis	Obs	F-Statistic	Prob.
IMP does not Granger Cause IPI	133	0.02025	0.9800
IPI does Granger Cause IMP		4.13846	**0.0181****
REER does Granger Cause IPI	133	4.24195	**0.0165** **
IPI does not Granger Cause REER		2.41011	0.0939
IMP does Granger Cause EXP	133	4.85051	**0.0093****
EXP does Granger Cause IMP		3.06242	**0.0502****
REER does not Granger Cause EXP	133	2.67550	0.0727
EXP does not Granger Cause REER		0.23569	0.7904
REER does Granger Cause IMP	133	4.44775	**0.0136****
IMP does not Granger Cause REER		0.08074	0.9225
Wald Test			
Test Statistic	**Value**	**df**	**Probability**
F-statistic	22.32545	(10, 119)	0.0000**
Chi-square	223.2545	10	0.0000**
Note: *(**) Significant at the 5 % level.*			

Source: *Compiled by the author.*

In Table 5, the causality relationship between the variables arranged monthly from January 2013 to March 2024 was tested using Pairwise Granger Causality Tests and the Wald Test. When the results of Pairwise Granger Causality Tests with 2 delays are examined, in cases where the probability value is less than 0.05, Null Hypothesis (H_0: It is not a Granger cause.) is rejected and the alternative hypothesis (H_1: It is a Granger cause) is accepted. Thus, it is possible to talk about Granger causality between this variables. When the Wald test results were examined, the probability value was found to be significant at the 0.05 significance level. There is a short-term causality relationship between variables.

Two factors are crucial for testing the robustness of the ARDL model. First, the specification error in the established ARDL Model is tested using the RAMSEY Reset Test, and second, whether there is serial correlation in the residuals is tested. The test for serial correlation is conducted using the Breusch-Godfrey Serial Correlation LM Test.

Table 6: RAMSEY Reset Test

	Value	df	Probability
t-statistic	0.967372	117	0.3354
F-test summary:	**Sum of Sq.**	**df**	**Mean Squares**
Test SSR	9.321052	1	9.321052
Restricted SSR	1174.692	118	9.955015
Unrestricted SSR	1165.371	117	9.960434

Source: *Compiled by the author.*

The F-test statistic (0.3354) is found to be greater than the 0.05 significance level, indicating its significance. Therefore, the null hypothesis (H_0: There is no specification error in the model) is accepted, and the alternative hypothesis (H_1: There is a specification error in the model) is rejected. Thus, no specification error was encountered in the established ARDL (1,4,1,1) model.

Table 7: Breusch-Godfrey Serial Correlation LM Test

F-statistic	0.649797	**Prob. F(2,117)**	0.5240
Obs*R-squared	1.439115	**Prob. Chi-Square(2)**	0.4870

Source: *Compiled by the author.*

When examining the test results in Table 7, the probability value of Chi-Square(2) (0.4870) is found to be greater than 0.05. Therefore, the null hypothesis (H_0: There is no serial correlation in the residuals) is accepted, indicating that there is no serial correlation in the residuals.

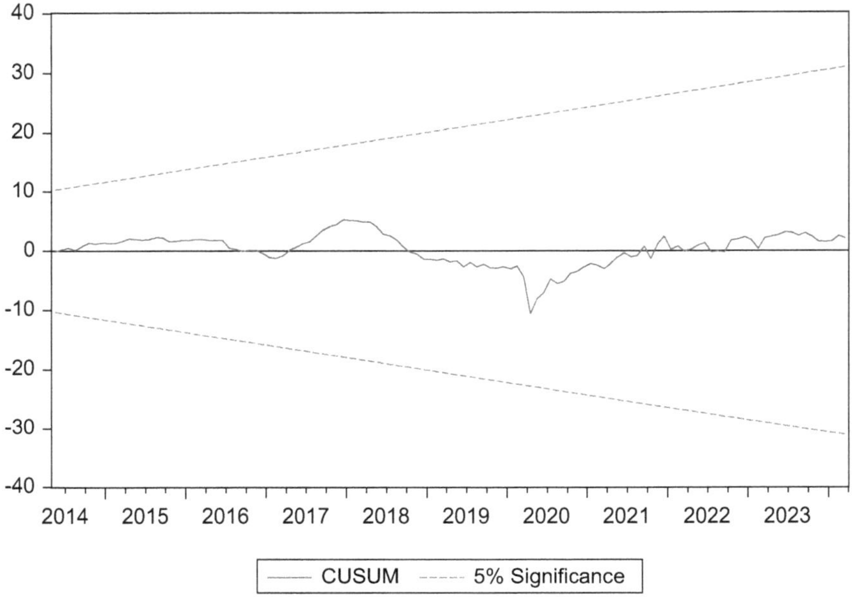

Graph 1: Cusum Test

CUSUM test was applied and graphical analysis was conducted to examine the stability of the long-term coefficients. When examining the lines in Graph 1 above, it can be observed that they fall within the confidence intervals representing the 5 % significance level. Therefore, the parameters are significant, the coefficients are stable, and the model is stable. Additionally, it indicates that the estimated regression analysis is stable.

Conclusion

It is an undeniable fact that foreign trade indicators are one of the most important indicators for a country's economy. Foreign trade is actually closely related to how efficiently national resources are utilized. Additionally, it provides important information about the extent to which external resources need to be accessed. Considering that virtually no country is managed through autarky nowadays, it is possible to emphasize the significance of foreign trade indicators in the national economic structure.

In this study, the aim was to uncover the relationship between the Industrial Production Index and various foreign trade indicators in Türkiye for the period between January 2013 and March 2024. The study was supported by econometric analysis and tested with monthly data sets. The findings indicate a short-term causality relationship among the variables. Furthermore, according to the results of the Granger Causality Test, it was determined that the Export and Import Unit Index provide statistically significant information about the future values of the Industrial Production Index and were identified as Granger causes. Additionally, it was found that the Real Exchange Rate Index is a Granger cause for the Industrial Production Index and imports, while the Export Unit Index is a Granger cause for imports, and finally, the Import Unit Index is a Granger cause for exports.

Like other developing countries, it can be said that production in the Turkish economy is dependent on Imports. Therefore, production inputs are sourced from other countries, leading to the importation of goods requiring high technology for production. It can be argued that products that can substitute for these imported goods need to be developed, thus enabling the production of high-value-added products. In line with various developments in global trade, it is possible to review our foreign trade policies and develop various strategies to achieve global competitive advantage. Considering the impact of exchange rate volatility, one of the indicators of foreign trade, on exports and imports, measures can be taken to ensure exchange rate stability.

References

Alancıoğlu, E., & Bayraktutan, Y. (2023). Dış Ticaret İktisadi Büyüme Etkileşimine İlişkin Hipotezlerin Geçerliliği: Türkiye Örneği. *Akademik Araştırmalar ve Çalışmalar Dergisi (AKAD), 15*(29), 401–410. https://doi.org/10.20990/kilisiibfakademik.1334618.

CBRT. (2024).AvailableAt: https://www.tcmb.gov.tr/wps/wcm/connect/tr/tcmb+tr (16.05.2024)

Gövdere, B., & Can, M. (2016). Türkiye'de Enerji Tüketimi, Dışa Açıklık, Finansal Gelişme, Sabit Sermaye Yatırımları ve Dış Ticaretin Ekonomik Büyümeye Etkisi: Sınır Testi Yaklaşımı. *Niğde Üniversitesi İktisadi ve İdari Bilimler Fakültesi Dergisi, 9*(1), 209–228.

Hayaloğlu, P. (2015). Kırılgan Beşli Ülkelerinde Finansal Gelişme ve Ekonomik Büyüme İlişkisi: Dinamik Panel Veri Analizi. *Ekonomik ve Sosyal Araştırmalar Dergisi, 11*(1), 131–144.

Kazaz, A. A., & Uzunoğlu, S. Türkiye'de İmalat Sanayi, Reel Döviz Kuru ve Dış Ticaret İlişkisi. *Marmara Üniversitesi İktisadi Ve İdari Bilimler Dergisi, 45*(2), 160–179.

Koç, E., Kaya, K., & Şenel, M. C. (2016). Türkiye'de Sanayi Sektörü ve Temel Sanayi Göstergeleri – Sanayi Üretim Endeksi. *Mühendis ve Makina*, *57*(682), 42–53.

Korkmaz, S., & Aydın, A. (2015). Türkiye'de Dış Ticaret- Ekonomik Büyüme İlişkisi: Nedensellik Analizi. *Eskişehir Osmangazi Üniversitesi İİBF Dergisi, Aralık*, *10*(3), 47–76.

Kutlar, A. (2007). *Ekonometriye Giriş*. Nobel Yayın Dağıtım.

Küçük Karaman, S. (2023). İthalat Bağımlılık Oranının Hesaplanması ve Türkiye'de Sektörel İthalat Bağımlılık Oranını Etkileyen Faktörlerin Panel Veri Analizi. *İktisadi İdari ve Siyasal Araştırmalar Dergisi*, *8*(21), 686–711. https://doi.org/10.25204/iktisad.1225874

Pazarlıoğlu, M. V., & Gülay, E. (2007). Net Portföy Yatırımları ile Faiz Arasındaki İlişki: Türkiye Örneği 1922: I – 2005:IV. *Dokuz Eylül Üniversitesi Sosyal Bilimler Enstitüsü Dergisi*, *9*(2), 201–221.

Pesaran, M. H, Shin, Y., & Smith, R. J. (2001). Bounds Testing Approaches to the Analysis of Level Relationships. *Journal of Applied Econometrics*, *16*(3), 289–326.

Şahin, H. (2016). Türkiye Ekonomisi. Ezgi Yayınevi, 1-746. 16. Basım. ISBN: 9786054484041.

Şerefli, M. (2016). Dış Ticaretin Ekonomik Büyüme Üzerine Etkisi: Türkiye Örneği. Kastamonu Üniversitesi, İİBF Dergisi, Temmuz, Sayi: 13, 136– 143.

Terzi, H., & Oltulular, S. (2004). Türkiye'de Sanayileşme ve Ekonomik Büyüme Arasındaki Nedensel İlişki. *Doğuş Üniversitesi Dergisi*, *5*(2), 219–226. https://dergipark.org.tr/en/download/article-file/2151852

TURKSTAT. (2024), Available At: https://www.tuik.gov.tr/ (12.03.2024)

Yılmaz, Ö., & Albayrak, M. (2023). Türkiye'de Dış Ticaretin Ekonomik Büyüme Üzerindeki Etkisinin Ampirik Analizi. *Uluslararası İktisadi ve İdari İncelemeler Dergisi*, *38*, 89–108. https://dergipark.org.tr/tr/download/article-file/2540714

Gökhan Güven[1]

Two-Country Model of International Trade with Different Pollution Industries and the Amount of Labor Force

1. Introduction

This paper explores the influences of free trade on resource stock functions and household utility expressions in the face of concurrent environmental pressures such as overharvesting and transboundary pollution. Transboundary pollution presents a challenge in constructing a model, as industrial damage occurring in one country can impact the actions of one or more countries.

Our model extends Copeland and Taylor (CT) (1999), Benarroch and Thille (2001), and Unteroberdoerster (2001) and others by including another negative externality on resource stock in addition to cross-border pollution, which is one point of originality of our analysis. Their groundbreaking Copeland and Taylor (1999) study examined the implications of trade within a framework where production in one sector has a detrimental impact on another. Their work explores a Ricardian model in which countries specialize in dirtier or environmentally sensitive industries, with findings complemented by Acharyya (2013). Copeland and Taylor suggested that international trade consistently leads to improved welfare for a small, open economy by circumventing the detrimental effects of pollution. This implies that the resource-depleting activities resulting from dirtier industries in one region can only affect the productivity of environmentally sensitive activities within the same region. Benarroch and Thille (2001) and Unteroberdoerster (2001) are seen extensions of this influential paper because they incorporate cross-border pollution. Consequently, their findings align closely (refer to the empirical work by Gibson and Carnovale, 2015). They conclude that an economy specializing solely in environmentally sensitive goods (resource goods) tends to suffer post-trade welfare losses due to the significant impact of transboundary pollution on other countries.

1 Ph. D., Faculty of Political Sciences, Department of Economics, Sakarya University, 54050, Serdivan, Sakarya, Turkey; e-mail: gokhan.guven@ogr.sakarya.edu.tr. ORCID: 0000-0003-2604-301X.

This analysis explores the impact of potential overharvesting as an external factor affecting resource supply, creating a distinct model feature. A related study by Takarada (2009) examines the influence of inter-industry pollution on trade benefits, assuming a diversified post-trade economic equilibrium similar to our model. Takarada's findings highlight the consequences of environmental policy interventions and suggest that in cases where pollution taxes are set below optimal levels, smaller economies can benefit from trade by subsidizing industries with higher pollution levels to promote economic growth.

This model puts forward a new approach to managing countries' open-access systems, which differs from the literature. Our model explores how this diversity affects trade patterns, including the post-trade diversified case, a topic that has yet to be extensively covered in previous research. Our findings demonstrate that countries with lower pollution intensity coefficients and smaller labor forces may experience a decline in utility and resource stock depletion under free trade, as spatial separation is not an option. Merely regulating polluting industries within national borders does not sufficiently mitigate pollution costs. This is because trade partners' polluting industries can impact the resource stock through the diffusion of transboundary pollution. Despite differences in the model constructions, our results align with the findings of Benarroch, Thille, and Unteroberdoerster to some extent.

This paper presents a theoretical finding that suggests that trade can improve welfare and exhibits characteristics of stock rebuilding for countries with poorly defined pollution management regulations, assuming a diversified steady-state production pattern post-trade. In essence, if a country's natural resource stock is more adversely affected by the manufacturing sector of a foreign country than its own domestic production, the negative transboundary externality becomes crucial in determining efficiency. This might incentivize the country to maintain production in a more damaging sector to reflect its pollutants to a trading partner, rebuild its resource stock, and secure favorable long-term trade terms. Consequently, the country may prefer to sustain production in a more polluting sector, increasing global pollution exposure. Interestingly, in our model, trade improves welfare and resources when the economy exports the industry's output, which is more damaging to the resource stock. However, welfare may decrease if a country exports environmentally friendly goods. This is due to the transboundary externality impacting another country more and the assumption of diversified production post-trade.

The remainder of this paper is organized as follows: To construct the basic components of the model and to present the intuition, the paper investigates the two-country setting in the second part, in which countries with different forms of environmental pollution standards are analyzed in terms of the stability of trade patterns and welfare implications, and provides the post-trade results. The last section briefly summarizes our main findings.

2. The Model

The fundamental framework we employ (including welfare expressions, renewable resource functions, production functions for various goods, etc.) is frequently utilized to examine the connection between trade and renewable resources. This is evidenced by the works of prominent scholars such as Brander and Taylor (1997a, 1997b, 1998), Emami and Johnston (2000), Rus (2012), Takarada et al. (2013), and Rus (2016).

Following Unterobersdoerster's model with a linear steady-state resource stock evolution function, we use the acronyms and symbols in the following table:[2]

Exogenous	Endogenous	Definition
r		Intrinsic growth rate
	S	Renewable resource stock
K		Maximum carrying capacity
	S^A	Resource level at equilibrium
q		Productivity parameter
	H	Harvesting level
L		Total supply of labor
	D	Total detrimental level of both activities
	L_M	Labor allocated to manufacturing industry
	M	Production of the manufacturing industry
	Z	Total amount of pollution
	γ	Pollution intensity parameter
	θ	The rate of pollution remaining in the original
	u	Individual utility, Cobb-Douglas function
I		Representative consumer's income
β		Elasticity, the taste parameter for resource good
p^T		World market price of the resource good
	p^A	Domestic price of the resource good
	w	Wage of a representative labor
	Π^M	Profit, manufacturing industry
p^M		Numeraire, price of manufacturing good
	p^H	Price of resource good

Table 1: Symbols and acronyms

2 Asterisks are used to refer to the foreign country.

Trade between the two countries assumed to be identical except for pollution intensity parameters and the labor force are assumed in this setting in which the domestic dirtier industry M is less polluting than the polluter industry in foreign, which implies that $\gamma < \gamma^*$. We also know that a stable trade pattern exists only under the assumption that manufacturing is more harmful in both countries under the assumption $\theta > \underline{\theta} > \frac{1}{2}$. This means that $\left[\overline{\gamma} < \gamma < \gamma^*\right]$ must hold. Unlike the reference models, we differentiate the amount of labor in each country to consider the harvesting activity as well, i.e., $L < L^*$. Since domestic has a lower pollution intensity parameter and the labor force enjoys more enhanced technologies or more restrictive environmental regulations on dirty industries, it can be considered more developed than a foreign to some extent. Note that the analysis here explaining the trade in the existence of transboundary pollution is similar to that of the South-South trade model with different levels of environmental awareness and labor force. That is, it can be assumed that both countries can be identified as less-developed to a certain extent. We also presume that both countries have diverse post-trade production features. For otherwise-identical countries, we also demonstrate that a domestic country's autarky steady-state resource stock is greater than that of a foreign country, provided that $\overline{\gamma} < \gamma < \gamma^*, L < L^*$ and $\theta > \underline{\theta} > \frac{1}{2}$. The following proposition is proposed[3]:

Proposition 1. *Given the conditions* $\overline{\gamma} < \gamma < \gamma^*, L < L^*$ and $\theta > \underline{\theta} > \frac{1}{2}$ *.the total level of externalities in domestic is lower than that of externalities in foreign, which means that* $D < D^*$. **Proof.** See the Appendix.

From Proposition 1, the domestic country in autarky has a higher steady-state equilibrium and a lower relative price of resource goods based on the equation $\frac{p_H}{p_M} = \frac{1}{qS}$.[4] When trade opens, the pattern of trade is determined by the

3 We assume the parameters ensure each country has a positive steady state in autarky.

4 We know $S_A = \frac{rK - \left[\theta\gamma(1-\beta)L + \left(1-\theta^*\right)\gamma^*(1-\beta)L^*\right]}{r + q\beta L}$ and $S_A^* = \frac{rK - \left[\theta^*\gamma^*(1-\beta)L^* + (1-\theta)\gamma(1-\beta)L\right]}{r + q\beta L^*}$. As $L < L^*$ and also

differing pollution coefficients and the amount of labor force in each country. Then, domestic production will shift toward the H good because of the comparative advantage, whereas foreign country will tend to produce more of the manufacturing good.

Different from the benchmark paper, we focus on the diversified case, which has not been analyzed thus far in the literature[5]. The fully specialized cases were investigated in Unteroberdoerster's paper in a more detailed and formal analysis but were less empirically relevant for determining real-life situations (Brander and Taylor, 1998, p.191). In addition, this section contributes to the literature by improving the assumptions in Brander and Taylor's model in which they do not consider the effects of transboundary pollution as another harmful activity on a nationally controlled open-access resource. However, we here presume transboundary pollution, which impacts the open-access resource stock through the activities that have emerged in both countries.

Given space constraints, allowing transboundary pollution as an extension to the Brander and Taylor model does not alter the main findings, such as factor price equalization $\left[w = w^{*} = 1\right]$ and resource stock equalization $\left[S = S^{*}\right]$ if both countries remain diversified after trade (we refer to Proposition 1 in Brander and Taylor for more formal and detailed analysis).

Proposition 2. *In a post-trade diversified steady state, given the conditions* $\bar{\gamma} < \gamma < \gamma^{*}$, $L < L^{*}$ and $\theta > \underline{\theta} > \frac{1}{2}$, ***i)*** the country that has a lower pollution intensity can export H goods. ***ii)*** the country that has a higher pollution intensity can export the M goods. **Proof.** See Appendix.

This finding indicates that a country with lax pollution regulations may have a greater incentive to export goods that deplete natural resources in exchange for exporting environmentally sustainable products. This result aligns with Proposition 2 in BT's model. In this model, trade patterns are influenced by resource abundance, but they also account for variations in environmental

---

$\left[\theta\gamma(1-\beta)L + (1-\theta^{*})\gamma^{*}(1-\beta)L^{*}\right] < \left[\theta^{*}\gamma^{*}(1-\beta)L^{*} + (1-\theta)\gamma(1-\beta)L\right]$ satisfies the following $S_A > S_A^{*}$.

5 The conditions that must be satisfied for the post-trade pattern of diversified production in both countries were formally analyzed in proof of Proposition 4 in the Brander and Taylor model. The main thrust of the logic in the Brander and Taylor model is valid here, and then we refer to the reader Proposition 4 in the BT model and its proof.

awareness between countries. The differing levels of environmental regulation in industries impact domestic prices and trade directions. We have established that the disparity between pollution intensity coefficients and the labor force impacts trade direction and the model's ability to redistribute production between countries, ultimately equalizing post-trade resource stock and relative prices.

Proposition* 3.** *In a diversified trading steady-state, given the conditions* $\bar{\gamma} < \gamma < \gamma^*, L < L^*$ and $\theta > \underline{\theta} > \frac{1}{2}$, ***i) *the common post-trade renewable resource stock, referred to in* S_T, *takes a value between autarkic and steady-state stock sizes* $\left[S_A^* < S_T < S_A\right]$, *and* ***ii)*** *the world relative price,* p_T, *lies between the autarkic prices of both countries.* **Proof.** See the Appendix.

Proposition 3 indicates that even though manufacturing is considered a more resource-depleting activity, provided that the condition $\bar{\gamma} < \gamma < \gamma^*$ and the domestic resource stock are supposed to benefit from a rise in the production of *H* goods, the marginal impact of total pollution exceeds the marginal impact of harvesting, which results in lower post-trade steady-state equilibrium. The foreign economy with diversified production reflects the pollution via the transboundary pollution mechanism to the domestic. Then, it experiences a lower post-trade total externality level and higher resource stock than autarky. The remaining diversified production after trade claims that free trade can increase the total externality level in domestic shifting of production toward less stock-reducing activity and decrease the total externality level in foreign country that supplies more manufacturing good that is more detrimental to stock, which is rather interesting compared to Unteroberdoerster's findings. Rus (2016) argues that if the pollution externality is strong enough and the economy is diversified in both goods, the country can experience welfare gains and rebuild its resource stock because it specializes in less distorted industry. However, the effects here are different from those of Rus (2016) because the present paper allows transboundary pollution, reverses stock rebounding, and does not permit spatial separation of incompatible industries. For this reason, if a country is exposed to more industrial pollution than domestic pollution, even though trade brings about the expansion of the less distorting sector, it cannot gain from trade in terms of resource conservation.

Proposition 4. *Comparing a diversified post-trade steady-state with autarky, given the conditions* $\overline{\gamma} < \gamma < \gamma^{*}, L < L^{*}$ and $\theta > \underline{\theta} > \frac{1}{2}$, ***i)*** *trade liberalization* causes the welfare at the steady-state to decrease in a country that has a lower pollution parameter and labor force. ***ii)*** *trade liberalization* causes the steady-state welfare to improve in the country that has a higher pollution parameter and a larger labor force.

First, consider the domestic country with a lower pollution intensity parameter and the lower amount of labor. In autarky, the domestic resource stock is greater than the foreign one. For this reason, the domestic PPF at autarky is a straight line with a higher slope. After the trade, the domestic consumption bundle must be on the budget line running through the post-trade production point with the resource stock S_T and with the slope $\frac{1}{P_T}$, which is identified as the post-trade PPF in which $S_T < S_A$. Therefore, post-trade PPF lies below the autarkic production and consumption possibility frontier. Taking into nominal income consideration, it remains constant at L and, while the price of manufacturing good is constant (with diversified production, $w = p_M$ before and after trade), the price of H goods is higher compared to that in autarky.

Consequently, the consumption bundle of a representative consumer in a domestic economy must be lower than the autarkic one, implying that the post-trade steady-state utility must shrink from U^A to U^T. The logic is similar except for $S_A^{*} < S_T$. Therefore, the foreign gains from trade in terms of the welfare perspective are shown in Figure 1.[6]

Our findings are different from the results of Rus (2016). Rus's model implies that when a comparative advantage in resource goods is strong under the condition that the pollution intensity parameter is high enough, the country initially reaches full specialization in H-goods. This leads to stock rebuilding, and the PPF expands outward, which results in long-run welfare gains due to the spatial separation of different economic activities. However, the present paper eliminates full specialization H and thus, the country cannot benefit from a possible

6 Nominal income in foreign is the same as in autarky, and also with the diversified production assumption, the price of M good is the same before and after the trade, whereas the price of resource good is lower than the autarky as it is proven in Proposition 6-ii. As a result, the consumption bundle in the foreign country for an arbitrarily choosen consumer must rise in terms of the real purchasing power index.

initial specialization in H good. Moreover, with the impact of transboundary pollution, the resource good exporter country suffers more from total detrimental impacts on the resource stock, so the vertical intercept of the PPF decreases, trade leads to welfare losses, and industrial pollution problems worsen in the domestic country.

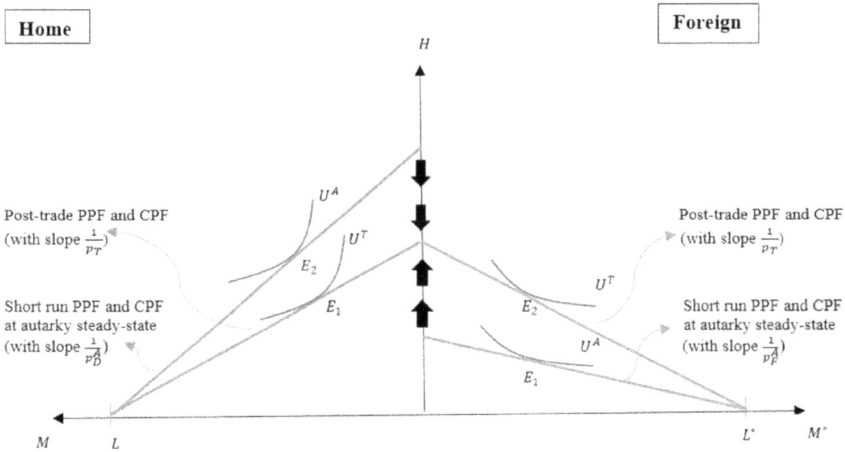

Figure 1: Welfare implications of trade on each country with the diversified trading equilibrium.

3. Conclusion

This research article expands on Unterobersdoerster's model by incorporating a two-country framework that differentiates between environmental regulation systems on resource stock and the labor force. We assume that manufacturing activities are more resource-depleting than harvesting activities for stability. The variance in environmental regulations serves as a source of comparative advantage. A nation with stricter resource management and a relatively small labor force exports resource goods, while a country with more lenient resource regulations and a larger labor force exports manufacturing goods. Our study also assumes that the post-trade production pattern varies across each country, eliminating the impact of industrial pollution that is limited to a single nation. Based on these assumptions, we can identify a specific trade scenario between two less developed or developing countries. Our results indicate that international trade tends to enhance the resource stock level and overall welfare in a nation with more lenient environmental standards and a larger labor force when its pollution level exceeds the threshold level of the pollution intensity parameter in the situation of diversified production at equilibrium.

In contrast, countries with more stringent environmental standards and smaller territorial sizes tend to experience a decline in resource stocks and diminished welfare levels when the transboundary pollution rate is assumed to exceed the threshold level, as in the current model. Interestingly, our findings suggest that the total externality resulting from a domestic shift in the labor force to produce more resource goods due to comparative advantage in resource goods is greater than the autarky level, based on the assumptions of a stable trade pattern. Even if a country implements effective regulations to reduce excessive harvesting issues, exposing the industrial externality in the form of transboundary pollution generated in other countries can lead to post-trade welfare losses and stock degradation as a result of a diversified production equilibrium after trade.

References

Acharyya, R. (2013). *"Trade and Environment"*. Oxford University Press, New Delhi.

Benarroch, M., & Thille, H. (2001). "Transboundary Pollution and the Gains From Trade". *Journal of International Economics, 55*(1), 139–159.

Brander, J. A., & Taylor, M. S. (1997a). "International Trade and Open Access Renewable Resources: The Small Open Economy Case". *Canadian Journal of Economics, 30*(3), 526–552.

Brander, J. A., & Taylor, M. S. (1997b). "International Trade between Consumer and Conservationist Countries." *Resource and Energy Economics, 19*(4), 267–297.

Brander, J. A., & Taylor, M. S. (1998). "Open Access Renewable Resources: Trade and Trade Policy in a Two-Country". *Journal of International Economics*, 44(2), 181–209.

Copeland, B. R., & Taylor, M. S. (1999). "Trade, Spatial Separation, and the Environment". *Journal of International Economics, 47*(1), 137–168.

Emami, A., & Johnston, R. S. (2000). "Unilateral Resource Management in a Two-Country General Equilibrium Model of Trade in a Renewable Fishery Resource". *American Journal of Agricultural Economics, 82*(1), 161–172.

Gibson, M., & Carnovale, M. (2015). "The Effects of Road Pricing on Driver Behavior and Air Pollution". *Journal of Urban Economics*, 89, 62–73.

Rus, H. A. (2012). "Transboundary Marine Resources and Trading Neighbours". *Environmental and Resource Economics, 53*, 159–184.

Rus, H. A. (2016). "Renewable Resources, Pollution and Trade". *Review of International Economics*, *24*(2), 364–391.

Unteroberdoerster, O. (2001). "Trade and Transboundary Pollution: Spatial Separation Reconsidered". *Journal of Environmental Economics and Management*, *41*(3), 269–285.

Takarada, Y. (2009). "Transboundary Renewable Resource and International Trade". Discussion papers 09041. Research Institute of Economy, Trade and Industry (RIETI).

Takarada, Y., Dong, W., & Ogawa, T. (2013). "Shared Renewable Resources: Gains from Trade and Trade Policy". *Review of International Economics*, *21*(5), 1032–1047.

Appendix

Proof of Proposition 1.

The total externality level in autarky for both countries is as follows: $D = q\beta LS + \left[\theta\gamma(1-\beta)L + (1-\theta^*)\gamma^*(1-\beta)L^*\right]$ and, $D^* = q\beta L^* S + [\theta^*\gamma^*(1-\beta)L^* + (1-\theta)\gamma(1-\beta)L]$. Assuming $D < D^*$ implies that $q\beta LS + \left[\theta\gamma(1-\beta)L + (1-\theta^*)\gamma^*(1-\beta)L^*\right] < q\beta L^* S + \left[\theta^*\gamma^*(1-\beta)L^* + (1-\theta)\gamma(1-\beta)L\right]$. Rearranging the last equation and then we have a more simplified equation as follows: $q\beta S\left[L - L^*\right] < [2\theta - 1](1-\beta)\left[\gamma^* L^* - \gamma L\right]$. The left-hand side of the last equation takes negative value based on $L < L^*$ and, the right-hand side is positive; thus, we conclude that $D < D^*$ is true.

Proof of Proposition 2.

We need to show that $H_D - H_S < 0$ for the domestic country with a lower pollution intensity parameter. From expression (6), the demand for H good in each country is $H_D = q\beta LS$ and $H_D^* = q\beta LS^*$ respectively. Recall that resource stock levels in both countries must be equal after trade, implying that $S = S^*$ and then, $\frac{H_D}{H_D^*} = \frac{L}{L^*} < 1$. We know that at steady-state equilibrium $H_S = G\left[S(t)\right] - D(t)$ and $H_S^* = G^*\left[S^*(t)\right] - D^*(t)$

must be held in which $H_s = r\left[K-S\right]-\left[\theta\gamma(1-\beta)L+(1-\theta^*)\gamma^*(1-\beta)L^*\right]$ and $H^{S^*} = r\left[K-S^*\right]-\left[\theta^*\gamma^*(1-\beta)L^*+(1-\theta)\gamma(1-\beta)L\right]$. It then yields that

$$\frac{H_D/H_D^*}{H_S/H_S^*}=\frac{H_D/H_S}{H_D^*/H_S^*}=$$

$$\frac{L/L^*}{r\left[K-S\right]-\left[\theta\gamma(1-\beta)L+\left(1-\theta^*\right)\gamma^*\left(1-{}^2\right)L^*\right]/r\left[K-S^*\right]-\left[\theta^*\gamma^*\left(1-\beta\right)L^*+\left(1-\theta\right)\gamma\left(1-\beta\right)L\right]}<1$$

(2.1)

because $r\left[K-S\right]-\left[\theta\gamma(1-\beta)L+\left(1-\theta^*\right)\gamma^*\left(1-{}^2\right)L^*\right]$ is higher than $r\left[K-S^*\right]-\left[\theta^*\gamma^*\left(1-\beta\right)L^*+\left(1-\theta\right)\gamma\left(1-\beta\right)L\right]$ and the numerator L/L^* is lower than 1 depending on the assumptions identified above. The material balance condition expressed by $H_D+H_D^*=H_S+H_S^*$ must also be satisfied at steady-state equilibrium. Taking these two conditions into account together, it follows that $H_D<H_S$. Therefore, a country that has a lower pollution intensity parameter and a labor force can produce more resource goods than it can consume, and then the conclusion is that it can export these goods, and import M goods.

Proof of Proposition 3.

(i) Considering the supply-demand balance condition when the economy reaches steady-state equilibrium after trade, $H_D\left(S_T\right)+H_D^*\left(S_T\right)=H_S\left(S_T\right)+H_S^*\left(S_T\right)$ and by substituting Eq. (3) and Eq. (6), we obtain

$$\begin{aligned}q\beta LS_T+q\beta L^*S_T=&\left[r\left[K-S_T\right]-\left[\theta\gamma(1-\beta)L+\left(1-\theta^*\right)\gamma^*\left(1-{}^2\right)L^*\right]\right]\\&+\left[r\left[K-S_T\right]-\left[\theta^*\gamma^*\left(1-\beta\right)L^*+\left(1-\theta\right)\gamma\left(1-\beta\right)L\right]\right]\end{aligned}\qquad(3.1)$$

The relevant solution after making the necessary arrangements is $S_T=\dfrac{rK-(1-\beta)\dfrac{\left[\gamma L+\gamma^*L^*\right]}{2}}{r+q\beta\left[\dfrac{L+L^*}{2}\right]}$. The stable autarkic steady-state stock equilibrium is $S_A = \frac{rK-(1-\beta)\left[\theta\gamma L+(1-\theta^*)\gamma^* L^*\right]}{r+q\beta L}$ and $S_A^* = \frac{rK-(1-\beta)\left[\theta^*\gamma^* L+(1-\theta)\gamma L\right]}{r+q\beta L^*}$ for both countries. Recall that $L < \left[\frac{L+L^*}{2}\right] < L^*$ and comparing the autarkic equilibria with the post-trade, equilibrium yields

i) Assume that $\left[\theta\gamma L+(1-\theta^*)\gamma^* L^*\right] < \frac{\left[\gamma L+\gamma^* L^*\right]}{2}$. This expression decreases to $\left[\theta-\frac{1}{2}\right]\left[\gamma L-\gamma^* L^*\right]$, which is certainly negative. Therefore, $S_T < S_A$.

ii) Assume that $\left[\theta^*\gamma^* L+(1-\theta)\gamma L\right] > \frac{\left[\gamma L+\gamma^* L^*\right]}{2}$. This expression decreases to $\left[\theta-\frac{1}{2}\right]\left[\gamma^* L^* -\gamma L\right]$, which is certainly positive. Therefore, $S_A^* < S_T$.

(ii) Depending on the diversified production after trade implies that $p_M = w = w^*$ and then, the autarkic relative prices and post-trade world price are $p_A = \frac{1}{qS_A}$, $p_A^* = \frac{1}{qS_A^*}$ and $p_T = \frac{1}{qS_T}$ in which $S_A^* < S_T < S_A$. Using the resource stock relation proven above, we reach the conclusion that $p_A < p_T < p_A^*$.

Leman Erdal[1]

The New Concept of Energy Security

1 Introduction

As a source of wealth and power, energy has been the main cause of geopolitical rivalry between states that want to conduct and exploit energy resources. For decades, technological innovations, globalization, and the interdependent relationship between the world economies have made cooperation urgent, rather than conflicts between both producer and consumer countries. on the one hand, there has been increasing global energy demand with the persistent dominance of fossil fuels; on the other hand, environmental degradation and climate change, anticipated that energy security issues will continue to be the main source of conflict and also cooperation for many more years.

The study is based on a method of descriptive analysis for the concept of energy security. The conventional definition of energy security and its components has been formulated by some factors of energy- geopolitics. It has been changing due to a variety of factors such as growing energy demand and new terms like energy efficiency, equity, and environmental concerns together with climate change. The need for "reconceptualization" of energy security concerns has ranged from government policies aimed at ensuring the security of energy such as the strategy of diversified energy supply, the energy markets, and clean technologies related to the transition of energy systems.

This part is structured under three sections. The following section briefly deals with a history of global energy security concerns as a result of transitions from coal to oil and natural gas in energy demand. The second section includes the conventional definition of the concept of energy security and its new components stressing uninterrupted energy supply, interdependence as a security issue, and sustainable economic development

The strategic nature of energy sources and the dynamic factors influencing energy security will be discussed from both traditional and modern perspectives, particularly considering the role of climate change in energy security. Key

1 Ph.D., Assoc. Prof., Istanbul Medeniyet University, Faculty of Political Sciences, Department of International Relations, Kadikoy/Istanbul –Turkiye, ORCID:0000-0001-9388-659X.

concepts such as supply and demand security, dependency on imported energy sources, geopolitical risks associated with energy routes, energy equity, and resilience will be defined. The final section will focus on energy policy goals related to energy security, with a specific emphasis on supply security and sustainable development goals, especially highlighting environmental concerns.

2 Global Energy Trends in Energy Demand / Energy Sources Classification

The security of primary energy supply has consistently been a focal point on the global agenda. It remains a top energy security concern worldwide due to several factors: *the persistent dominance of fossil fuels in the energy sector, liberalization of energy markets, escalating energy demand in developing economies, ongoing political unrest, environmental disasters, and the increasing likelihood of fossil fuel depletion* (WEC, 2024); (EU, 203).

Recent global trends in total primary energy consumption, encompassing both fossil fuels and renewable sources, are depicted in Graphic 1. As of 2023, fossil fuels account for 82 % of total primary energy consumption, down from 87 % in 2010. While natural gas and coal demand has stabilized, oil consumption has nearly returned to pre-pandemic (COVID-19) levels. Despite significant growth in renewables such as solar and wind, the world remains heavily reliant on fossil fuels (IEA, 2024: 185).

Although renewable energy capacity expanded at record rates, global total primary energy consumption continued to be dominated by fossil fuels, maintaining an 82 % share in 2023. The global demand for renewables increased by 5.6 % in 2018, reaching a total share of 15 % according to IEA statistics. Renewable-based electricity generation rose by almost 450 TWh, growing by 7 % and now accounting for over 25 % of global power generation. According to the IEA, solar PV, hydropower, and wind each contributed about a third to this growth, with bioenergy making up the remainder (Ember, 2024; IEA, 2024). In 2022, fossil fuels maintained their dominance in global electricity generation, comprising 62 % of the total, as illustrated in Graphic 1 (Ember, 2024; IEA, 2024).

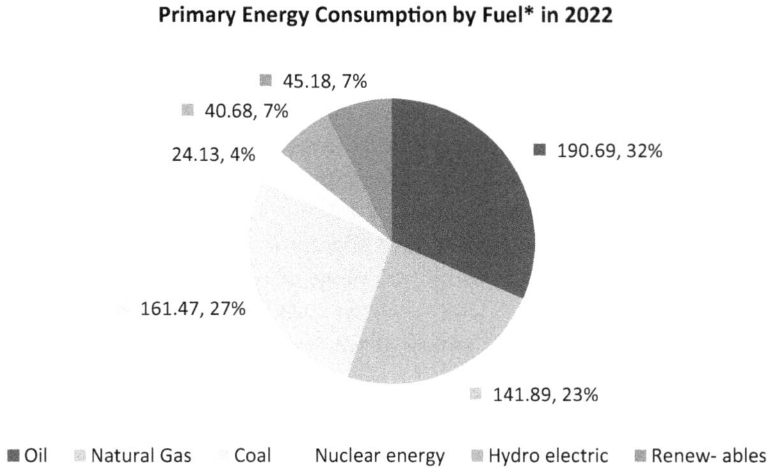

Grafic 1 World Primary Energy Consumption by Fuel[2]
Source: (EMBER, Energy Institute, 2024).

Disruptions at Ukraine's Zaporizhzhia nuclear power plant and closures of French nuclear units due to safety issues resulted in a 4 % decrease in global nuclear power use. France, which experienced a 22 % decline in electricity generation (to 294.7 TWh in 2022), transitioned from being the world's largest electricity exporter to a net importer. Globally, electricity generation increased by 2.3 % from the previous year, reaching 29,165.2 terawatt hours (TWh) in 2022. Coal remains the leading global source for electricity generation, accounting for *35.6 %*, followed by natural gas *at 22.7 %*, and hydroelectric power at *14.9 %* (Ember, 2024).

Nuclear plants contributed 9.11 % to global electricity generation with 440 power reactors operational, and an additional 48 planned for construction from 2023 to 2027, as reported by the World Nuclear Association. These developments are expected to generate around 420 TWh annually, with a capacity of 60 GW, across *nineteen plants in China, eight in India, six in South Korea, four in Turkey, and three in Russia.* According to the IEA's *Net Zero Emissions scenario,* nuclear generation is projected to increase significantly between 2021 and 2030, alongside substantial growth in solar and wind energy, which are expected to account for 75 % of all clean electricity growth through 2040 (Ember, 2024); (WEC, 2024).

---

2 *Exajoules in OECD, Non-OECD countries and European Union.

Renewable power growth surpassed electricity demand growth in 2023, marking the beginning of a new era in the power sector expected to lead to reduced use of fossil fuels and lower emissions. According to prediction by Ember (2024) the first decline in fossil fuel generation occurred in 2023. This is based on the assumption that electricity demand would grow by an average of 2.5 % annually (+726 TWh), while wind and solar generation would increase by 19 % (+641 TWh), and other clean power sources by 1.7 % (+132 TWh) over the year. As a result, fossil fuel generation decreased slightly to 47 TWh (-0.3 %) in 2023. Although this reduction falls within the range of uncertainty, Ember suggests that "*if fossil fuel generation did not decline in 2023, it would likely decrease starting in 2024, given the continued growth of wind and solar power*" (Ember, 2024).

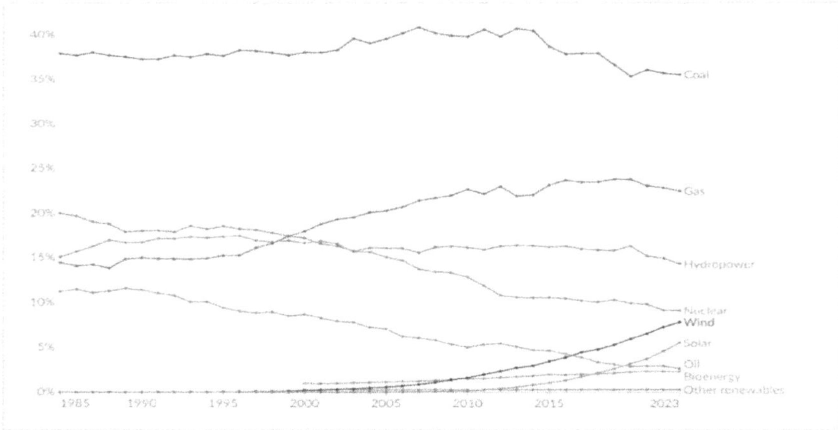

Grafic 2: Global Electricity Generation by Source (%)
Source: (Ember, 2024).

In 2022, the world marked a significant turning point in its transition to clean power. The invasion of Ukraine by Russia prompted governments to reconsider their energy strategies amidst rising fossil fuel prices and concerns over energy supply security related to fossil fuel imports. This geopolitical event also accelerated the adoption of electrification technologies such as heat pumps, electric vehicles, and electrolyzers. These advancements are expected to anable emission reductions across different sectors and increase the urgency to accelerate the deployment of clean power infrastructure.

Energy security has evolved significantly throughout history, from its early use in prehistoric times to its critical role in modern society's infrastructure and economy. Although not explicitly defined in ancient times, the concept of ensuring sustainable energy supply gained strategic importance just before World War I with Winston Churchill's decision to transition the British Navy from stable and reliable coal torisky but cheaper oil sourced from Persia. The two world wars referring to World War I and II, underscored the essential need for consistent oil supplies, leading to increased dependence on Middle Eastern oil post-WWII[3]. In the 1970s, deliberate oil embargoes by OPEC underscored vulnerabilities in global energy supply chains and catalyzed efforts to diversify energy sources and promote conservation. International organizations like the IEA were subsequently established to address energy security through policy frameworks. Recent challenges such as *the COVID-19* pandemic and geopolitical tensions, like the Ukraine crisis, have renewed global focus on energy transition and climate action. Initiatives like the EU's REPowerEU plan reflect current efforts to reduce fossil fuel dependence and accelerate renewable energy adoption amidst concerns about resource depletion, geopolitical instability, economic impacts, and environmental sustainability, highlighting the ongoing evolution and importance of energy security in shaping global policy and strategic agendas. (Novikau, 2021)

3 Historical and Conceptual Evolution of Energy Security

The concept of energy security has improved significantly over time, rooted in the fundamental role of energy in human civilization as a source of "*heat, light, and power*". It initially focused on ensuring the "*uninterrupted supply of energy to support economic functions and societal needs, the definition has expanded to encompass broader dimensions, including environmental sustainability and geopolitical stability*" (IEA, 2024).

Historically, energy security has been closely tied to oil supply due to high global oil dependency, particularly highlighted by events like the 1973 Arab oil embargo, which underscored vulnerabilities in global energy markets (Cherp & Jewell, 2011).

The modern understanding of energy security integrates multiple dimensions (Table 1), such as "*the physical availability of energy resources, the economic cost*

3 refers to the geopolitical and economic significance of oil resources in the Middle East after World War II, marking a period of increased global dependency on and competition for oil from countries in the region.

of energy, and the reliability of energy infrastructure". Sustainability has emerged as a critical component, emphasizing the long-term sustainability of energy sources for economic growth without compromising environmental cohesion (Vakhshouri, 2011).

Geopolitical considerations have also become central, reflecting concerns over energy transmission routes and infrastructure security (Cherp & Jewell, 2014). The concept of energy security, as defined by the International Energy Agency (IEA) through the well-known "*Four A's*" framework – *availability, affordability, accessibility, and acceptability* – comprehensively summarize the criteria for evaluating the reliability of energy sources (Tufail et al., 2019).

Energy security influenced by a complex network of interconnected factors, covering advance technology, social impacts, military considerations, and international relations. The current evolution of energy security frameworks highlights the necessity for comprehensive strategies that harmonize economic development, environmental sustainability, and political stability in global energy markets (Minke & Merel, 2019).

Table 1: The New Dimensions of Energy Security

Dimensions	**Components**
Availability of Energy Supply	*Sufficient supplies of energy* *Reliable energy system and energy mix of the system* *Using domestically available energy sources* *Dependency of imported energy supply* *Diversification of energy sources and suppliers* *Prudent reserve to production ratio*
Affordability of Energy Services	*Affordable and stable energy services prices* *Accessible and equitability to all population*
Technology Development and Efficiency	*Promoting innovation and research into energy system* *Resilience and adaptive capacity* *Sufficient investment and employment* *Promoting energy efficient among consumers*
Environmental and Social Sustainability	*Mitigation efforts for controlling pollutions and climate change* *Balance of energy spending budget for consumers*
Governance of Regulations, Policies, and International Cooperation	*Clear energy regulations and policies* *A stable and transparent regulations and policies* *Participant from authorities in making regulations and policies* *Promoting educating public* *Commitment toward international cooperation*

Source:(Tufail et al., 2019; WEC, 2024).

The "energy security" term is composed of two main components: "*energy*" and "*security*". Energy is a core element in nature for a "*physical system to perform work*" which produce goods or provide the required services. It can be *harnessed, converted, stored, transmitted, and reconverted into light and heat* to meet specific needs (Narula, 2019). The concept of Energy security has roots as ancient as fire itself, serving as "*a fundamental resource for heating, cooking, and protection*" throughout human history. Security, in this context, implies "*the state of being free from disruptions, encompassing threats to energy supply from technical, geopolitical, or natural causes*" (Ang et al., 2015; Narula, 2019; Paravantis & Nikoletta, 2020). The concept has evolved in parallel with human civilization, reflecting changes in social structures and international relations.

Historically, energy security has been associated with "*the reliable availability of energy sources at an affordable price, ensuring that the national economy can function without interruption*". This concept, while ancient in its origins, has evolved considerably, reflecting the complexity of *modern energy systems and the diverse risks they face* (Narula, 2019; Tufail et al., 2019).

Traditionally, energy security has been closely tied to the security of oil supply due to historical dependence, aiming to *mitigate risks of supply disruptions.* Thus, the classic definition of energy security which underscored the vulnerabilities associated with over-reliance on a single energy source or supplier. It was focusing "*on the assurance of sufficient energy supplies to support national economies within a politically stable environment*" (Cherp & Jewell, 2014; Narula, 2019; Paravantis & Nikoletta, 2020). Ensuring a sustainable physical supply of fossil fuels to meet economic demands remains a primary concern in energy policies. The supply chain disruptions and vulnerabilities, the threat of terrorist attacks, and the integration of large emerging economies, such as china and India, into the global market are new challenges for both producers and consumers in global energy trade

The modern definition of energy security shaped by, obviously during the 20th century, particularly following serial oil shocks of the 1970s. It has been address a wider range of concerns, including "international *energy trade, supply chain vulnerabilities, the geopolitical integration of emerging economies and, environmentel and climate issues*". Modern definitions (Table 1), emphasize "*the availability, affordability, accessibility, and acceptability of energy sources*, matching with the framework set by International Energy Agency's (IEA). This comprehensive approach includes factors related to the sustainability and resilience of energy systems against various risks (Vakhshouri, 2011; Tufail et al., 2019; IEA, 2024).

Energy security definitions, are relatively new in academic literature and initially, they focused on access to fossil fuels, although sustainability of energy supply has proven more crucial for economic wellbeing (Vakhshouri, 2011). By the 2010s, the concept of energy security expanded to include various *new dimensions* such as *technological, social, military, economic, political, and geographic aspects*. Geopolitics, particularly concerning energy's political geography and other international security concerns like energy transition or pipeline security, have become crutial policy goals for sustainable development (Azzuni & Breyer, 2018).

The new global energy security framework largely emerged shortly after the oil shocks triggered by the 1973 Arab oil embargo. It recognized energy security both as a policy stance and as a critical condition amid deliberate threats such as embargoes and price manipulations by energy producers (Cherp & Jewell, 2011).

The International Energy Agency (IEA) defines energy security as "*the uninterrupted availability of energy sources at an affordable price.*" This famous definition is often known as "*Four A's: as availability, affordability, accessibility, and acceptability*" "for energy sources". This framework emphasizes not only "*the physical availability*" of energy resources but also *the economic aspects of affordability and accessibility, ensuring that energy* is both *obtainable and priced at reasonable level for end users*. Moreover, it includes *the acceptability of energy sources*, which considers "*environmental and social impacts*", particularly in relation to fossil energy sources, thus synchronizing energy security with wider sustainability goals (Vakhshouri, 2011).

Today, energy security encompasses new challenges such as "*data-empowered energy consumers and producers, necessitating technological and innovative transitions from fossil fuels to renewable energy sources*". This global transition is supported by both private firms and public policies, particularly in the transportation sector, requiring robust institutional policy frameworks and domestic financial markets to enable large-scale investments in renewable energy infrastructures (Soto & Martinez-Cobas, 2024).

The new evolved dimensions of energy security include policy priorities such as "*resilience, innovations, adaptability, and a strong sense of initiative*" to ensure sustainability of energy security. Policies, promoted by international private corporations and governments, emphasize the transition of energy systems mainly from fossil fuels to renewable energy sources, enhancing energy efficiency, and advocating environmental sustainability. These policy initiatives seek to *reduce dependency on fossil fuels, mitigate carbon emissions, enforce security and sustainability* of energy systems. Moreover, strong institutional policy and domestic financial markets play a crucial role in facilitating substantial investments in renewable energy infrastructure. This proactive approach is essential for fostering economic growth and social development, as it boosts the emergence of innovative products, services, and industries, while also driving significant societal and technological advancements (Ang et al., 2015).

Energy security policies set by governments, aims mitigation efforts for "*regulating pollutions and climate change, includes environmental concerns which are crucial component of social acceptability dimensions*", which gained prominence during the 1980s environmental awakening. Recent studies highlight challenges in oil supply alongside concerns for *energy equity and climate change, emphasizing nuclear energy's impact on air quality and land use requirements* (Minke & Merel, 2019; Soto & Martinez-Cobas, 2024).

National energy strategies focus on "*increasing energy access, reducing import dependencies through diversification, enhancing energy storage, optimizing domestic resources, promoting energy efficiency, and reducing carbon and energy intensities*". These strategies vary based on specific country values, priorities, risks, and historical contexts, often influenced by economic fluctuations and energy market conditions. The EU has restricted, *seaborne imports*[4]of crude oil, refined petroleum products, and coal from Russia, resulting in a significant decrease in gas imports from 45 % in 2021 to 15 % in 2023. By March 2024, the EU increased its gas imports from alternative suppliers to compensate for the reduction, with Norway and the U.S. providing 34 % and 20 % of the EU's gas imports, respectively (EU, 2024; WEC, 2024).

Energy security concept maintains its dynamic structure and also is an evolving concept as its crucial role for *economic stability, environmental sustainability, and geopolitical relations* ınternaitonal in arena. Its definitions have expanded over time to encompass new and multiple dimensions, reflecting "*the complex interplay of technological advancements, economic policies, environmental concerns, and international cooperation in ensuring a resilient global energy system*" (WEC, 2024).

4 Conclusion

The sustainable energy supply is a vital for economic growth, social developement and sustainable enviroment for future generations. It is a fact that the supply chain pieces of equipment for energy resources ranges from "*resource*

4 the transportation of Russian crude oil by sea, typically via tankers or other maritime vessels.

exploration, extraction, power generation, conversion, transportation, transmission, storage facilities, distribution to end-user equipment". It is stated that; *increasing environmental awareness, uncontrollable rise of digitalisation are driving forces* that is changing the style of production, consumption and trade for energy. The contemporary energy policy agenda has been dominated by energy security aspects of "*affordability*" and "*reliability*" of energy supplies and climate among three energy security aspects; the "*availability of energy supply*", the "*economic affordability*", and the "*social acceptability*".

Sustainable development is the final target associated with security of nation-states, a group of integrated-states to some extent politically and economically such as EU, and interdependent global economy. The EU's active efforts to decrease reliance on Russian fossil fuels while advancing towards a more sustainable energy future and supporting Ukraine have significant global implications. Initiatives, such as *REPowerEU d*isplays the EU's commitment to reducing Russian dependence and promoting renewable energy adoption. These policies advocate for structural changes that reinforce energy security and provide a framework for global climate action. By reducing Russian gas imports, enhancing industrial decarbonization, accelerating renewable energy deployment, investing in energy infrastructure, and improving energy efficiency, the EU aims *to reinforce energy security, meet climate targets, and strengthen geopolitical stability* in the region and beyond (EC, 2024).

The international energy trade provides diversification of both energy mix and import location, enhanced infrastructure, innovations and R&D on energy-tech that increase energy efficiency. These are also cobbles on the way for sustainable economic development backed by public (energy/economic and foreign) policies which are set up for national, international and global security. Resolving all the problems associated with the energy security will take enormous efforts and a long lasting efforts

References

Ang, B. W., Choong, W. L., & Ng, T. S. (2015). Energy security: Definitions, dimensions and indexes. *Renewable and Sustainable Energy Reviews, 42*, 1077–1093. https:// doi.org/10.1016/ j.rser.2014.10.064.

Azzuni, A., & Breyer, C. (2018). Definitions and dimensions of energy security: A literature review. (Wiley Interdisciplinary Reviews: Energy and Environment, Du.) *Wiley Interdisciplinary Reviews: Energy and Environment, 7(1)*. https:// doi.org/10.1002/ wene.268.

Cherp, A., & Jewell, J. (2011). The three perspectives on energy security: Intellectual history, disciplinary roots the potential for integration. *Current Opinion in Environmental Sustainability*, *3*(4), 202–212. doi: 10.1016/j.cosust.2011.07.001.

Cherp, Aleh & Jewell, Jessica (2014), The concept of energy security:Beyond the four As, Energy Policy, 75, pp. 415–421, journal homepage: "http://www.elsevier.com/locate/enpol" www.elsevier.com/locate/enpol.

EC. (2024). *REPowerEU at a glance.* Commision, European, The European Green Deal. European Commision. https://commission.europa.eu/strategy-andpolicy/priorities-2019-2024/european-green-deal/repowereu-affordablesecure-and-sustainable-energy-europe_en#repo.

EMBER. (2024). *European Electricity Review 202Energy data.* Creative Commons ShareAlike Attribution Licence (CC BY- SA 4.0, Lead writers-Sarah Brown and Dave Jones. https://ember-climate.org/insights/research/europeanelectricity-review-2024/#supporting-material.

Eurostat, (2024), Eurostat database (Comext) and Eurostat estimates, EU imports of energy products, 2021-2024, EU imports of energy products - https://ec.europa.eu/eurostat/statistics-explained/index.php?title=EU_imports_of_energy_products_-_latest_developments.

IEA, I. E. (2024). *Strategies for Affordable and Fair Clean Energy Transitions, IEA,.* Paris: Licence: CC BY 4.0. 06 06, 2024 tarihinde https://www.iea.org/reports/strategies-for-affordable-and-fair-clean-energy-transitions

Minke, M., & Merel. (2019). *Economic Security Methodology and Approach.* Clingendael- the Netherlands Institute of International Relations – . the Netherlands Institute of International Relations. www.clingendael.org.

Narula, K. (2019). *Energy Trade – The Backbone of Sustainable Energy Security. In The Maritime Dimension ofSustainable Energy Security,.*

Novikau, A. (2021). onceptualizing and Redefining Energy Security: A comprehensive Review. J. Ren icinde, *China's Energy Security Analysis, Assessment and Improvement*" (s. Chapter 2, 248). Hong Kong: The Hong Kong Polytechnic University. doi: https://doi.org/10.1142/q0270.

Paravantis, J., & Nikoletta, K. (2020). Energy Security and Renewable Energy: A Geopolitical Perspective. Mansour Al Qubeissi & Ahmad El- Kharouf & Hakan Serhad Soyhan (ed.) icinde, *Renewable Energy – Resources, Challenges and Applications.* In. doi: https://doi.org/10.5772/intechopen.91848.

Soto, G. H., & Martinez-Cobas, X. (2024). Nuclear energy generation's impact on the CO2 emissions and Ecological footprint among European Union countries Author links open overlay panel. *Science of the Total Environment, 945*(173844).

Statista. (2024). *Global oil and gasoline prices – Statistics & facts*. statista.

Tufail, M. M., Ibrahim, J. A., Melan, M., & Nawi, M. N. (2019). Novel approach of quantifying energy security in terms of economic, environmental and supply risk factors. *Journal of Advanced Research in Fluid Mechanics and Thermal Sciences*, 16(12), 4979–4984. doi: https://doi.org/10.1166/jctn.2019.8551.

Vakhshouri, S. (2011, October 11). Measuring the Effect of Political Instability in Middle East and North Africa on Global Energy Security. (Paper No. 30,). Available at SSRN: https://ssrn.com/abstract=1944628.

WEC. (2024). *Energy Trilemma Index*. The World Energy Council. https://trilemmaworldenergy.org/.

İlkay Güler[1]

Investigation for Effects of Selected Macroeconomic Factors on the Energy Demand: Evidence from Türkiye

1 Introduction

Modern society's social and economic welfare has resulted from various human-oriented efforts such as technological advancement, tools, and innovation. However, energy resources have been irreplaceable, though scientific leaps have promoted less reliance on natural resources. Concerning human history, various political and economic events such as discovering steam power achieved from hydro and fossil energy resources, the 1970s energy crisis orchestrated by OPEC members, and political disturbances that occurred in Middle-East countries are some pivotal experiences confirming the importance of energy resources in society. Moreover, some undesirable events and experiences, such as climate change and global warming, especially dominating the political agendas since the 1960s, have presented other addressing properties and effects of energy resources. Since the industrial revolution, fossil energy resources have served to fulfill the required energy, and usage of fossil energy resources leads to greenhouse emissions, driving forces of climate change and global warming. Therefore, finding accurate solutions and transforming the recent economic and social patterns from high energy intensity to energy efficiency have been the challenges that policymakers and researchers encounter.

The empirical evidence becomes a pivotal guideline in this context because of the tight connection between energy and economic patterns. Regarding the literature, the volume of research endeavored to find the relationship between energy and various kinds of financial and social-related indicators, which has reached a massive number. Concerning the literature, Kraft and Kraft (1978) made a pioneering study addressing the relationship between energy consumption and economic growth. Following Kraft and Kraft's (1978) seminal research,

1 Ph.D., Asst. Prof., Ankara Hacı Bayram Veli University, School of Land Registry and Cadastre, Department of Land Registry and Cadastre, Ankara, Türkiye, ilkay.guler@hbv.edu.tr, ORCID:0000-0003-1289-6652.

various researchers have strived to explore the connection between energy consumption and economic growth and the drivers and forces of economic growth, such as exports and financial development. The four theoretical frameworks have been established to explore the relationship between energy consumption and economic-related indicators. Suppose economic growth and other variables are induced by energy consumption. In that case, the growth hypothesis is presented while the economic forces drive the energy consumption, which approves the conservation hypothesis. However, the feedback hypothesis is valid when the bidirectional connection between the variables is detected, whereas the connection's absence means the neutrality hypothesis. If the growth or the feedback hypothesis is valid, any attempts aiming to reduce energy consumption lead to a shrinking in economic performance; while the conservation or the neutrality hypothesis is held, the policymakers can implement the policies addressing protecting the environment without the cost of decreasing economic performance (Apergis, 2010; Apergis & Tang, 2013; Shahbaz et al., 2022)

Regarding these four hypotheses, when considering economic growth, some authors prove the growth hypothesis Fan & Hao, 2020; Kahouli, 2019 while the conservation hypothesis is also supported by Akarca & Long, 1980; Jaraite et al., 2017; Jobert & Karanfil, 2007 support the neutrality hypothesis, whereas Pao & Fu, 2013; Yasmeen et al., 2023 confirm the feedback hypothesis. As for the financial development Kassi et al., 2020; Raza et al., 2020; Zhe et al., 2021 present the evidence detecting the growth while Dimnwobi et al., 2022; Lin et al., 2016 prove the conservation hypothesis. Besides, the neutrality hypothesis is verified by Burakov & Freidin, 2017 while the feedback hypothesis is disclosed by Brunnschweiler, 2010; Vo et al., 2022. The growth hypothesis concerning the export and energy consumption is found by Erkan et al., 2010, while the conservation hypothesis is shown by Sharma et al., 2021, Odhiambo, 2021. In addition, Shakeel et al., 2014; Shakeel & Ahmed, 2020; Lean & Symth, 2010; promote the evidence supporting the neutrality hypothesis while Sadorsky, 2012; Shahbaz et al., 2019 prove the feedback hypothesis.

All in all, the study on the relationship between energy consumption and economic-related indicators has protected their importance. Within this scope, the study has been organized to find the connection between energy consumption, economic growth, exports, and financial development in Türkiye from 1965 to 2023. The contribution of the study is that focusing on three used economy-related indicators in the literature is considered together, and Türkiye is an excellent example because of countries' high imported energy reliances and the dominant share of fossil energy resources in the total energy and the endowment with various renewable energy resources. Within this scope, various time-series methods such as Johansen and FADL cointegration analysis, The VAR models, the Granger, the Toda-Yamamoto, the Fourier Granger, and Fourier Toda-Yamamoto analysis are applied.

The structure of the remaining paper is arranged as follows. The data, methodology, and empirical results are presented in Section 2, and Section 3 discloses the concluding remarks.

2 Data, Methodology and Empirical Results

The study aims to address the determinants of the energy demand in Türkiye over the period from 1965 to 2023 by performing various time-series methods involving the cointegration analysis, VAR model, and causality analysis. The properties of the considered variables are shown in Table 1, and the descriptive analysis is posed in Table 1.

Table 1: Abbreviations and Sources

Variables	**Abbreviation**	**Description**	**Log transformation**	**Data Sources**
Energy Consumption	Energy	Primary energy consumption per capita	lnENERGY	Ourworldindata
Exports of goods and services	Export	Exports of goods and services (% of GDP)	LnEXPORT	World Development Indicators (World Bank Data Stream, 2024)
Per Capita Real Gross Domestic Product (Constant 2015 US dollar)	GDP	GDP per capita is gross domestic product divided by midyear population.	lnGDP	World Development Indicators (World Bank Data Stream, 2024)
Financial Development	FD	Domestic credit to private sector by banks (% of GDP)	lnFD	World Development Indicators (World Bank Data Stream, 2024)

According to Table 2, the highest mean and maximum value belongs to energy consumption, followed by economic growth and financial development, and the exports contain the highest standard deviations.

Table 2: Descriptive Analysis

	LNENERGY	LNEXPORT	LNFINANCIAL	LNGDP
Mean	9.245150	2.618017	3.199722	8.690634
Median	9.344508	2.934437	3.000856	8.648639
Maximum	10.04796	3.652843	4.261222	9.590855
Minimum	7.977369	1.168768	2.609215	7.928562
Std. Dev.	0.575863	0.740627	0.506132	0.454487
Skewness	-0.449881	-0.685092	0.973061	0.321794
Kurtosis	2.209885	1.999380	2.444790	2.037645
Jarque-Bera	3.524890	7.076673	10.06847	3.294985
Probability	0.171625	0.029062	0.006511	0.192532
Sum	545.4638	154.4630	188.7836	512.7474
Sum Sq. Dev.	19.23385	31.81467	14.85785	11.98040
Observations	59	59	59	59

Investigation of the stochastic properties of the time series or stationary analysis is vital for the econometric methods because the following econometric methods are determined as a result of the stationary analysis. Besides, inconsistent and biased results are encountered when ignoring the stationary analysis due to fallacy regression. Within this scope, the conventional traditional unit root tests composed of the ADF and KPSS unit root tests and the Fourier unit root tests, allowing for multiple, sharp, smooth, and unknown, involving the Fourier ADF and KPSS unit root tests. All unit root tests with constant and constant&trend models are employed.

Table 3: Unit Root Test Results for the Variables at the Level

Variables	ADF		KPSS		FADF		FKPSS	
	Constant	Constant and Trend	Constant	Constant and Trend	Constant	Constant and Trend	Constant	Constant and Trend
lnENERGY	-2.906392 (0.0507)	-2.788689 (0.2073)	0.938781	0.222313	-1.190206 [1,5, 34.22113]	-3.475799 [1,5, 45.72627]	0.461961 [1,5, 34.22113]	0.050864 [1,5, 45.72627]
lnEXPORT	-1.135853 (0.6958)	-2.062425 (0.5553)	0.854120	0.179049	-1.587913	-3.599212	0.401565 [1,5, 41.57377]	0.044568 [1,4, 21.37987]
lnGDP	0.563411 (0.9875)	-1.566454 (0.7941)	0.943191	0.224132	-0.454035	-2.910259	0.468130 [1,5, 34.89194]	0.036529 [1,2, 53.80471]
lnFD	-1.179548 (0.6774)	-1.984525 (0.5971)	0.606324	0.189687	-2.383880 [1,5, 36.45069]	-3.736259 [1,4, 57.53209]	0.341467 [1,5, 36.45069]	0.050985 [1,4, 57.53209]

Values in the parenthesis related to the ADF unit root tests mean the probability while three values in the brackets about the FADF and FKPSS unit root tests donate frequencies, Bandwidth, and F-test statistics, respectively. The table critical values of the FADF and FKPSS unit root tests are shown in the study of Christopoulos and León-Ledesma (2010: 1083) and Becker et al. (2006: 389), respectively. In contrast, the critical values of the F-test are shown in Becker et al.'s study (2006: 389).

The null hypothesis of the ADF and FADF unit root tests shows the series is not stationary, while the KPSS and FKPSS unit root tests' null hypothesis presents that the series is stationary. However, when interpreting the outcome of the Fourier unit root tests, the significance of the trigonometric terms becomes important. When the values of F-test statistics are greater than the table critical values, the FADF and FKPSS unit root tests can be examined. However, the results of the conventional unit root tests are interpreted in the case of the insignificance of the trigonometric terms. Within this scope, the result of the unit root tests for the variables at the level is presented in Table 3, and the finding of the unit root tests for the variables at the first difference is disclosed in Table 4. As for the variables at the level, all considered unit root tests can be considered, while the ADF and KPSS unit root tests are only regarded. As a consequence of the outcome of all unit root tests concerning the level and first difference, it is concluded that all variables become stationary at the first difference (Adalı et al., 2023)

Table 4: Unit Root Test Results for the Variables at the First Differences

Variables	ADF		KPSS		FADF		FKPSS	
	Constant	**Constant and Trend**	**Constant**	**Constant and Trend**	**Constant**	**Constant and Trend**	**Constant**	**Constant and Trend**
lnENERGY	-7.962186 (0.000)	-8.589156 (0.000)	0.575026	0.073708	- [2.501125]	- [1.459964]	- [2.501125]	- [1.459964]
lnEXPORT	-6.721297 (0.000)	-6.680006 (0.000)	0.069616	0.050726	- [5,2, 4.007851]	- [5,2,3.910763]	- [5,2, 4.007851]	- [5,2,3.910763]
lnGDP	-7.423185 (0.000)	-7.524929 (0.000)	0.143903	0.072492	- [1,6,1.164017]	- [3,6, 0.932672]	- [1,6,1.164017]	- [3,6, 0.932672]
lnFD	-5.520984 (0.000)	-5.475141 (0.0002)	0.107896	0.088045	- [5,2,2.533328]	- [1,6,3.383199]	- [5,2,2.533328]	- [1,6,3.383199]

Values in the parenthesis related to the ADF unit root tests mean the probability while three values in the brackets about the FADF and FKPSS unit root tests donate frequencies, Bandwith, and F-test statistics, respectively. The table critical values of the FADF and FKPSS unit root tests are shown in the study of Christopoulos and León-Ledesma (2010: 1083) and Becker et al. (2006: 389), respectively. In contrast, the critical values of the F-test are shown in Becker et al.'s study (2006) shown in the FADF and FKPSS unit root tests means the insignificance of the trigonometric terms at a 1 % significance level.

With respect to the evidence achieved from the unit root tests, it is confirmed that all considered series contain the same integration order, I(1); in other words, they become stationary at their first differences. Then, the cointegration tests are employed to detect whether the variables are cointegrated and move together in the long run. Within this objective, the analysis of Johansen (1991) and the FADL cointegration proposed by Banerjee et al. (2017) is performed, and the outcome of the tests is presented in Table 4. When examining the results of Table 5, it is indicated that the null hypothesis of tests, meaning the absence of the cointegrated relationship, can not be rejected, and the considered variables do not move together in the long run

Table 5: Results of the Cointegration Tests

Johansen Cointegration Analysis			**FADL Cointegration Analysis**					
Trace Statistics		**Maximum Eigenvalue statistics**	-1.952425					
r=0	74.51666 (0.0049)	30.23406 (0.0834)	**Table Critical Values**					
r=1	44.28261 (0.03629)	21.66028 (0.1614)	%1			%5	%10	
r=2	22.62233 (0.1205)	14.49434 (0.2226)	-4.75			-4.03	-3.65	
r=3	44.28261 (0.2414)	8.127993 (0.2414)	Lnenergy lags	lnGDP lags	Lnexport lags	Lnfinancial lags	Frequency	Min_AIC
			2	3	1	1	4	-3.446677

Due to the absence of the long-run movement among the variables proved by Johansen and Fourier-Shin cointegration analysis, the VAR models are utilized, and the impulse-response and the Variance Decomposition analysis are performed within this scope. Optimal lags are selected in light of the diagnostic tests involving inverse roots of AR characteristic polynomial, residual correlation LM, and heteroskedasticity tests. The impulse-response function discloses that any variable responds against any shocks emerging in any variables. The 500 replication Monte-Carlo stimulation is used for confidence, and the generalized impulse-response methods are used. The outcome of the impulse-response function analysis is posed in Figure 1. According to Figure 1, energy demand positively responds for two periods when any shocks occur in financial development and economic growth

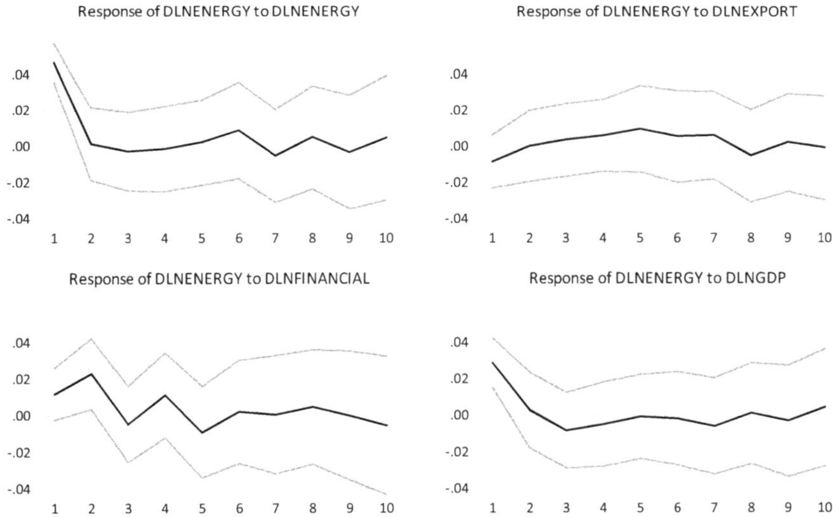

Figure 1: Results of The Impulse-Response Function Analysis

In addition to the impulse response function analysis, the Variance Decomposition analysis is also executed to detect the dynamic structure of variables, expressing the parts of a variable's variability in the model emerging from itself and other variables as percentages. The finding of the Variance Decomposition analysis is presented in Figure 1. As a result of Figure 1, it is identified that the proportion of the forecast error variance of energy demand is predicted by financial development, with the second period calculated as %20.

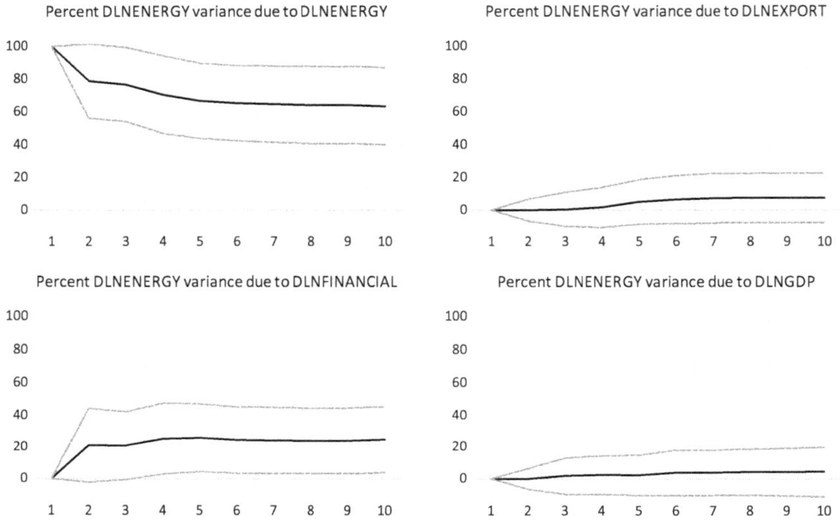

Figure 2: Results of The Variance Decomposition Analysis

Following the dynamic effects of the explanatory variables on the energy demand, various causality analyses are applied to find the causality nexus between the explanatory variables and the energy demand. Employed causality analyses are classified into their assumptions on the presence of the structural breaks. Granger (1969) and Toda-Yamamoto (1995) are ranked the first group ignoring the structural breaks, while the Fourier Granger (Enders & Jones, 2016) and the Fourier Toda-Yamamoto causality (Nazlioglu et al., 2016) analysis allow for considering multiple, smooth and unknown structural breaks are ranked as the second group. The series must be stationary when performing Granger and Fourier Granger Causality analysis. In contrast, the series' level values are used for the Toda-Yamamoto and Fourier Toda-Yamamoto causality analysis by adding maximum integration values plus optimal lags. The results of the causality tests are disclosed in Table 6. As a consequence of Table 6, all employed tests prove that financial development is a significant driver of the energy demand while energy demand is induced by export, concerning the finding of the Toda-Yamamoto causality analysis.

Table 6: Results of the Causality Tests

Model	**Granger**	**Toda-Yamamoto**	**Fourier Granger**	**Fourier Toda-Yamamoto**
lngdp=/>lnenergy	1.903736 (5, 0.8623)	1.877184 (6, 0.8659)	0.037714191 [1,1, 0.8434]	4.1875378 [1,1, 0.9950]
lnenergy=/>lngdp	1.891340 (5, 0.8640)	2.468625 (6, 0.7812)	0.36842011 [1,1, 0.5485]	0.000731039 [1,1, 0.9786]
lnexport=/>lnenergy	3.876787 (5,0.5673)	13.30499** (6, 0.0207)	0.65970388 [1,1, 0.4254]	4.7152197 [1,1, 0.9981]
lnenergy=/>lnexport	3.478623 (5,0.6266)	2.014813 (6, 0.8471)	1.3836179 [1,1, 0.2467]	1.1852598 [1,1, 0.2875]
lnfinancial=/>lnenergy	12.54919** (5,0.0280)	17.62540*** (6, 0.0035)	9.3433468 ** [1,1, 0.0032]	8.8707289 ** [1,1, 0.0050]
lnenergy=/>lnfinancial	3.926880 (5,0.5600)	3.903966 (6, 0.5633)	1.8027556 [1,1, 0.1928]	1.7842317 [1,1, 0.18830]

, and * mean 5 % and 1 % significance level. Number of boostrap replication 10000. SIC selects the optimal lag for Fourier Granger and Fourier Toda-Yamamoto analysis, while the process of optimal lag selection on VAR models presented in the appendix is considered for Granger and Toda-Yamamoto analysis. Values in the parenthesis donate used lags and probability values, respectively, while values in the bracket represent lag, structural break, and bootstrap probability values.

3. Conclusion

The study is organized to address the determinants of energy demand in Türkiye between 1965 and 2023 under the light of the energy and economic forces-related hypothesis composing the growth, conservation, neutrality, and feedback hypothesis. The GDP per capita, exports, and financial development are considered variables as the economic forces indicators, and various time-series methods are applied to reveal the evidence guiding the policy actions.

Initially, the unit root tests covering the traditional and Fourier-based tests confirm that the variables are integrated as order I (1). Johansen and FADL cointegration tests present the exception of the null hypothesis, which presents the absence of the long-run movement among the variables. Following the cointegration tests, the dynamic connection between the variables is explored with the help of the impulse-response function and the Variance decomposition analysis, and as a result of the estimations, it is found that financial development promotes a significant effect on the energy demand. In order to test the energy-economic forces linked hypothesis, Granger, Toda-Yamamoto, Fourier Granger, and Fourier Toda-Yamamoto tests are performed. When checking all outcomes of the tests, it is identified that the neutrality hypothesis concerning economic growth, export, and energy demand is verified. In contrast, the energy demand is induced by financial development, which presents the finance-growth hypothesis. Moreover, verifying the shreds of evidence of the finance-growth hypothesis and the neutrality hypothesis on economic growth and export also supports other detected proof from the VAR models. The finance-growth hypothesis is also approved by some studies; Kassi et al., 2020; Raza et al., 2020; Zhe et al., 2021 while Akarca & Long, 1980; Jobert & Karanfil, 2007; Jaraite et al., 2017, Lean & Symth, 2010; Shakeel et al., 2014; Shakeel & Ahmed, 2020 also confirm the neutrality hypothesis.

Focusing on the evidence revealed by the study, the conservation hypothesis, which aims to mitigate energy demand, protect the environment, reduce fossil fuels, and stimulate renewable energy resources, can be implemented without worrying about the cost of shrinking economic growth and exports. However, an expansion in domestic credits viewed as financial development causes higher energy demand, leading to higher environmental degradation. Within this scope, financial institutions should reorganize and change the perspective of the loan conditions, supporting enterprises to invest in renewable energy resources and upgrading firms through energy efficiency.

References

Adalı, Z., Toygar, A., & Yıldırım, U. (2023). Assessing the stochastic behavior of fishing grounds footprint of top ten fishing countries. *Regional Studies in Marine Science*, *63*, 103015.

Akarca, A. T., & Long, T. V. (1980). On the relationship between energy and GNP: A reexamination. *The Journal of Energy and Development*, 326–331.

Apergis, N., & Payne, J. E. (2010). The emissions, energy consumption, and growth nexus: evidence from the commonwealth of independent states. *Energy policy*, 38(1), 650–655.

Apergis, N., & Tang, C. F. (2013). Is the energy-led growth hypothesis valid? New evidence from a sample of 85 countries. *Energy Economics*, *38*, 24–31.

Banerjee, P., Arčabić, V., & Lee, H. (2017). Fourier ADL cointegration test to approximate smooth breaks with new evidence from crude oil market. *Economic Modelling*, *67*, 114–124.

Becker, R., Enders, W., & Lee, J. (2006). A stationarity test in the presence of an unknown number of smooth breaks. *Journal of Time Series Analysis*, *27*(3), 381–409.

Brunnschweiler, C. N. (2010). Finance for renewable energy: An empirical analysis of developing and transition economies. *Environment and Development Economics, 15*(3), 241–274.

Burakov, D., & Freidin, M. (2017). Financial development, economic growth and renewable energy consumption ın russia: A vector error correction approach. *International Journal of Energy Economics and Policy, 7*(6), 39–47.

Christopoulos, D. K., & León-Ledesma, M. A. (2010). Smooth breaks and non-linear mean reversion: Post-Bretton Woods real exchange rates. *Journal of International Money and Finance, 29*(6), 1076–1093.

Dimnwobi, S. K., Madichie, C. V., Ekesiobi, C., & Asongu, S. A. (2022). Financial development and renewable energy consumption in nigeria. *Renewable Energy, 192*, 668–677.

Enders, W., & Jones, P. (2016). Grain prices, oil prices, and multiple smooth breaks in a VAR. *Studies in Nonlinear Dynamics & Econometrics, 20*(4), 399–419.

Erkan, C., Mucuk, M., & Uysal, D. (2010). The impact of energy consumption on exports: The Turkish case. *Asian Journal of Business Management, 2*(1), 17–23.

Fan, W., & Hao, Y. (2020). An empirical research on the relationship amongst renewable energy consumption, economic growth and foreign direct investment in China. *Renewable Energy, 146*, 598–608.

Granger, C. W. (1969). Investigating causal relations by econometric models and cross-spectral methods. *Econometrica: Journal of the Econometric Society*, 424–438.

Jaraite, J., Karimu, A., & Kazukauskas, A. (2017). Policy-induced expansion of solar and wind power capacity: Economic growth and employment in EU countries. *The Energy Journal, 38*(5), 197–222.

Jobert, T., & Karanfil, F. (2007). Sectoral energy consumption by source and economic growth in Turkey. *Energy Policy, 35*(11), 5447–5456.

Johansen, S. (1991). Estimation and hypothesis testing of cointegration vectors in Gaussian vector autoregressive models. *Econometrica: Journal of the Econometric Society*, 1551–1580.

Kahouli, B. (2019). Does static and dynamic relationship between economic growth and energy consumption exist in OECD countries? *Energy Reports, 5*, 104–116.

Kassi, D. F., Sun, G., & Ding, N. (2020). Does governance quality moderate the finance- renewable energy-growth nexus? Evidence from five major regions in the world. *Environmental Science and Pollution Research, 27*, 12152–12180.

Kraft, J., & Kraft, A. (1978). On the relationship between energy and GNP. *The Journal of Energy and Development*, 401–403.

Lean, H. H., & Smyth, R. (2010). Multivariate Granger causality between electricity generation, exports, prices and GDP in Malaysia. *Energy*, *35*(9), 3640–3648.

Lin, B., Omoju, O. E., & Okonkwo, J. (2016). Factors influencing renewable electricity consumption in China. *Renewable and Sustainable Energy Reviews*, *55*, 687–696.

Nazlioglu, S., Gormus, N. A., & Soytas, U. (2016). Oil prices and real estate investment trusts (REITs): Gradual-shift causality and volatility transmission analysis. *Energy Economics*, *60*, 168–175.

Odhiambo, N. M. (2021). Trade openness and energy consumption in sub-Saharan African countries: A multivariate panel Granger causality test. *Energy Reports*, *7*, 7082–7089.

Pao, H. T., & Fu, H. C. (2013). Renewable energy, non-renewable energy and economic growth in Brazil. *Renewable and Sustainable Energy Reviews*, *25*, 381–392.

Raza, S.A., Shah, N., Qureshi, M.A., Qaiser, S., Ali, R., & Ahmed, F. (2020). Non-linear threshold effect of financial development on renewable energy consumption: Evidence from panel smooth transition regression approach. *Environmental Science and Pollution Research*, *27*, 32034–32047.

Sadorsky, P. (2012). Energy consumption, output and trade in South America. *Energy Economics*, *34*(2), 476–488.

Shahbaz, M., Gozgor, G., & Hammoudeh, S. (2019). Human capital and export diversification as new determinants of energy demand in the United States. *Energy Economics*, *78*(C), 335–349.

Shahbaz, M., Song, M., Ahmad, S., & Vo, X. V. (2022). Does economic growth stimulate energy consumption? The role of human capital and R&D expenditures in China. *Energy Economics*, *105*, 105662.

Shakeel, M., & Ahmed, A. (2020). Economic growth, exports, and role of energy conservation: Evidence from panel co-integration-based causality models in South Asia. *Energy & Environment*, *32*(1), 3–24.

Shakeel, M., Iqbal, M. M., & Majeed, M. T. (2014). Energy consumption, trade and GDP: A case study of South Asian countries. *The Pakistan Development Review*, *53*(4), 461–476.

Sharma, R., Shahbaz, M., Kautish, P., & Vo, X. V., (2021). Analyzing the impact of export diversification and technological innovation on renewable energy consumption: Evidences from BRICS nations. *Renewable Energy*, *78*(C), 1034–1045.

Toda, H. Y., & Yamamoto, H. (1995). Statistical inference in vector autoregressions with possibly integrated processes. *Journal of Econometrics*, Vol. 66/ 1995, 225–250.

Vo, D. H., Tran, Q., & Tran, T. (2022). Economic growth, renewable energy and financial development in The CPTPP Countries. *Plos One*, *17*(6).

Yasmeen, R., Zhang, X., Sharif, A., Shah, W. U. H., & Dincă, M. S. (2023). The role of wind energy towards sustainable development in top-16 wind energy consumer countries: Evidence from STIRPAT model. *Gondwana Research*, *121*, 56–71.

Zhe, L., Yüksel, S., Dinçer, H., Mukhtarov, S., & Azizov, M. (2021). The Positive influences of renewable energy consumption on financial development and economic growth. *SAGE Open*, *11*(3), 1–10.

Fatih Volkan Ayyıldız[1]

Could New Zealand be the First Country to Use All Renewable Energy for Electricity Generation? Future Prediction with Artificial Neural Networks for the Year 2030

1 Introduction

The use of fossil-based energy based on traditional production techniques harms the ecological balance on a large scale. The fact that energy is an indispensable resource has caused the need for clean energy in this field to gain increasing importance. Along with technological developments, especially with the influence of the Kyoto Protocol and the Paris Climate Agreements, the importance given to renewable energy is felt tangibly in every field at the international level, which has led the countries to compete in the production and consumption of renewable energy (Lund et al., 2016, Şahin & Ayyıldız, 2020). One of these areas is the use of renewable energy in the generation of electrical energy.

New Zealand is the country with the highest use of renewable energy in electricity generation in the world with 87 % in 2022. In this respect, it is the country with the highest potential to use completely renewable energy in electrical energy production. This situation also coincides with the target of using 100 % renewable energy sources in electrical energy generation in 2030, made by New Zealand official sources.

1 Ph.D., Asst. Prof., Ordu University Unye Faculty of Economics and Administrative Sciences Department of Economics, f.volkanayyildiz@odu.edu.tr, ORCID: 0000-0001-5991-3574.

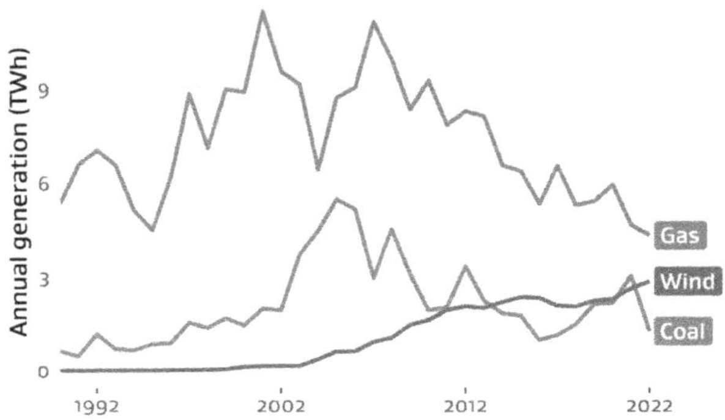

Figure 1: Distribution of renewable energy sources in total electricity production in New Zealand 1990–2022
Source Ministry of Business, Innovation & Employment 2024

According to Figure 1, the increase in the share of wind energy in electrical energy production in New Zealand in the period 1990–2022 is noteworthy. Electricity generation from wind energy, which was only close to 1 % in 2004, exceeded 5 % in 2019, thus increasing by over 500 %. According to 2022 official data, the share of hydroelectric energy in electricity production is 60 %, the share of geothermal energy is 18 %, the share of natural gas energy is 9.9 %, the share of wind energy is 6.5 %, and the share of coal energy is 2.9 % (Ministry of Business, Innovation & Employment 2024, New Zeeland Wind Energy Association, 2024). In addition, according to official records, the winter of 2022 in New Zealand will be the rainiest and warmest winter season in the history of New Zealand, causing renewable electrical energy production to reach the highest rate due to the effect of climate change.

The aim of this study is to determine whether New Zealand has achieved its goal of using completely renewable energy in electrical energy production in 2030 by future modeling using artificial neural networks. Therefore, using renewable energy, population and inflation data for the period between 1971 and 2021 in New Zealand, electricity use from renewable energy for 2030 was estimated with the Artificial Neural Networks methodology. Accordingly, in the modeling, the supply side was determined as electrical energy production and the demand side as total population and inflation. This study is original because it is the first study to apply future prediction using the artificial neural networks method of renewable energy in electricity production in the case of New Zealand. In addition, it is thought that the study will raise awareness by contributing to the realization of this gap in the literature with its original dimension.

2 Statistical Model and Datas

There is a broad consensus in the literature that the two main determinants of renewable energy are population and inflation. In this context, the model representation explaining the basic function of renewable energy-based electricity production is as follows:

$$lnRen_t = \alpha_0 + \delta_1 lnPop_t + \delta_2 lnInf_t + \varepsilon_t \quad (1)$$

From the notations in the model (1) α_0, to the constant coefficient; δ_1 *ve* δ_2, parameter coefficients; ε_t, to the error term; Inrefers to values whose logarithms have been taken against Ren, excessive differentiation operations. renewable energy; Pop, its population; are the parameters that indicate Inf, inflation . Descriptions of the parameters are presented in Table 1. subnotation *t* It refers to the period 1971–2021.

Table 1: Descriptions of parameters

Abbreviation	Parameter	Description	Source
lnRen	Renewable energy	Share of renewable resources in total electricity production (GWh)	British Petroleum Corporation (BP 2024 data)
lnPop	Population	Total population	World Bank 2024 (world development indicators data)
lnInf	Inflation	Consumer prices (annual %)	World Bank 2024 (world development indicators data)

The analysis operations were carried out through the functions included in EViews and SPSS econometric package programs

2.1 Results

Table 2 presents descriptive statistics of the renewable energy variable and its determinants in the country of New Zealand. It is seen that other variables, except inflation, which varies more, exhibit similar average values. It also explains that the difference between maximum and minimum values usually indicates a small band gap. For standard deviation values that provide reliability in data consistency, it is seen that the standard deviations of all variables are in the +3 and -3 band range (Özdamar & Ersoy, 2016: 144). The specified finding is a desired situation in statistical data analysis. Skewness coefficients show that renewable energy and inflation variables are skewed to the right; It indicates the left-skewed distribution of the population variable. Kurtosis coefficients draw attention to a sharp distribution in all variables.

Table 2: Descriptive statistics

Statistics	lnRen	lnPop	lnInf
Average	1.898492	6.571942	0.504743
Maximum	1.967833	6.708540	1.234278
Minimum	1.805631	6.455302	-0.533570
Standard Deviation	0.041386	0.072842	0.440451
Skewness	-0.326376	0.231608	-0.079754
Kurtosis	2.519485	1.839589	2.216960
Jarque-Bera	1.396083 (0.4975)	3.317387 (0.1903)	1.330406 (0.5141)

Note: Values in parentheses are probability (p) values.

Correlation analysis is the analysis of two or more variables that are related to measure how closely the variables are related to each other. In order to perform correlation analysis, there must be a certain relationship or probability between the variables. In statistics, correlation analysis generally expresses the degree to which binary variables are linearly related. Kendall's Tau correlation analysis was performed to determine the strength and direction of the sequential relationship between the variables in the established model. Kendall's tau_b correlation coefficients are summarized in Table 3. In this direction, a weak correlation and negative linear relationship is observed between renewable energy and population (-0.283). However, it explains the weak and positive linear relationship between renewable energy and inflation (0.232). The ordered pairwise correlation coefficients between the explanatory variables are considered to indicate that there is no multicollinearity problem in the model.

Table 3: Kendall's correlation analysis

Statistics			lnRen	lnPop	lnInf
Kendall's tau_b	lnRen	Correlation coefficient	1.000	-0.283[a]	0.232[b]
		Sig. (2-tailed)	.	0.003	0.017
	lnPop	Correlation coefficient	-0.283[a]	1.000	-0.506[a]
		Sig. (2-tailed)	0.003	.	0.000
	lnInf	Correlation coefficient	0.232[b]	-0.506[a]	1.000
		Sig. (2-tailed)	0.017	0.000	.

Note:e [(a)] The correlation is significant at the 0.01 level (2-tailed). [(b)] The correlation is significant at the 0.01 level (2-tailed).

Artificial Neural Networks (ANN) model consists of input layers, hidden layers and output layers. The hidden layer can have one or more layers. In ANN models used to estimate electricity production from renewable energy, three different input types are usually included in the model: time, population and inflation. However, the bias effect is also added to the model. Figure 2 shows the topology of the ANN model used in this study. Accordingly, the model consists of three input, three hidden and one output layers.

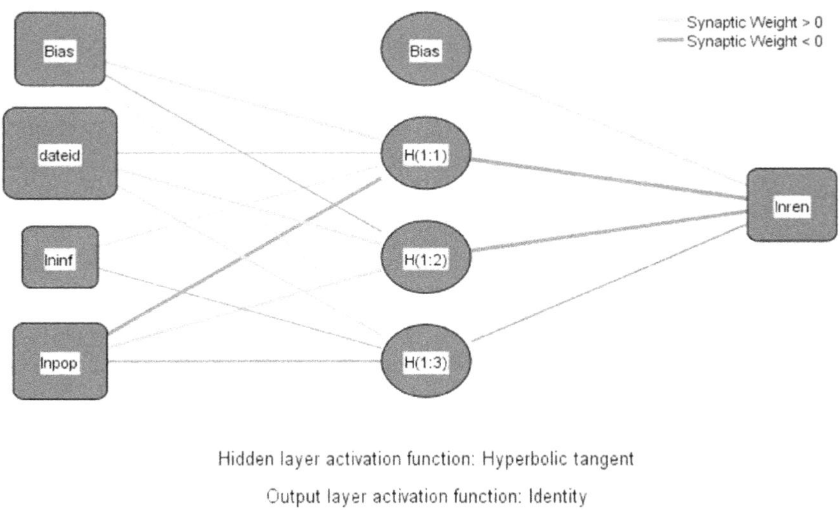

Figure 2: Artificial neural networks model topology

70 % of the data set in the established model is determined as training data and 30 % as test data. Table 4 informs that the model corresponds to approximately 71 % of the training data and approximately 29 % of the test data.

Table 4: Sample Processing Summary

		Observation	Percentage
Sample	Training	36	%70.6
	Test	15	%29.4

The values regarding the importance levels of the independent variables are reported in Table 5. When the findings are examined, it is seen that the importance levels of the variables are time, population and inflation parameters, respectively.

Table 5: Importance levels of independent variables

	Importance level	Normalized significance level
Time	0.415	%100
lnInf	0.223	%53.6
lnPop	0.362	%87.4

Success evaluation metrics preferred in regression models offer the opportunity to determine the differences between real values and predicted values. These metrics reflect average error measurements. Among these metrics, MAE (Mean Absolute Error), RMSE (Root Mean Square Error) and MAPE (Mean Absolute Percentage Error) are frequently used in the literature as model success evaluation techniques. In the study, the MAPE technique was adopted for the performance evaluation metric of the ANN model. The calculation formula for the MAPE metric is given in Equation (2) (Wang et al., 2018: 11):

$$MAPE = \frac{1}{P}\sum_{p=1}^{P}\frac{lnRenA_p - lnRenF_p}{lnRenA_p} \tag{2}$$

P in equation (2) is the number of observations; lnRenAp represents the actual values of renewable energy in electricity generation; lnRenF_p indicates the estimated values of renewable energy in electricity generation. Actual (lnRenA), estimated (lnRenF) and difference (Δ) values of the renewable energy variable in electricity generation are presented in Table 6.

Table 6: Renewable Energy Values in Electricity Generation: 1971–2021 period

Year	*lnRenA*	*lnRenF*	*Δ*	Year	*lnRenA*	*lnRenF*	*Δ*
1971	1.9678326	1.9234814	0.0443512	1997	1.8473195	1.8736508	-0.0263313
1972	1.9453940	1.9249799	0.0204141	1998	1.8803271	1.8660013	0.0143257
1973	1.9240828	1.9241553	-0.0000725	1999	1.8446539	1.8702008	-0.0255469
1974	1.9232753	1.9223051	0.0009702	2000	1.8568493	1.8458587	0.0109906
1975	1.9535474	1.9197785	0.0337688	2001	1.8056314	1.8409344	-0.0353030
1976	1.8859352	1.9178980	-0.0319628	2002	1.8437297	1.8354719	0.0082578
1977	1.8674824	1.9192600	-0.0517775	2003	1.8212679	1.8394794	-0.0182114
1978	1.9020824	1.9207923	-0.0187098	2004	1.8605648	1.8365127	0.0240521
1979	1.9589848	1.9196423	0.0393425	2005	1.8104591	1.8333521	-0.0228931
1980	1.9561305	1.9164740	0.0396565	2006	1.8129177	1.8344582	-0.0215404
1981	1.9513840	1.9173086	0.0340754	2007	1.8179592	1.8441559	-0.0261968
1982	1.8923042	1.9149737	-0.0226696	2008	1.8096619	1.8373339	-0.0276719
1983	1.9139486	1.9222940	-0.0083454	2009	1.8568871	1.8545749	0.0023122
1984	1.8984915	1.9225613	-0.0240697	2010	1.8656812	1.8594409	0.0062403
1985	1.8842168	1.9086254	-0.0244086	2011	1.8826334	1.8519408	0.0306926
1986	1.9123226	1.9110504	0.0012722	2012	1.8585772	1.8762594	-0.0176822
1987	1.9101938	1.9026081	0.0075857	2013	1.8732799	1.8804941	-0.0072142
1988	1.9192584	1.9181484	0.0011099	2014	1.9004149	1.8896789	0.0107360
1989	1.8927645	1.9175562	-0.0247917	2015	1.9055841	1.9064259	-0.0008417
1990	1.9043044	1.9140484	-0.0097440	2016	1.9271792	1.9147171	0.0124621
1991	1.8888615	1.9112057	-0.0223441	2017	1.9112331	1.9176023	-0.0063692
1992	1.8623526	1.9120137	-0.0496611	2018	1.9224756	1.9249142	-0.0024387
1993	1.8878931	1.9058653	-0.0179722	2019	1.9139368	1.9294413	-0.0155045
1994	1.9150094	1.8969919	0.0180175	2020	1.9054229	1.9341235	-0.0287006
1995	1.9243205	1.8778155	0.0465050	2021	1.9108291	1.9293640	-0.0185348
1996	1.8990128	1.8756131	0.0233996	MAPE	-	-	1.605781

Figure 3 shows the trend and model success evaluation metrics of the forecast values for the renewable energy variable in electricity generation in the period 1971–2021. In this regard, it is understood that RMSE, MAE and MAPE measurement statistics are approximately 0.04, 0.03 and 1.61, respectively.

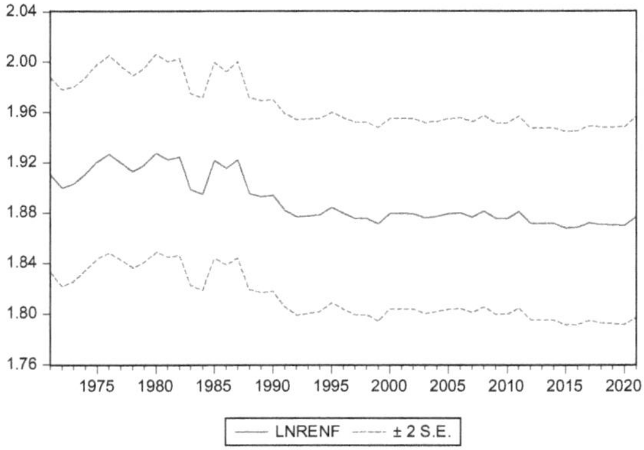

Forecast: lnRenF	
Actual: lnRenA	
Forecast sample: 1971 2021	
Included observations: 51	
Root Mean Squared Error	0.036197
Mean Absolute Error	0.030217
Mean Abs. Percent Error	1.605781
Theil Inequality Coefficient	0.009575
Bias Proportion	0.000000
Variance Proportion	0.361686
Covariance Proportion	0.638314
Theil U2 Coefficient	1.276275
Symmetric MAPE	1.602577

Figure 3: Success metrics in the model of renewable energy use in electricity generation 1971–2021 period

Statistical values of the model success evaluation measurements of the renewable energy model in electricity generation in the 2022–2030 period are visualized in Figure 4. The RMSE, MAE, and MAPE metrics are approximately 0.00, 0.00, and 0.18, respectively. It is stated in the literature that model predictions with a MAPE value below 0.10 have a high degree of accuracy. In this context, the established model indicates a high degree of accuracy. It can be seen that the coefficient of determination value (R2) of the dependent variable, which expresses the proportion of variance determined by the independent variables, is 0.94. When examined together with the F statistic value [50.16(p=0.0001)], it is found that the model is significant and its explanatory power is quite high.

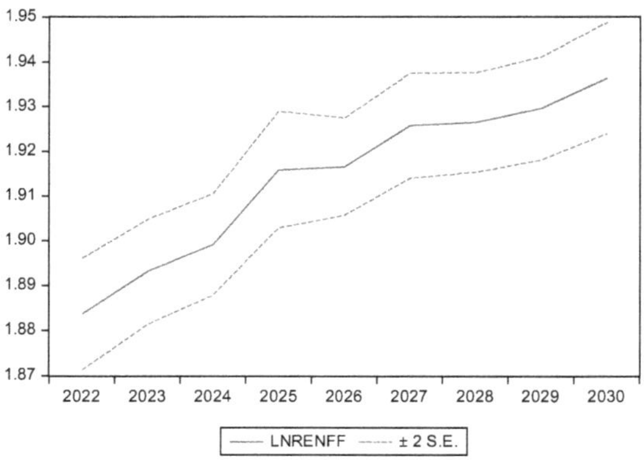

Forecast: lnRenFF	
Actual: lnRenF	
Forecast sample: 2022 2030	
Included observations: 9	
Root Mean Squared Error	0.004178
Mean Absolute Error	0.003502
Mean Abs. Percent Error	0.182925
Theil Inequality Coefficient	0.001091
Bias Proportion	0.000000
Variance Proportion	0.014520
Covariance Proportion	0.985480
Theil U2 Coefficient	0.508440
Symmetric MAPE	0.182940

Figure 4: Success metrics in the model of using renewable energy in electricity generation: 2022–2030 period.

Figure 5 shows the predicted values of renewable energy use in electricity generation in New Zealand over the period 2022–2030. In this regard, according to the evidence obtained from the empirical analysis, it is predicted that the share of renewable energy sources in the total electricity production, which is expected to be 1.85 (GWh) in 2022, will be 1.94 (GWh) in 2030. When these values are eliminated from logarithmic transformation, it is assumed that 76 % and 85 % will be realized, respectively.

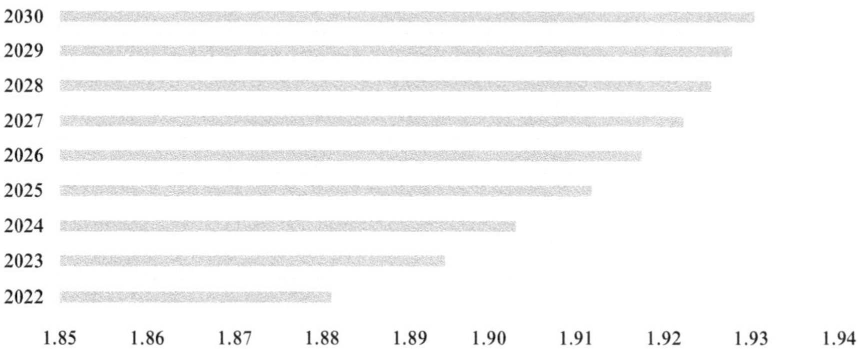

Figure 5: Estimated values of renewable energy use in electricity generation: 2022–2030 period

4 Conclusion

With the increasing importance given to renewable energy, efforts to increase the use of renewable energy have led to diversification of energy policies. Although it is important to implement these policies in the field, projections compatible with the targets in this field should also be taken into account. Therefore, in this study, it was investigated whether the goal of being the first country to use completely renewable energy in electricity production in 2030, set out by New Zealand official sources, could be achieved. In the analysis part of the study, a forecast was made for 2030 by taking into account the renewable energy values in electricity production in the 1971–2022 period. While the modeling creates the supply side and dependent variable of the renewable energy production model in electricity, the demand side of the model is determined as population and inflation variables, taking into account the studies in the literature. As a result of the future prediction made with the artificial neural networks method, it is estimated that the use of renewable energy in electricity production in New Zealand will be 85 % in 2030.

As a result of the artificial neural networks and future projection method used in this study, it was determined that although New Zealand could not achieve the goal of using completely renewable energy in electricity generation, it was very close to achieving the goal. As a result of the study, it was determined that New Zealand's policies on the use of renewable energy, especially in electricity generation, were generally successful and these policies should be continued.

References

British Petroluem. (2024). British Petroleum Datas. Available at: https://www.bp.com/en/global/corporate/energy-economics.html (12.04.2024)

Lund, H., Østergaard, P. A., Connolly, D., Ridjan, I., Mathiesen, B. V., Hvelplund, F., Thellufsen J.c., & Sorknæs, P. (2016). "Energy Storage and Smart Energy Systems". *International Journal of Sustainable Energy Planning and Management*, *11*, 3–14.

Ministry of Business, Innovation & Employment. (2024). Available at: https://www.mbie.govt.nz/ (10.04.2024)

New Zeeland Wind Energy Association. (2024). Available at: https://www.windenergy.org.nz/ (20.04.2024)

Özdamar, İ. H., & Ersoy, G. (2016). "Standart Skorlar Kullanılarak Ürün Kalite Performansının Belirlenmesi ve Bir Çalışma". *Süleyman Demirel Üniversitesi Fen Bilimleri Enstitüsü Dergisi*, *20*(1), 142–147.

Şahin, G., & Ayyıldız, F. V. (2020). Climate Change and Energy Policies: European Union-Scale Approach to a Global Problem. In: Qudrat-Ullah, H., Asif, M. (Eds.), *Dynamics of Energy, Environment and Economy*. Lecture Notes in Energy, vol. 77. Springer, Cham, pp. 295–320.

Wang, D., Liu, Y., Wu, Z., Fu, H., Shi, Y. ve Guo, H. (2018). "Scenario Analysis of Natural Gas Consumption in China Based on Wavelet Neural Network Optimized by Particle Swarm Optimization Algorithm". *Energies*, *11*(825), 1–16.

WorldBank. (2024). World Development Indicators Data. Available at: https://data.worldbank.org/indicator (09.04.2024)

Ömer Akçayır[1]

Asymmetrical Effects of Macroeconomic Indicators on GHGs Emissions: Analysis of Türkiye

1 Introduction

Although we face many macroeconomic problems on a national, regional and global scale, the biggest problem of the last century is undoubtedly environmental degradation and the gigantic dangers it brings with it. What makes this problem the biggest is that environmental problems have reached irreversible limit. In particular, the unavoidable increase in greenhouse gases (GHGs) reaches limits that cannot allow the environment to renew itself and brings about the destructive situation called global warming, which threatens future generations (IPCC, 2023). Diminishing of trees from our forests day by day, lakes tending to disappear, unsustainably decreasing potable water resources, disappearing some species and decreasing biodiversity are only some of the main headings of the harmful consequences of global warming. The extraordinary environmental degradation of the last few decades in particular has increased the responsibilities of executives at all levels, through the efforts of The Club Of Rome, has led to marginal decisions (TCOR, 2019). Both the decisions taken and the decisions postponed necessarily involve foregone costs. Moreover, each decision has positive or negative, direct or indirect effects on other macroeconomic variables.

In the literature, the relationship between environmental degradation and many macroeconomic variables has been examined, but there are no studies that deal with the direct relationship with inflation and unemployment. In the industrial production perspective, it is possible to reveal the direct relationship between GDP, imports and exports and GHGs (Haug & Ucal, 2019). However, when evaluated on the basis of production, inflation and unemployment variables may also be indirectly related to GHGs.

1 Ph.D., Lecturer, Nevşehir Hacı Bektaş Veli University, Health Services Vocational, omerakcayir@nevsehir.edu.tr, ORCID: 0000-0002-1645-5312.

Furthermore, the effect of an increase or a decrease in a macroeconomic variable may not always be as expected. Example for the last 3 years of Turkiye, while decreasing interest rate only has significant negative effects on many variables such as inflation, increasing interest rate alone is not sufficient for the disinflation process (Egilmez, 2023). Therefore, asymmetric effects are not always expected to be in the same direction.

In research, the asymmetric effects of imports, exports, inflation and unemployment on GHGs in the Turkish sample are investigated with the non-linear ARDL method. The model investigated by Shpak et al. (2022) for the USA and the Asia-Pacific countries, which is the region with the highest environmental pollution, was investigated for Turkiye with data for 1990–2022. It aims to contribute to the literature with the findings of the study.

2 Theoretical Background & Literature Review

In the literature on the impact of the environment on the economy, there are many different types of studies in many different fields of science. Anthropogenic environmental degradation is increasing day by day, consumption frenzy supported by hedonistic feelings, profit motive of producers mobilises states for growth and development. With the demand for energy and food triggered by the increasing population, GHGs (especially carbondioxide), are increasing and global warming accelerates environmental degradation (Bothun, 2018). Undoubtedly, environmental degradation and climate changes affect every agents directly or indirectly. The main impact of environmental degradation is on future generations (Attfield, 2010).

Because the rapidly depleting resources and the nature that cannot renew itself as a result of global warming reduce the welfare of generations yet to be born. A disturbed balance has the power to affect everything from agriculture to livestock, from energy demand to consumption habits. Considering the many macroeconomic variables that affect each other directly or indirectly, all macroeconomic variables originating from production and consumption affect GHGs emissions. Environmental pollution and GHGs statistics are now one of the most important macroeconomic variables that governments and all economic agents should follow. There is a huge literature on GHGs and its determinants. We limit our literature review due to page limitations. The study is theoretically based on 4 basic hypotheses.

1st Hypothesis [H_1]. Exports directly affect GHGs, In Turkiye.

2nd Hypothesis [H_2]. Imports directly affect GHGs, In Turkiye.

The need for energy triggered by the increase in the volume of economic activities leads to rapid depletion of resources, leading to deterioration of the ecological balance and ultimately to an increase in GHGs (IPCC, 2023). Each increasing in GDP means an increase in GHGs emissions due to production. Therefore, increase in foreign trade volume, FDI, exports and economic growth are expected to increase GHGs values due to energy demand and many studies in the literature examine these issues.

Al-Mulali Usama and Sheau-Ting (2014) found statistically significant and long-run relationship between import and export variables on CO_2 emission in their study. The study of Haug and Ucal (2019) on asymmetric effects for Turkiye, found that decreases in exports significantly reduce CO_2 emissions, while no statistically significant results were found for increases. While increases in imports increase CO_2 emissions, no statistically significant results were found for decreases. Mahmood et al. (2020) found that exports reduce CO_2 emissions but have spillover effects to neighbouring countries, while imports increase emissions but do not spill over to neighbouring countries.

3rd Hypothesis [H_3]. Inflation indirectly affect GHGs, In Turkiye.

The number of studies on inflation and GHGs is quite a few in the literature and there are no studies that directly address this relationship. Musarat et al. (2021) concluded in their study for the Malaysian construction industry that a decrease in the inflation rate increases CO_2 emissions. Setyadharma et al. (2021) concluded that inflation has a positive effect on air pollution and that high inflation can reduce air pollution. According to the findings of Ronaghi et al. (2020), as inflation increases, purchasing power weakens and CO_2 emissions decrease as falling demand reduces production.

4th Hypothesis [H_4]. Unemployment indirectly affect GHGs, In Turkiye.

The number of studies on unemployment and GHGs is quite low in the literature, and there are no studies that directly address this relationship. Mrabet and Jarboui (2017) concluded that labour inputs have a positive impact on CO_2 emissions for Arab countries. Liu and Feng (2022) showed that unemployment has negative effects on CO_2 emissions at the global level; however, the situation looks different at the regional level.

Shpak et al. (2022) confirmed these relationships in their study based on the hypothesis that GDP, import and export variables directly and inflation and unemployment variables indirectly affect GHGs for the most polluted countries, the two regions with the highest carbon emissions in the world. Parameter coefficients differ for regions and models.

As can be seen, the intensity of studies in the literature on environmental degradation consists of combinations of economic growth, foreign trade, FDI and energy variables. Very few studies have addressed the inflation and unemployment variables, albeit indirectly. We think that the study conducted in Turkiye sample will contribute to the literature.

3 Data, Methodology and Results

In this study, imports, exports, inflation rate, unemployment rate and GHGs (logarithmic) time series variables of Turkiye for the years 1990–2022 are considered. The data set is obtained from the World Bank database. In the econometric model, GHGs is the dependent variable and the others are determined as independent variables. The closed form of the model constructed in the research using the nonlinear ARDL method is as follows. The model investigated by Shpak et al. (2022) for where the dirtiest countries in the world, was analysed for Turkiye.

Linear Function Form: GHGs= f(INF, UNEMP, EX, IM)

Nonlinear Function Form: GHGs= f(INF^+, INF^-, $UNEMP^+$, $UNEMP^-$, EX^+, EX^-, IM^+, IM^-)

The model in its explicit form is as in Eq.1.

$$LnGHG_t = \beta_0 + \beta_1 INF_t + \beta_2 LnUNEMP_t + \beta_3 EX_t + \beta_4 IM_t + \varepsilon_t \quad \text{(Eq.1)}$$

LnGHG: Greenhouse gas emissions (*CO_2 equivalent- CO_2, NH_4, N_20, F-Gas*)
INF: Inflation, consumer prices (*annual %*) **EX**: Exports of goods and services *(% of GDP)*
UNEMP: Unemployment, total *(% of total labor force)* **IM**: Imports of goods and services *(% of GDP)*

Important descriptive statistics of the model variables are shown in Table 1.

Graph 1: Changes Model Variables in Turkiye Between 1990 and 2022.

Table 1: Describtive Statistics

	LNGHG	UNEMP	INF	EX	IM
Mean	5.890964	9.807030	36.44365	23.42366	25.77689
Median	5.900526	10.22700	16.33246	23.08351	25.50033
Max.	6.349118	14.02600	105.2150	38.58420	42.58593
Min.	5.429354	6.495000	6.250977	13.36510	16.63488
Std. Dev.	0.293820	1.950777	32.29887	5.704668	5.523879
Jarque-Bera	2.650836	0.411946	3.936981	2.517967	3.725925
Prob.	0.265692	0.813855	0.139668	0.283943	0.155212
Sum	194.4018	323.6320	1202.641	772.9809	850.6374
Sum Sq. Dev.	2.762567	121.7769	33382.94	1041.384	976.4236

The correlation matrix for the direction and intensity of the relationship between the variables is shown in Table 2. In particular, it can be said that there is a strong relationship between dependent variables and independent variables. In terms of the direction of the relationship, there is only an negative relationship between inflation and GHGs, while there is positive relationship between the other independent variables and the dependent variable.

Table 2: Correlation Matrix and Graphs

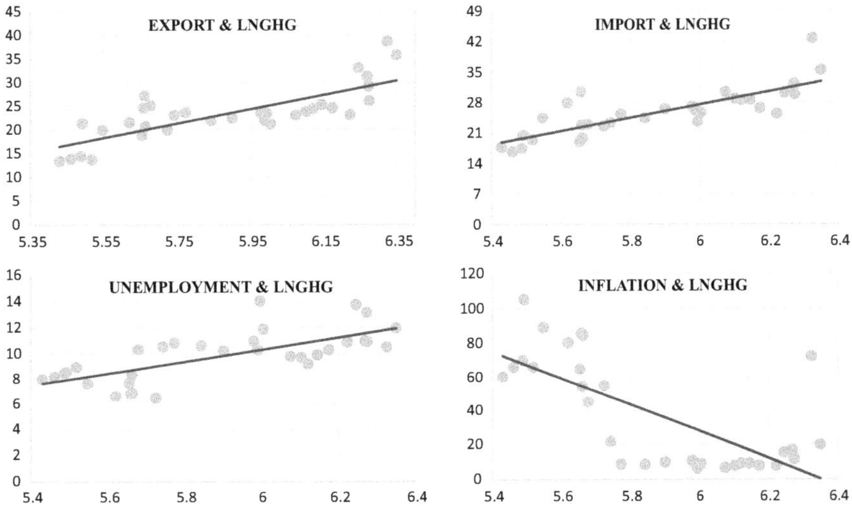

	LNGHG	UNEMP	XM	INF	EX	IM
LNGHG	1					
UNEMP	0.698	1				
XM	0.817	0.497	1			
INF	-0.716	-0.736	-0.341	1		
EX	0.778	0.532	0.970	-0.316	1	
IM	0.807	0.430	0.968	-0.345	0.879	1

In the first step of econometric analysis, whether the series contain unit roots or not is investigated with the Augmented Dickey Fuller [ADF] (1979) and Phillips-Perron [PP] (1988) standard unit root tests and Zivot and Andrews [ZA] (1992) tests that take into account structural breaks. ADF and PP unit root test results are presented in Table 3 and ZA structural break unit root test results are presented in Table 4.

Table 3: ADF & PP Test Results

Series	ADF Test Statistics		Phillips-Perron Test Statistics	
	H_0: Series is non-stationary.		H_0: Series is non-stationary.	
	Level	Δ (1st Difference)	Level	Δ (1st Difference)
LNGHG	-2.78 [0]^{it}	-5.80* [0]^{it}	-2.84 [1]^{it}	-6.67* [9]^{it}
UNEMP	-1.84 [0]^{i}	-4.73* [0]^{i}	-1.82 [4]^{it}	-5.67* [16]^{it}
INF	-0.60 [0]^{none}	-3.59* [0]^{none}	-0.67 [2]^{none}	-3.59* [1]^{none}
IM	-2.16 [0]^{it}	-4.49* [1]^{it}	-2.08 [4]^{it}	-4.78* [10]^{it}
EX	-2.24 [0]^{it}	-5.40* [1]^{it}	-2.28 [1]^{it}	-6.67* [6]^{it}

^{it} : intercept & trend ^{i}: intercept, ^{none}: no intercept & no trend *: stationary (%1)
[]; values in brackets denote lag length. Optimum lag lengths (max. lag length is 8) are determined by Schwartz Information Criterion. []; In PP tests, it refers to the bandwidth determined using the Newey West criterion.

Table 4: ZA (1992) Test Results

	Model A (intercept)		Model B (trend)		Model C (intercept & trend)	
	H_0: Series is non-stationary.					
Series	ADF-Stat.	Break Date	ADF- Stat.	Break Date	ADF- Stat.	Break Date
LNGHG	-4.21 [0]	2006	-3.10 [0]	2018	-4.13 [0]	2005
UNEMP	-4.31 [1]	2000	-3.31 [1]	2010	-6.78* [8]	2008
INF	-1.86 [0]	2002	-2.23 [2]	2007	-2.15 [2]	2010
IM	-4.10 [1]	2021	-4.21 [1]	2021	-4.28 [1]	2020
EX	-3.27 [1]	2021	-3.66 [1]	2017	-3.53 [1]	2015
Δ LNGHG	-6.36* [0]	2003	-5.98* [0]	2008	-6.26* [0]	2003
Δ UNEMP	-5.59* [8]	2011	-5.07* [0]	2020	-7.74* [0]	2011
Δ INF	-5.30** [0]	2021	-5.28* [0]	2022	-5.11* [0]	2020
Δ IM	-6.13* [1]	1998	-5.24* [1]	2017	-6.06* [0]	1998
Δ EX	-6.80* [0]	1997	-6.53* [0]	2000	-7.67* [0]	2001
	Critical Values (Zivot & Andrews, 1992)					
	%1		%5		%10	
Model A	-5.34		-4.85		-4.60	
Model B	-5.06		-4.52		-4.26	
Model C	-5.57		-5.08		-4.82	

Δ is the difference operator. *: Denotes stationarity at 1 % and **: at 5 % significance level. []; values in brackets denote lag length. Optimum lag lengths (max. lag length is 8) are determined by Schwartz Information Criterion (SIC). []; In PP tests, it refers to the bandwidth determined using the Newey-West criterion.

The cointegration relationship between the series is investigated with the NARDL method developed by Shin et al. (2014), which also takes into account the asymmetric changes of the ARDL method first proposed by Pesaran and Shin (1999) and Pesaran et al. (2001).

NARDL models allow the investigation of both short-run and long-run asymmetry. The method is also useful in revealing the hidden cointegration relationship that cannot be detected with the linear approach by accepting asymmetry instead of linearity assumption (Shahzad et al., 2017). In the bounds test approach, whether the integrated levels of the series to be included in the model are I(0) or I(1) is not a precondition, only the precondition of not being I(2) is in question. In Table 5, the statistics and critical values of the bounds test are reported and it is concluded that the series are cointegrated.

Table 5: Bound Test Results

F-Statistic H_0: There is no cointegration.	**Critical Values Pesaran et al. (2001)**			**Critical Values Narayan (2005)**			**Decision**
	Significance Level	**I(0) (Lower Limit)**	**I(1) (Upper Limit)**	**Significance Level**	**I(0) (Lower Limit)**	**I(1) (Upper Limit)**	
7.591679 (k=8) [N=33]	10 %	2.20	3.09	10 %	2.46	3.46	H_0 Reject.
	5 %	2.56	3.49	5 %	3.05	4.22	
	1 %	3.29	4.37	1 %	4.28	5.84	
Diagnostic Statistics							
$\chi^2(1)_{BGAC}$=2.40 [0.02] χ^2_{BPG}= 0.77 [0.99] $JB_{norm.}$ = 0.67 [0.70] DW_{ist} = 2.71							

Note: k, represents the number of independent variables and N, represents the number of observations. Pesaran et al. (2001: 300) critical values are taken from Tablo CI (iii). Narayan (2005: 1988) critical values are taken from *Case III.* χ^2_{BGAC} ; Breusch-Godfrey autocorrelation LM test statistics, χ^2_{BPGDV} : Breusch-Pagan-Godfrey heteroskedasticity test statistic, JB_{norm}; Jarque-Bera normallity test statistic, DW_{ist}; Durbin Watson test statistic. []; values in brackets are p-probability.

Schwarz information criterion was used to select the model with a maximum lag of 2. NARDL Dynamic Asymmetric Estimation Results are reported in Table 6.

Table 6: NARDL Dynamic Asymmetric Estimation Results

Dependend Variable: LNGHG			
Variables	**Coefficient**	**t-Statistic**	**Prob.**
constant	1.949258	3.415510	0.0033
LNGHG(-1)*	-0.349515	-3.351119	0.0038
$INF^{+}(-1)$	0.002130	1.555050	0.1384

Table 6: Continued

Dependend Variable: LNGHG			
Variables	**Coefficient**	**t-Statistic**	**Prob.**
INF^{-**}	0.001773	2.214709	0.0407
IM^{+**}	0.000406	0.077285	0.9393
$IM^{-}(-1)$	0.006643	1.464541	0.1613
EX^{+**}	-0.005424	-1.810697	0.0879
$EX^{-}(-1)$	-0.026299	-2.223459	0.0400
$UNEMP^{+}(-1)$	0.015676	1.543704	0.1411
$UNEMP^{-**}$	0.005881	0.601305	0.5556
$\Delta(INF^{+})$	-0.001344	-2.073371	0.0536
$\Delta(IM^{-})$	-0.000258	-0.058349	0.9542
$\Delta(EX^{-})$	0.002418	0.227007	0.8231
$\Delta(UNEMP^{+})$	-0.009613	-1.174772	0.2563
Diagnostic Statistics			
$R^2=0.995$ $Adj\text{-}R^2=0.992$		$JB_{norm.}=0.67\ (0.71)$	
$F_{ist.}=291.4\ (0.000)$ $DW_{ist}=2.709$		$Ç^2_{AC}(1)=2.40\ (0.12)$ $Ç^2_{BPG}=0.77\ (0.67)$	

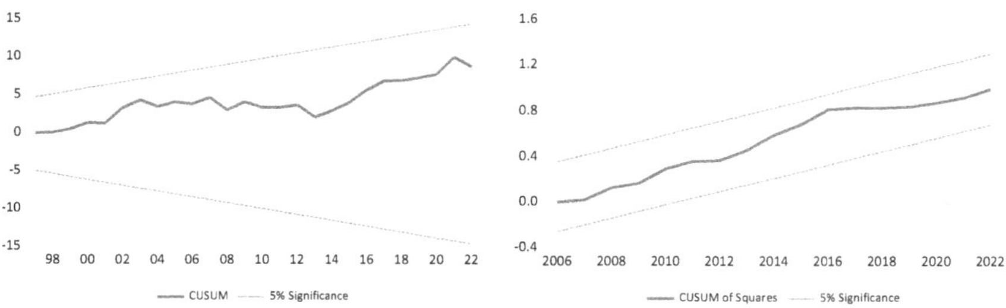

Graph 2: Cusum and Cusum of Squares

Cusum and Cusum Of Squares plots of the long-run analysis results are also presented Graph 2. Since the lines are within the confidence intervals, it can be stated that the regression coefficients are stable. The findings of the Short Run Analysis (Error Correction Model) are reported in Table 7. The coefficients of the error terms are negative and statistically significant. Deviations are self-correcting after a certain period[2].

2 Because of 1/1,3495≅2,86 the deviations disappear completely after approximately 2.86 years.

Table 7: Error Correction Regression

Variables	Coefficient	t-Statistic	Prob.
Δ (INF^+)	-0.001344	-4.237330	0.0006
Δ (IM^-)	-0.000258	-0.123584	0.9031
Δ (EX^-)	0.002418	0.684931	0.5026
Δ ($UNEMP^+$)	-0.009613	-2.359388	0.0305
CointEq(-1)*	**-0.349515**	**-10.77534**	**0.0000**

The coefficients and significance levels of the long-run model are reported in Table 8.

Table 8: Long-Run Asymmetric Coefficient Estimation Results

Variables	Coefficient	t-Statistic	Prob.
INF^+	0.006093	1.845927	0.0824*
INF^-	0.005072	2.254216	0.0377*
IM^+	0.001162	0.088172	0.9308
IM^-	0.019007	1.214970	0.2410
EX^+	-0.015520	-2.453874	0.0252*
EX^-	-0.075244	-1.849593	0.0818*
$UNEMP^+$	0.044850	1.698829	0.1076
$UNEMP^-$	0.016826	0.695309	0.4963
Constant	5.577035	99.45376	0.0000*

LNGHG= 0.0061***INF^+** + 0.0051***INF^-** + 0.0012***IM^+** + 0.0190***IM^-** - 0.0155***EX^+** - 0.0752***EX^-** + 0.0448***$UNEMP^+$** + 0.0168***$UNEMP^-$** + 5.5770

According to the long-run coefficient results, the positive and negative effects of imports and unemployment are not statistically significant. However, the relationship between inflation rate and exports and GHGs is significant.

- A 1 unit (percentage point) increase in the inflation rate increases GHGs by 0.6 %, while a 1 percentage point decrease decreases GHGs by 0.5 %.
- While a 1 unit increase in the share of exports in GDP decreases GHGs by 1.5 %, a 1 percentage point decrease increases GHGs by 7.5 %.

Graphical representation of asymmetric long-run and short-run coefficients (Dynamic Asymmetric Multiplier Impulse Response Functions) is given in the Graph 3.

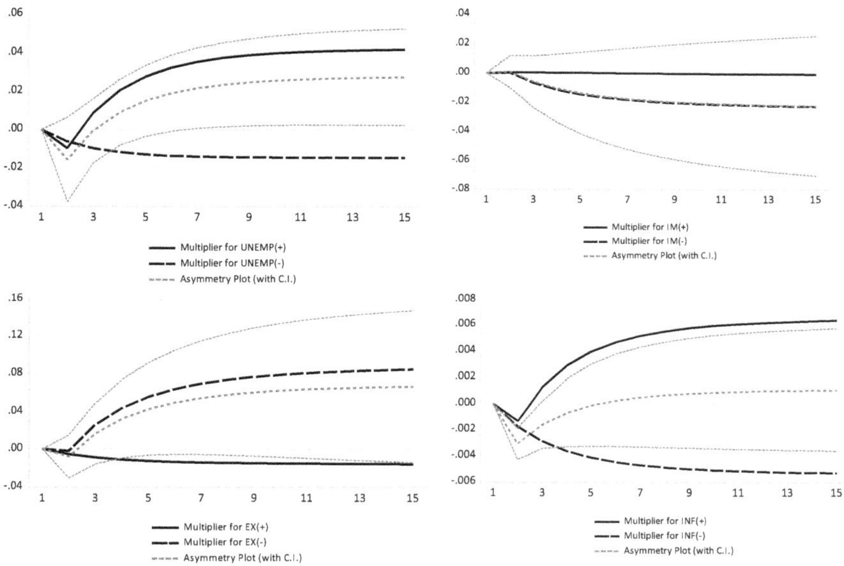

Graph 3: Dynamic Asymmetric Multiplier Impulse Response Functions

4 Conclusion

Nowadays, where anthropogenic environmental damage is a very important threat to sustainable growth, GHGs emissions (especially CO_2), is a variable that should be monitored with the effect of legal sanctions. While there are many studies on GHGs in the literature, there are very few studies on key macroeconomic variables such as unemployment and inflation. In research, the asymmetric effects of imports, exports, inflation and unemployment on GHGs emissions are investigated with the non-linear ARDL method with the data of Turkiye for 1990–2022. The model builded by Shpak et al. (2022) for which is the region with the highest environmental pollution, is analysed for Turkiye.

According to the long-run findings of the NARDL method, which also provides information about the effects of positive and negative movements of independent variables, the positive and negative effects of imports and unemployment are not statistically significant, while the relationship between inflation rate and exports and GHGs is significant. While an increase of 1 unit in the inflation rate increases GHGs by 0.6 %, a decrease of 1 percentage point decreases GHGs by 0.5 %. While a 1 unit increase in the share of exports in GDP decreases GHGs by 1.5 %, a 1 percentage point decrease increases GHGs by 7.5 %. As stated in the hypothesis, the indirect relationship between inflation and GHGs has been confirmed econometrically. H_1 and H_3 are confirmed, but H_2 and H_4 are not confirmed. The low coefficient indicates that this indirect relationship is not very strong.

This is because the cause of inflation (demand or cost induced), which causes a significant erosion in purchasing power and the inflation-protection behaviour of citizens exposed to inflation (consumption or saving preference) vary significantly from country to country. Turkiye, which currently has the 4th highest inflation rate in the world, is the 15th highest country in terms of carbon emissions. As a result of the analysis, it is seen that inflation has a low impact on GHGs both in terms of increase and decrease. Moreover, the increase in export data of Turkiye affects GHGs emissions as a negative output of production, especially energy demand. Contrary to the expected empirical findings, the amount of GHGs increases when exports decrease and decreases when exports increase.

References

Al-mulali, U., & Sheau-Ting, L. (2014). "Econometric analysis of trade, exports, imports, energy consumption and CO2 emission in six regions". *Renewable and Sustainable Energy Reviews, 33*, 484–498.

Attfield, R. (2010). "Global warming, equity and future generations". *Human Ecology Review*, 102–105.

Bothun, G. D. (2018). "Earth systems thinking: global consumerism, climate change, and the spiritual value of the earth". *Environmental Management and Sustainable Development, 7*(3), 35.

Dickey, D., & Fuller, W. A. (1979). "Distribution of the Estimates for Autoregressive Time Series with a Unit Root".*Journal of the American Statistical Association, 74*, 427–431.

Egilmez, M. (2023, September 25). "Faiz Gerçek Bir Beladır." Kendime Yazılar. https://www.mahfiegilmez.com/2023/09/faiz-gercek-bir-beladr.html

Haug, A. A., & Ucal, M. (2019). "The Role of Trade and FDI for CO2 Emissions in Turkey: Nonlinear Relationships". *Energy Economics, 81*, 297–307.

Intergovernmental Panel on Climate Change (IPCC). (2023). Climate Change Synthesis Report. http://www.ipcc.ch/S.

Liu, Y. Q., & Feng, C. (2022). "The effects of nurturing pressure and unemployment on carbon emissions: Cross-country evidence". *Environmental Science and Pollution Research*, 1–20.

Mahmood, H., Alkhateeb, T. T. Y., & Furqan, M. (2020). “Exports, imports, foreign direct investment and CO2 emissions in North Africa: Spatial analysis”. *Energy Reports, 6*, 2403–2409.

Mrabet, A., & Jarboui, S. (2017). “Do institutional factors affect the efficiency of GDP and CO2 emission? Evidence from Gulf and Maghreb countries”. *International Journal of Global Energy Issues, 40*, 259.

Musarat, M. A., Alaloul, W. S., Liew, M. S., Maqsoom, A., & Qureshi, A. H. (2021). “The effect of inflation rate on CO2 emission: A framework for Malaysian Construction Industry”. *Sustainability, 13*, 1562.

Narayan, P. K. (2005). “The saving and investment Nexus for China: Evidence from cointegration tests”. *Applied Economics, 37*(17), 1979–1990.

Pesaran, M. H., Shin, Y., & Smith, R. J. (2001). “Bound testing approaches to the analysis of level relationships”. *Journal of Applied Econometrics, 16*, 289–326.

Pesaran, M. H., & Shin, Y. (1999). “An autoregressive distributed lag modeling approach to cointegration analysis”. In: Strom, S. (Ed.), *Econometrics and economic theory in the 20th century: The Ragnar Frisch Centennial Symposium*. Cambridge University Press, Cambridge.

Phillips, P. C. B., & Perron, P. (1988). “Testing for a unit root in time series regression”. *Biometrika, 75*(2), 335–346.

Ronaghi, M., Reed, M., & Saghaian, S. (2020). “The impact of economic factors and governance on greenhouse gas emission”. *Environmental Economics and Policy Studies, 22*, 153–172.

Setyadharma, A., Oktavilia1, S., SriWahyuningrum, I. F., Indah Nikensari, S., & Mei Saputra, A. (2021). “Does Inflation Reduce Air Pollution? Evidence from Indonesia. In Proceedings of the the 6th International Conference on Energy”, *Environment, Epidemiology, and Information System (ICENIS)*, Semarang, Indonesia, 4–5 August.

Shahzad, S. J. H., Nor, S. M., Ferrer, R., & Hammoudeh, S. (2017). “Asymmetric determinants of CDS spreads: US industry-level evidence through the NARDL approach”. *Economic Modelling, 60*, 211–230.

Shin, Y., Yu, B., & Greenwood-Nimmo, M. (2014). “Modelling asymmetric cointegration and dynamic multipliers in a nonlinear ARDL framework”. *Festschrift in honor of Peter Schmidt: Econometric Methods and Applications*, 281–314.

Shpak, N., Ohinok, S., Kulyniak, I., Sroka, W., Fedun, Y., Ginevi˘cius, R., & Cygler, J. (2022). “CO2 emissions and macroeconomic indicators: Analysis of the most polluted regions in the world”. *Energies*, 15, 2928. https://doi.org/10.3390/en15082928

The Clup of Rome (TCOR). (2019). "Activity Report", https://www.clubofrome.org/wp-content/uploads/2021/10/2019-Annual-Report.pdf

Zivot, E., & Andrews, D. W. K. (1992). "Further evidence on the Great Crash, the Oil Price shock, and the unit-root hypothesis". *Journal of Business & Economic Statistics*, *10*(3), 251–270.

Ahmet Tiryaki[1] & Çağlar Karaduman[2] & Harun Demir[3]

The Impact of Institutional and Macroeconomic Factors on Environmental Technology Diffusion in OECD Countries*

1 Introduction

Classification and optimal application of environmental technologies have dominated environmental studies' theoretical and empirical literature. A few studies have examined the effects of socioeconomic factors on environmental technology diffusion. Since environmental technology development is crucial to green transformation and requires global commitment, an OECD report suggests that practitioners, academics, regulators, and international agencies can learn from its determinants (OECD, 2012). Environmental technology development should be part of a global agenda to produce, provide, and distribute such technologies efficiently. For this reason, identifying the main determinants of environmental technologies using all 37 OECD countries would provide many insights and inputs for firms producing this technology, governments, and international agencies working to improve the environment.

Environmental technology drivers should include country-specific institutional and macroeconomic factors. Töbelmann and Wendler (2020) and Meirun et al. (2021) show that environmental patents cut carbon dioxide emissions and boost economic growth. Horbach et al. (2012) indicate that government regulation is crucial to decreasing harmful emissions, while Ghisetti and Pontoni (2015) discover that "only certain types of policy have proven to affect environmental innovation" –particularly regulatory stringency. Lindman and Soderholm (2016) state that technological advancements need practical applications. To determine

---

1 Ph.D., Professor, Anadolu University, Open Education Faculty, Economics, ahmettiryaki@anadolu.edu.tr, ORCID: 0000-0002-9527-7736.
2 Ph.D., Researcher, caglarkaraduman@anadolu.edu.tr, ORCID: 0000-0002-4956-6684.
3 Ph.D., Asst. Prof., Dogus University, Vocational School, Department of Foreign Trade, hdemir@dogus.edu.tr, ORCID: 0000-0001-7778-2897.
* The early-draft version of this study is verbally presented in "14th SCF International Conference on "Economic, Social, and Environmental Sustainability in the Post Covid-19 World", Antalya, Türkiye, 13 – 16 Ekim 2022".

the causes of environmental technology development, a study must contain a proxy for the practical application of learning-by-doing and theoretical breakthroughs in R&D. This study uses the Two-step system GMM and Prais-Winsten/Panel Corrected Standard Errors methods to identify the factors of environmental technology diffusion in 37 OECD countries from 2000 to 2017. Based on the theoretical and empirical literature, the study selects two groups of variables that may affect environmental patent spread. The first group includes macroeconomic variables like GDP/capita, human capital, and trade volume (openness). In contrast, the second group includes institutional quality indicators like government effectiveness, regulatory quality, corruption control, and rule of law.

As expected, both estimating approaches reveal that human capital and GDP/capita positively affect environmental patents. There is negative statistically significant relationship between openness and environmental patents. The empirical data show that government efficacy, regulatory quality, corruption control, and rule of law do not statistically affect environmental patents.

This analysis contributes to the literature by presenting cross-sectional dependency and homogeneity consistent empirical evidence for laggard institutions in most OECD nations. The findings also corroborate and underscore macroeconomic factors' deterministic impact on environmental technology dissemination. These findings can guide policymakers, practitioners, and academics in understanding the factors that influence the diffusion of environmental technology, thereby informing the development of effective strategies and policies.

2 Literature Review

Few research has examined the determinants of environmental technology, according to Brunnermeier and Cohen (2003) and Fujii and Managi (2016). We selected GDP/capita to represent total economic activity, human capital, and openness to reflect total trade volume. Most research and development (R&D) efforts that lead to innovations and adaptation occur in high-income countries. Therefore, economic activity and per capita income are significant. Education improves human capital, which drives national progress. According to Bartel and Lichtenberg (1987), highly educated people adapt to new technology more efficiently, which affects dissemination and demand. Another critical factor for the diffusion of technology is the country's trade volume.

Many studies show how macroeconomic variables affect technology diffusion. Total economic activity, measured by GDP, GDP/capita, employment, etc., is a crucial determinant in technological diffusion. For this reason, Pece et al. (2015) and Zhou and Luo (2018) stress economic growth. Mitra (2020) and Zhu et al.

(2021) report employment. Wang and Wang (2020), Murshed and Alam (2021) explore the consequences of socioeconomic factors like energy, while Saudi et al. (2019), Zameer et al. (2020) analyze environmental pollution. Human capital is a well-known factor of technological diffusion (Mayer, 2001; Keller, 2004; Fedderke, 2005; Banhabib & Spiegel, 2005) and economic growth. Zhou et al. (2021) show that education boosts environmental innovation. According to Jänicke and Jacob (2004), Jaffe et al. (2005), Popp (2006), Horbach (2008), Veugelers (2012), Quitzow (2015), and Halkos and Skouloudis (2021), institutions are crucial for the diffusion of environmentally friendly technologies. Porter's (1991) theory states that environmental laws boost economic performance by enhancing company competition. Porter and van der Linde (1995) also claim that environmental laws will boost enterprises' innovation and economic performance. However, numerous researchers stress the "inhibitory effect." This effect states that corporations can fully or partially cover their environmental costs; thus, they will maintain their technology level. Cost rises reduce efficiency.

3 Data and Methodology

This study employs an annual panel data of 37 OECD member countries over the period of 2000–2017 to examine the effects. For this purpose, we identify the selected environmental patents, obtained from OECD Database, as dependent variable based on Lanjouw and Mody (1996). The explanatory variables used in the study are the GDP/capita, the human capital index, the level of openness measured as trade to GDP ratio and four institutional quality indicators of the government effectiveness, regulatory quality, control of corruption and rule of law. These four (4) indicators of institutional quality are the country percentile ranking data, showing the relative positioning of each country in the world, from World Bank World Governances Indicators (WGI). Data on human capital index extracted from the latest Penn World Table (Feenstra et al., 2015). The index is based on the years of schooling and returns to education. This provides us with the advantage of including both a theoretical and a practical part for human capital in a single variable, fulfilling the perspective of Lindman and Soderholm (2016). Also, the level of openness is included in our model as it shows the level of integration with global markets and thus directly affects innovative capacity and tendency through technology diffusion. Similarly, the GDP per capita directly represents the effective demand while being a coarse proxy for macroeconomic performance and economic activity. The data on level of openness and GDP per capita were obtained from World Development Indicators. Table 1 lists this study's variable data sources.

Table 1: Variables used in the study

Dependent Variable	**Environmental Patents Count**	**OECD Patents Database**
Independent Variables	Trade as a percentage of GDP Human Capital Index GDP (Per Capita, PPP Constant 2017 International $)	WDI (databank.worldbank.org) PWT 10.0 (Feenstra et al., 2015) WDI (databank.worldbank.org)
	Government Effectiveness Index (Percentile Rank) Regulatory Quality Index (Percentile Rank) Control of Corruption Index (Percentile Rank) Rule of Law Index (Percentile Rank)	WGI (databank.worldbank.org) WGI (databank.worldbank.org) WGI (databank.worldbank.org) WGI (databank.worldbank.org)

In order to estimate the stated models, the estimation methodologies of (i) Two-step system GMM and (ii) Prais-Winsten Correlated Panels Corrected Standard Errors are utilized. The latter one uses panel OLS estimation and standard error correction to control for autocorrelation and cross-sectional dependency. Despite its benefits, this method cannot handle endogeneity and lacks dynamic modeling. In contrast, the GMM technique addresses the drawbacks by utilizing dynamic modeling and instrumental variables to account for endogeneity. However, GMM does not address cross-sectional dependencies. Thus, a post-estimation test for cross-sectional independence is needed to verify its robustness. System GMM is preferable over difference GMM because it preserves an additional observation by not differing. Besides, system GMM approach does not require strict exogeneity of the explanatory variables. Also, the two-step approach produces more efficient results and the bias occurring due to the utilization of two-step system is corrected by Windmeijer (2005)'s method. Lastly, our GMM approach also accounts for the year dummies and excludes any other dummies to let the STATA command xtabond2 correctly calculate the number of instruments. We estimate the relations using four models:

Model I: $lnEnvPAT_{it} = \beta_1 + \beta_2 lnTRADEg_{it} + \beta_3 lnHC_{it} + \beta_4 lnGE_{it} + \beta_5 lnGDP_{it} + \beta_6 X'_{it} + \varphi_t + u_{it}$

Model II: $lnEnvPAT_{it} = \beta_1 + \beta_2 lnTRADEg_{it} + \beta_3 lnHC_{it} + \beta_4 lnRQ_{it} + \beta_5 lnGDP_{it} + \beta_6 X'_{it} + \varphi_t + u_{it}$

Model III: $lnEnvPAT_{it} = \beta_1 + \beta_2 lnTRADEg_{it} + \beta_3 lnHC_{it} + \beta_4 lnCoC_{it} + \beta_5 lnGDP_{it} + \beta_6 X'_{it} + \varphi_t + u_{it}$

Model IV: $lnEnvPAT_{it} = \beta_1 + \beta_2 lnTRADEg_{it} + \beta_3 lnHC_{it} + \beta_4 lnRoL_{it} + \beta_5 lnGDP_{it} + \beta_6 X'_{it} + \varphi_t + u_{it}$

EPAT, TRADE, HC, and GDP represent all environmental patents, trade openness, human capital index, and GDP per capita. GE, RQ, CoC, and RoL index government effectiveness, regulatory quality, corruption control, and rule of law.

The system-GMM approach must estimate these panel models with fewer instruments than countries (Blundell & Bond 1998). Each equation had one institutional quality variable due to this constraint. Time dummies are included in estimations but not group dummies due to the same constraint. We had to assume that sample countries were homogenous for the sample period. Remember that we must exclude common factors in panel member nations. Knowing that ignorance of these standard components can lead to skewed and insignificant coefficients, we carefully picked our explanatory variables to avoid omitting essential variables and model specification errors. Next, we examined variable pairwise correlations and significance. According to pairwise correlations, multicollinearity is unlikely to be a big issue for our models, indicating they are technically feasible. We utilize Pesaran (2004) to evaluate our panel for cross-sectional dependence. A significantly cross-sectionally dependent panel is shown by rejecting the test hypothesis of cross-section independence. Pesaran's (2004) cross-sectional dependence test results are in Table 2.

Table 2: Pesaran (2004)'s Test for Cross-Sectional Dependence and Test Results of Slope Homogeneity

Test for Cross-Sectional Dependence			**Test for Homogeneity**			
CD-Test	**lnEPAT**	63.58***		**Type**	**Delta**[a]	**p-Value**
	lnTRADE	48.54***	Model I	Non-adjusted	-0.260	0.795
	lnHC	93.19***		Adjusted	-0.381	0.703
	lnGE	-0.75	Model II	Non-adjusted	-0.477	0.633
	lnRQ	-0.48		Adjusted	-0.701	0.484
	lnCoC	-1.03	Model III	Non-adjusted	0.133	0.894
	lnRoL	-0.38		Adjusted	0.195	0.845
	lnGDP	77.02***	Model IV	Non-adjusted	-0.494	0.621

(continued on next page)

Table 2: Continued

Test for Cross-Sectional Dependence	Test for Homogeneity
*** indicates a 1 % significance level. Negative values for the CD-Test are not abnormally far from 0, meaning strong independence.	Cross-sectional lags are included.

As seen in Table 2, all explanatory variables except institutional factors have high cross-sectional interdependence. This is significant because cross-sectionally dependent variables can cumulatively alter institutional elements. Thus, we must evaluate cross-sectional independence after GMM estimation to eliminate cross-sectional dependence in our analysis.

After finding substantial cross-sectional dependencies in selected models, homogeneity must be checked. First, we depict the distributions of our major institutional quality factors following log transformation. Our sample boxplots reveal that Colombia, Mexico, and Turkey have much lower institutional quality than the rest of the OECD. There are few outliers; hence, no outlier correction is needed.

As the boxplots show a discrepancy between these three nations and the rest of the OECD, we must ensure our panels are homogenous. Implementing a heterogenous method on an actual homogenous panel reduces efficiency, and imposing a homogenous approach biases estimation results. Pesaran and Yamagata's (2008) slope homogeneity results are in Table 2. According to Table 2, all four models cannot reject the null hypothesis of homogenous slope coefficients. Thus, we can ignore heterogeneity and maximize efficiency in our study.

3.1 Testing the Relationship with the Two-step System GMM

We choose system-GMM since Arellano and Bond (1991) recommend it for efficient estimation when the number of countries in the sample exceeds the number of observations (N>T). GMM estimation can be done by differencing the individual series to use the moment conditions from the estimated first or second differences of the error term (Arellano & Bond, 1991) or by using the residual levels (Arellano & Bover, 1995) and (Blundell & Bond, 1998). The difference is that the GMM approach does not assume instruments are uncorrelated with the error term, while the system GMM approach uses residuals to prevent data loss. Thus, the type of GMM to use can be determined by comparing the estimated coefficient values of the lagged dependent variable. Since the panel structure provides better explanatory power for common effects among groups, system-GMM becomes a better choice as the estimated value from a different GMM perspective approximates fixed effects (FE) estimation or diverges from pooled OLS (POLS) estimation. Thus, we tested our four pre-specified models using POLS, FE, and difference GMM estimation. The results are in Table 3.

Table 3: Lagged LogEnvPAT Coefficient Comparison for FE, POLS, and Diff. GMM

Estimator	Model I	Model II	Model III	Model IV
POLS	.9464396***	.947645***	.9492475***	.9481206***
	(.0177524)	(.0178791)	(.0177334)	(.0177957)
FE	.4297096***	.4252017***	.4321658***	.4143076***
	(.0785783)	(.0780561)	(.0782687)	(0770732)
Diff. GMM	-.3303627	-.3425977	-.3572373	-.1843974
	(.8396376)	(.8351844)	(.8354289)	(1.153141)

*** $p < 0.01$; ** $p < 0.05$; * $p < 0.10$
Standard errors are in parentheses.

Since all FE models estimate the lagged dependent variable coefficient almost the same, we used system-GMM. Using system-GMM improves estimation results and prevents data loss by differencing. Because it produces more efficient coefficients, we employed the two-step technique to estimate the model using system-GMM. According to Arellano and Bond (1991) and Blundell and Bond (1998), the two-step technique is asymptotically more efficient but produces downward biased coefficients. We use Windmeijer's (2005) corrective method to fix this.

3.2 Estimation Results from the Two-step System GMM

We estimated the models using STATA 15.1's xtabond2 function (Roodman, 2009). Tables 4 and 5 show the estimation results for models I-II and III-IV system-GMM.

Table 4: The system GMM estimation results for Models I&II

Model I Wald Chi²=17678.82 (P>Chi²=.000) Instruments/Groups: 36/37			Model II Wald Chi²=13828.87 (P>Chi²=.000) Instruments/Groups: 36/37		
Variable	**Coefficient**	z	**Variable**	**Coefficient**	z
EnvPAT(L1)	.732618*** (.056837)	12.89	**EnvPAT(L1)**	.7194748*** (.0651668)	11.04
Trade	-.626178 *** (.2086924)	-3.00	**Trade**	-.6684297*** (.2289669)	-2.92
HC	1.291742** (.6499432)	1.99	**HC**	1.596557** (.7470715)	2.14
GE	.2953502 (5755588)	.51	**RQ**	-.3882089 (.7590961)	-0.51
GDP	.5614269* (.3294733)	1.70	**GDP**	.7389719** (.3371147)	2.19
year	-.0028591*** (.0010062)	-2.84	**year**	-.0023457* (.0013871)	-1.69
Diagnostics					
Model I			**Model II**		
Arellano-Bond AR(1)		.002	Arellano-Bond AR(1)		003
Arellano-Bond AR(2)		.224	Arellano-Bond AR(2)		.223
Sargan Test of Rest.		.000	Sargan Test of Rest.		.000
Hansen Test of Rest.		.327	Hansen Test of Rest.		.317
Variance Explained by the Components		.984	Variance Explained by the Components		.984
KMO Measure of Sampling Adequacy		.883	KMO Measure of Sampling Adequacy		.883

1) *** $p < 0.01$; ** $p < 0.05$; * $p < 0.10$
2) Heteroskedasticity consistent standard errors are in parentheses.
3) Forward orthogonal deviations transformed

Table 5: The system GMM estimation results for Models III&IV

Model III **Wald Chi^2=18176.19** **(P>Chi^2=.000)** **Instruments/Groups: 36/37**			**Model IV** **Wald Chi^2=20888.81** **(P>Chi^2=.000)** **Instruments/Groups: 36/37**		
Variable	**Coefficient**	**z**	**Variable**	**Coefficient**	**z**
EnvPAT(L1)	.7226788*** (.0648855)	11.14	**EnvPAT(L1)**	.7254207*** (.0657813)	11.03
Trade	-.6659576*** (.238991)	-2.79	**Trade**	-.6476294*** (.2318786)	-2.79
HC	1.480509** (.6726294)	2.20	**HC**	1.42762** (.7102393)	2.01
CoC	-.1276676 (.317911)	-.40	**RoL**	.0077606 (.3371518)	.02
GDP	.7010228** (.3275127)	2.14	**GDP**	.6563909* (.3596081)	1.83
year	-.0026722** (.0010576)	-2.53	**year**	-.0027426** (.0010759)	-2.55
Diagnostics					
Model III			**Model IV**		
	Arellano-Bond AR(1)	002		Arellano-Bond AR(1)	.002
	Arellano-Bond AR(2)	.225		Arellano-Bond AR(2)	.224
	Sargan Test of Rest.	.000		Sargan Test of Rest.	.000
	Hansen Test of Rest.	.318		Hansen Test of Rest.	.318
	Variance Explained by the Components	.984		Variance Explained by the Components	.984
	KMO Measure of Sampling Adequacy.	.883		KMO Measure of Sampling Adequacy	.883

1) *** $p < 0.01$; ** $p < 0.05$; * $p < 0.10$
2) Heteroskedasticity consistent standard errors are in parentheses.
3) Forward orthogonal deviations transformed

The estimation diagnostics for all four models demonstrated no second-order serial correlations, making the results robust with 36 instruments. PCA was used to reduce the number of instruments, making explained variance relevant. All model PCA values (.984) confirmed the specifications. Finally, the KMO measure of sample adequacy showed merit. Since our models use percentile ranks to better capture relative changes among OECD members and percentile ranks are direct measures (higher values indicate normatively better conditions), we expect the institutional factor coefficients to be positive by hypothesizing that higher institutional quality improves environmental innovation. Table 3 shows that government effectiveness and regulatory quality were insignificant, around absolute – 50 z values. Control of corruption declined to 40 and the rule of law to 02. All four models showed a favorable and strong influence on human capital. While trade openness was negative, GDP per capita was continuously considerable and positive. Carefully analyze institutional variables' constant insignificance. Institutions are essential for a country to join the highly developed, technologically advanced nations. The findings may be due to the regulatory system's lag behind technology. GMM estimation results indicate that institutional variables are not influencing the development path of environmental technologies but rather adapting their internal structure to them as they grow. According to Glaeser et al. (2004), well-known macroeconomic causes of economic growth have significantly greater influence than institutions, and successful macroeconomic policies often lead to well-established institutions.

It could also be the "inhibitory effect." If corporations pay environmental expenses, they will maintain their technology level, reducing efficiency owing to cost increases. As regulatory costs rise, the benefit impacts of institutional quality increases may be negated, making test results meaningless. We found the opposite of Halkos and Skouloudis (2021), who used FE, RE, and GLS estimators. These methods work well under slope homogeneity assumption but fail miserably without cross-sectional dependence. We must evaluate the system-GMM estimation residuals for cross-sectional dependence to determine the robustness of our results. We use Pesaran's (2015) weak cross-sectional dependence technique, as shown in Table 6.

Table 6: Pesaran's (2015) Test for Weak Cross-Sectional Dependence

	Model I	**Model II**	**Model III**	**Model IV**
CD-Test	-1.571	-1.613	-1.634	-1.670*

* Indicates a 10 % significance level.

The derived test statistic for Models I, II, and III shows a weak cross-sectional dependence. However, Model IV's null hypothesis (errors are weakly cross-sectional dependent) can be rejected at 10 %.

3.3 Estimation Results from the Regression with the Prais-Winsten Regression/Panel Corrected Standard Errors (PCSEs)

Beck and Katz's Prais-Winsten estimation (1995) can also address cross-sectional dependence. According to Moundigbaye et al. (2018), the approach is optimal for hypothesis testing. The Prais-Winsten Regression/Panel Corrected Standard Errors (PCSEs) approach yields estimation results similar to GMM testing. Prais-Winsten estimation findings are in Table 7.

Table 7: Prais-Winsten Regression (PCSE)

	Model I		Model II		Model III		Model IV	
Variables	**Coef. (Std. Err.)**	**P > |t|**	**Coef. (Std. Err.)**	**P > |t|**	**Coef. (Std. Err.)**	**P > |t|**	**Coef. (Std. Err.)**	**P > |t|**
lnTRADE	-1.996551 (.1967133)	0.000	-2.060565 (.1838411)	0.000	-2.043669 (.1917595)	0.000	-2.010402 (.1941455)	0.000
lnHC	4.303986 (.8693968)	0.000	4.751826 (.8302818)	0.000	4.453575 (.8197321)	0.000	4.304986 (.8688493)	0.000
lnGE	.2604299 (.5297315)	0.623						
lnRQ			-.913165 (.6399334)	0.154				
lnCoC					-.0387163 (.3449672)	0.911		
lnRoL							.1726173 (.4120284)	0.675
lnGDP	2.430783 (.3088158)	0.000	2.657393 (.300634)	0.000	2.518826 (.2991099)	0.000	2.438236 (.3357479)	0.000

Common AR(1) implemented
Corrected SEs for correlated panels

4 Conclusion and Policy Implications

We observed that institutional measures do not statistically affect environmental innovation using two estimators for the OECD and controlled for trade openness, GDP per capita, and human capital. GDP per capita and human capital have statistically significant effects on environmental patents, but trade openness has a negative influence. Even when using PCSE to account for cross-sectional interdependence, these effects continue. Our findings support Glaeser et al. (2004)'s claim that institutions adapt to innovations. We believe that OECD nations, especially those with weaker institutional quality, should focus primarily on institutional aspects to reduce costs, channel economic resources, and enable technology adoption to spur innovation.

References

Arellano, M., & Bond, S. (1991). Some tests of specification for panel data: Monte Carlo evidence and an application to employment equations. *The Review of Economic Studies*, *58*, 277–297.

Arellano, M., & Bover, O. (1995). Another look at the instrumental variable estimation of error-components models. *Journal of Econometrics*, *68*, 29–51.

Bartel, A. P., & Lichtenberg, F. R. (1987). The comparative advantage of educated workers in implementing new technology. *The Review of Economics and Statistics*, 69, 1.

Beck, N., & Katz, J. N. (1995). What to do (and not to do) with time-series cross-section data. *American Political Science Review*, *89*, 634–647.

Benhabib, J., & Spiegel, M. (2005). Human capital and technology diffusion. *Handbook of Economic Growth*, *1*, 935–966.

Blundell, R., & Bond, S. (1998). Initial conditions and moment restrictions in dynamic panel data models. *Journal of Econometrics*, *87*, 115–143.

Brunnermeier, S. B., & Cohen, M. A. (2003). Determinants of environmental innovation in US manufacturing industries. *Journal of Environmental Economics and Management*, *45*(2), 278–293. https://doi.org/10.1016/S0095-0696(02)00058-X.

Fedderke, J. W. (2005). "Technology, Human Capital and Growth," Working Papers 027, Economic Research Southern Africa.

Feenstra, R. C., Inklaar, R., & Timmer, M. P. (2015). The Next Generation of the Penn World Table. *American Economic Review*, *105*(10), 3150–3182, available for download at www.ggdc.net/pwt

Fujii, H., & Managi, S. (2016). Research and development strategy for environmental technology in Japan: A comparative study of the private and public sectors. *Technological Forecasting and Social Change*, *112*, 293–302. https://doi.org/10.1016/j.techfore.2016.02.012.

Ghisetti, C., & Pontoni, F. (2015). Investigating policy and R&D effects on environmental innovation: A meta-analysis. *Ecological Economics*, *118*, 57–66.

Glaeser, E. L., la Porta, R., Lopez-de-Silanes, F., & Shleifer, A. (2004). Do institutions cause growth?. *Journal of Economic Growth*, *9*, 271–303.

Halkos, G., & Skouloudis, A. (2021). Environmental technology development and diffusion: panel data evidence from 56 countries. *Environmental Economics and Policy Studies*, 23, 79–92

Horbach, J. (2008). Determinants of environmental innovation: New evidence from German panel data sources. *Research Policy*, *37*, 163–173.

Horbach, J., Rammer, C., & Rennings, K. (2012). Determinants of eco-innovations by type of environmental impact: The role of regulatory push/pull, technology push and market pull. *Ecological Economics, 78*, 112–122.

Jaffe, A. B., Newell, R. G., & Stavins, R. N. (2005). A tale of two market failures: Technology and environmental policy. *Ecological Economics*, *54*(2–3), 164–174.

Jänicke, M., & Jacob, K. (2004). Lead markets for environmental innovations: A new role for the nation state. *Global Environmental Politics*, *4*, 29–47.

Keller, W. (2004). International Technology Diffusion. *Journal of Economic Literature*, 42(3), 752–782. https://doi.org/10.1257/0022051042177685.

Lanjouw, J. O., & Mody, A. (1996). Innovation and the International Diffusion of Environmentally Responsive Technology. *Research Policy*, *25*, 549–571.

Lindman, Å., & Söderholm, P. (2016). Wind energy and green economy in Europe: Measuring policy-induced innovation using patent data. *Applied Energy*, *179*, 1351–1359.

Mayer, J. (2001). Technology Diffusion, Human Capital and Economic Growth in. Developing Countries, No 154, UNCTAD Discussion Papers.

Meirun, T., Mihardjo, L. W., Haseeb, M., Khan, S. A. R., & Jermsittiparsert, K. (2021). The dynamics effect of green technology innovation on economic growth and CO_2 emission in Singapore: New evidence from bootstrap ARDL approach. *Environmental Science and Pollution Research*, *28*, 4184–4194. https://doi.org/10.1007/s11356-020-10760-w.

Mitra, A. (2020). Technological progress, innovation and employment: Firm-level evidence from India's manufacturing sector. *Innovation and Development*, *10*(3), 451–465. https://doi.org/10.1080/2157930X.2019.1689614.

Moundigbaye, M., Rea, W. S., & Reed, W. R. (2018). Which Panel Data Estimator Should I Use?: A Corrigendum and Extension, Economics: The Open- Access, Open-Assessment E-Journal, ISSN 1864-6042, Kiel Institute for the World Economy (IfW), Kiel, Vol. 12, Iss. 2018-4, pp. 1–31.

Murshed, M., & Alam, M. S. (2021). Estimating the macroeconomic determinants of total, renewable, and non-renewable energy demands in Bangladesh: the role of technological innovations. *Environmental Science and Pollution Research*, 28, 30176–30196.

Organisation for Economic Cooperation and Development (OECD). (2012). The future of eco-innovation: The role of business models in green transformation. OECD background paper, OECD, Copenhagen.

Pece, A. M., Simona, O. E. O., & Salisteanu, F. (2015). Innovation and economic growth: An empirical analysis for CEE countries. *Procedia Economics and Finance*, 26, 461–467. https://doi.org/10.1016/S2212-5671(15)00874-6.

Pesaran, H. M. (2015). Testing weak cross-sectional dependence in large panels. *Econometric Reviews, 34*, 1089–1117.

Pesaran, M. H. (2004). General diagnostic tests for cross section dependence in panels. CESifo Working Paper Series No. 1229; IZA Discussion Paper No. 1240. Available at SSRN: http://ssrn.com/abstract=572504.

Pesaran, M. H., & Yamagata, T. (2008). Testing slope homogeneity in large panels. *Journal of Econometrics*, *142*(1), 50–93.

Popp, D. (2006). International innovation and diffusion of air pollution control technologies: The effects of NOx and SO2 regulation in the US, Japan, and Germany. *Journal of Environmental Economics and Management*, *51*(1), 46–71.

Porter, M. (1991). America's green strategy. *Scientific American*, *264*, 168.

Porter, M., & van der Linde, C. (1995). Toward a new conception of the environment-competitiveness relationship. *Journal of Economic Perspectives*, *9*(4), 97–118. https://doi.org/10.1257/jep.9.4.97.

Quitzow, R. (2015). Assessing policy strategies for the promotion of environmental technologies: A review of India's National Solar Mission. *Research Policy*, *44*(1), 233–243.

Roodman, D. (2009). How to do xtabond2: An introduction to difference and system GMM in Stata. *The Stata Journal*, 9(1), 86–136.

Saudi, M. H. M., Sinaga, O., & Jabarullah, N. H. (2019). The role of renewable, non-renewable energy consumption and technology innovation in testing environmental Kuznets curve in Malaysia. *International Journal of Energy Economics and Policy*, *9*(1), 299–307. https://doi.org/10.32479/ijeep.7327.

Töbelmann, D., & Wendler, T. (2020). The impact of environmental innovation on carbon dioxide emissions. *Journal of Cleaner Production*, *244*, 118787. https://doi.org/10.1016/j.jclepro.2019.118787.

Veugelers, R. (2012). Which policy instruments to induce clean innovating? *Research Policy*, *41*(10), 1770–1778.

Wang, H., & Wang, M. (2020). Effects of technological innovation on energy efficiency in China: Evidence from dynamic panel of 284 cities. *Science of the Total Environment*, *709*, 136172. https://doi.org/10.1016/j.scitotenv.2019.136172.

Windmeijer, F. (2005). A finite sample correction for the variance of linear efficient two-step GMM estimators. *Journal of Econometrics*, *126*, 25–51.

Zameer, H., Yasmeen, H., Zafar, M. W., Waheed, A., & Sinha, A. (2020). Analyzing the association between innovation, economic growth, and environment: Divulging the importance of FDI and trade openness in India. *Environmental Science and Pollution Research*, *27*, 29539–29553. https://doi.org/10.1007/s11356-020-09112-5.

Zhou, G., & Luo, S. (2018). Higher education input, technological innovation, and economic growth in China. *Sustainability*, *10*(8), 2615. https://doi.org/10.3390/su10082615.

Zhou, M., Chen, F., & Chen, Z. (2021). Can CEO education promote environmental innovation: Evidence from Chinese enterprises. *Journal of Cleaner Production*, 297, 126725. https://doi.org/10.1016/j.jclepro.2021.126725.

Zhu, C., Qiu, Z., & Liu, F. (2021). Does innovation stimulate employment? Evidence from China. *Economic Modelling*, *94*, 1007–1017. https://doi.org/10.1016/j.econmod.2020.02.041

Cumali Marangoz[1] & Serap Coban[2]

Macro-Economic Determinants of FDI: A Comparative Literature Review of Türkiye & International Studies[3]

1 Introduction

Since the late 20th century, increasing globalization trends, along with the development of technology & communication tools, have tightened social, economic, & political ties between countries. In this context, some economies that position themselves as part of an open economic system in the global economy are extending their production capabilities beyond their national borders. Additionally, many economies aim to become attractive destinations for foreign investments to expand the production capacity they need to enhance global competitiveness or achieve economic development.

Indeed, in achieving these objectives FDI are of great importance, especially for economies with domestic savings deficits or limited machinery, equipment, & technology necessary for converting existing savings into investments. This is because FDI plays a crucial role in increasing employment, exports, and economic growth rates, as well as in the development of technological infrastructure and human capital in achieving sustainable development for host countries. Consequently, it is known that many regulatory policies are pursued to make the economy a magnet for FDI, particularly in developing countries.

As a result of these policies, according to data published by the United Nations Conference on Trade and Development (UNCTAD, 2019), investments in developing countries constituted 44 % of all FDIs in 2019. Despite this, the Turkish economy can attract only 1.23 % of these investments. With its demographic structure & strategic position in the global economy, Türkiye has the potential to make significant contributions to the world economy. For Türkiye to escape

1 Ph.D., Asst. Prof., Ağrı İbrahim Çeçen University, Faculty of Economics, Department of Economics, cmarangoz@agri.edu.tr, ORCID: 0000-0002-0216-3490.

2 Ph.D., Assoc. Prof., Nevşehir Hacı Bektaş Veli University, Faculty of Economics, Department of Economics, ORCID: 0000-0002-6622-0327.

3 This study is derived from a doctoral dissertation.

the middle-income trap & join the group of high-income countries, FDIs are critically important.

For this purpose, this study compares both domestic and international research on the determinants of foreign direct investment (FDI) to accurately identify these determinants for Türkiye, highlighting the differences and similarities.

2 Theoretical Framework: Determinants of Foreign Direct Investments

Early studies on the determinants of FDI are observed to examine FDIs within a limited theoretical framework without empirical results (Faeth, 2009; Lall, 1978). The first of these theories is the Heckscher-Ohlin model, which is based on the Neo-Classical trade theory. This model assumes that capital will move to countries where the return on capital is higher and the return on labor is lower due to changes in the relative factor intensities of commodities & the relative factor abundance of countries *Kılıçarslan (2018).*

Over time, although different theoretical versions of the Neo-classical approach, based on perfect competition market conditions, have emerged, such as market size, portfolio diversification and monetary area hypotheses, these versions are also seen to be inadequate in explaining the determinants of FDIs.Since the Neo-classical approach falls short in explaining the reasons for FDIs, Kindleberger (1969) & Hymer (1976) introduced the concept of imperfect competition to better explain the factors that firms consider when investing abroad. Theories of FDIs based on the assumption of imperfect competition include the Hymer-Kindleberger hypothesis, the product life cycle hypothesis, the oligopolistic reaction hypothesis, the internalization hypothesis and the eclectic paradigm. These theories assume that firms investing in a foreign country must have certain comparative advantages over local firms in the host country (Agarwal, 1980). According to Hymer, these advantages are "firm-specific advantages or monopolistic advantages in Kindleberger's terms, such as patent-protected superior technology, brand names, marketing & managerial skills, economies of scale, & cheaper sources of finance."These advantages are the main determinants of theories developed to explain FDIs under imperfect competition.

Although the above-mentioned theories have attempted to explain the determinants of FDIs, the clear emergence of development differences between countries over time, changes in stability environments depending on economic & political developments, & differences in basic input costs have diversified the factors determining FDIs (Mucuk, 2011).

3 Empirical Literature: Determinants of Foreign Direct Investments

In the literature, there are many studies focusing on the factors affecting the flow of FDIs to Türkiye and other countries or country groups. These studies examine the effects of macroeconomic variables such as market size, labor & energy costs, exchange rates on FDIs. Some of these variables are based on the theories explained above, while others are included in models because they are intuitively meaningful. A large portion of these variables, included in UNCTAD's (1998) classification of inward FDI determinants & frequently used in empirical studies, are explained below.

3.1 Market Size

The market size variable is one of the most frequently used variables in all applied studies on FDI determinants. This is because market size offers an opportunity for more effective use of economies of scale & potential demand (Dunning, 1980; Chakrabati, 2001; Demirhan & Masça, 2008; Ang, 2008; Bilgili et al., 2012; Blonigen, & Piger, (2018). Studies show that market size, measured by GDP, GDP per capita, or GNP, plays a positive role in explaining inward FDI for both different countries or country groups & Türkiye. In comparing the international and domestic literature on the impact of market size on FDI, both sets of studies demonstrate a significant correlation between market size and FDI. In the international literature, Dunning (1980), Chakrabarti (2001), Ang (2008), Masça and Demirhan (2008), Kök and Ersoy (2009), Vijayakumar et al. (2010), Ranjan and Agrawal (2011), Jadhav (2012), Castro et al. (2013), and Kumari and Sharma (2017) have all found positive results.In the literature specific to Türkiye, Erdal and Tatoğlu (2002), Kaya and Yılmaz (2003), Deichman et al. (2003), Yapraklı (2006), Açıkalın et al. (2006), Vergil and Çeştepe (2006), Batmaz and Tunca (2007), Kurtaran (2007), Kar and Tatlısöz (2008), Tarı and Bıdırdı (2009), Koyuncu (2010), Lebe and Ersungur (2012), Bilgili et al. (2012), Emir et al. (2013), Polat (2015), Çiftci and Yıldız (2015), Aydemir and Genc (2015), and Aytekin (2019) have all found positive results.

3.2 Labor Costs

One of the economic factors affecting FDI is labor cost. Foreign investors aim to minimize production costs; thus, relatively higher labor costs in the host country can deter FDI. However, studies examining the relationship between labor costs & FDI have produced conflicting results in the literature. Some studies that find a negative relationship between FDI & labor costs suggest that, all else being equal, high labor costs reduce the profitability of firms & thus negatively affect their FDI decisions (Lansbury et al., 1996; Galego et al., 2004; Johnson, 2006; Bilgili et al., 2012; Walch & Worz, 2012).

On the other hand, studies that find a positive or insignificant relationship between FDI & labor costs argue that, besides labor costs, other factors such as labor productivity, transportation costs, & energy prices also affect firms' profitability, and thus labor costs alone cannot determine firms' overall costs (Wheeler & Mody, 1992; Wang and Swain, 1995; Chakrabati, 2001; Bénassy & Lahrèche-Révil, 2005; Kinoshita & Campos, 2006).

In comparing the international and domestic literature on the impact of labor cost on FDI, the findings show varied correlations. In the international literature, Caves (1974) and Wheeler and Mody (1992) found positive results. Wang and Swain (1995), Bénassy et al. (2005), Kinoshita and Campos (2006), and Demirhan and Masça (2008) found neutral results. Lansbury et al. (1996), Galego et al. (2004), Vijayakumar et al. (2010), Ranjan and Agrawal (2011), and Walch and Worz (2012) found negative results. In the literature specific to Türkiye, Kaya and Yılmaz (2003), Yapraklı (2006), Batmaz and Tunca (2007), Kar and Tatlısöz (2008), Tarı and Bıdırdı (2009), and Bilgili et al. (2012) found negative results, whereas Aydemir and Genç (2015) found positive results. Moreover, studies specific to the Turkish economy generally find a negative relationship between FDI inflows & labor costs. An important reason for this is that firms operating in the Turkish economy are more labor-intensive rather than capital-intensive.

3.3 Economic Growth

When examining the empirical literature, it is evident that FDI & economic growth, defined as the annual increase in Gross Domestic Product (GDP), are two interrelated factors. The role of economic growth in attracting FDI is closely associated with the economic development level of countries. (Lunn, 1980; Schneider & Frey, 1985; Culem, 1988) found that growth has a significantly positive impact on FDI in developing countries. In contrast, Borenzstein et al. (1998) did not find a significantly positive impact of growth on FDI in developed economies, & Nigh (1985) found no significant impact in less developed countries. The primary reason for this is that a rapidly growing economy offers relatively better profit opportunities compared to slow-growing or stagnant economies (Charkrabarti, 2001; Lim, 1983). In comparing the international and domestic literature on the impact of economic growth on FDI, the findings are generally positive.In the international literature, Lunn (1980), Schneider and Frey

(1985), Culem (1988), De Angelo et al. (2010), Bhavan et al. (2011), and Saini and Singhania (2018) found positive results, while Nigh (1985) and Borenztein et al. (1998) found neutral results.In the literature specific to Türkiye, Erdal and Tatoğlu (2002) and Bilgili et al. (2012) found positive results.

3.4 Exchange Rate

The impact of exchange rate fluctuations on FDI is based on modern literature, rooted in Aliber's (1970) Monetary Area Hypothesis. The Monetary Area Hypothesis assumes capital markets are flawed, suggesting borrowers will face a premium for external borrowing. In this environment, foreign investors believe their relative wealth will increase, & input costs will decrease in terms of the host country's currency, potentially leading to depreciation of the host country's currency, which could increase FDI inflows & allow for more internal financing of investments. For example, if a foreign investor competes with local investors to acquire a local company, a depreciation of the source country's currency may increase the reservation price for the foreign investor, potentially outbidding local competitors. Froot and Stein (1991) were among the first to study this concept, finding that a decline in the value of the dollar reduces investment costs, prompting investors to prefer different countries.

Despite numerous studies supporting this hypothesis (see Blonigen, 1995; Blonigen & Feenstra, 1996), the assumption of market imperfections fails to fully explain how different types of FDIs are affected differently by exchange rate fluctuations. According to Blonigen (1997), investors may not have equal access to all markets, leading exchange rate fluctuations to affect the relative returns obtained by investors in the source and host countries to varying degrees. Additionally, activities involving the acquisition of firm-specific assets may be more sensitive to exchange rate movements.

Therefore, while the literature often argues that depreciation of the host country's currency supports FDI inflows (Ravenscraft, 1991; Blonigen, 1997; Buch & Kleinert, 2008), this hypothesis is not universally supported by some studies (Stevens, 1998; Healy & Palepu, 1993). In the international literature, Froot and Stein (1991), Blonigen and Feenstra (1996), and Stevens (1998) found negative results, while Harris and Ravenscarft (1991) and Buch and Kleinert (2008) found positive results.

In the literature specific to Türkiye et al. (2002), Yapraklı (2006), Batmaz and Tunca (2007), and Kar and Tatlısöz (2008) found negative results, while Vergil and Çeştepe (2006) found positive results. Emir et al. (2013) found neutral results. This indicates that international investors are sensitive to exchange rate volatility and prioritize economic stability.

3.5 Trade Openness

The positive relationship between FDI & the volume of foreign trade is considered an indicator of efforts by countries desiring more FDI to increase their foreign trade. The existing literature suggests that trade openness is directly associated with FDI in the host country, but its direct impact on FDI can vary depending on whether the investment is aimed at market-seeking or export-oriented activities (Mudiyanselage et al., 2021). According to the tariff avoidance hypothesis, economies with trade restrictions may have a positive impact on FDI (Market Seekers), while multinational companies oriented towards exports tend to prefer more open economies. This is because trade protectionism generally implies higher transaction costs associated with exports. Most of the FDI literature has used the share of trade in GDP as an indicator of openness (Demirhan & Masça, 2008; Ting & Tang, 2010; Leitao & Faustino, 2010; Leitao, 2010; Bhavan et al., 2011; Bilgili et al., 2012; Donghui et al., 2018). However, recent studies suggest that trade openness can lead to a better business environment through increased expectations of economic growth, thereby potentially encouraging more FDI in open economies (Lim, 2001). Nevertheless, some studies in the literature do not support the hypotheses mentioned above (Kumari & Sharma, 2017). In comparing the impact of openness on Foreign Direct Investment (FDI) between international and Turkish literature, several patterns emerge. International studies consistently show a positive correlation between openness and FDI. Masça and Demirhan (2008), Aw and Tang (2010), Leitao and Faustino (2010) across EU countries, Leitao (2010) across OECD countries, Ranjan and Agrawal (2011) for BRIC nations, Bhavan et al. (2011) for South Asian economies, Jadhav (2012) for BRIC nations, and Donghui et al., 2018 for selected Asian countries all find positive relationships. In contrast, Turkish studies also generally find a positive correlation between openness and FDI, as observed by Erdal and Tatoğlu (2002), Yapraklı (2006), Batmaz and Tunca (2007), Kurtaran (2007), Kar and Tatlısöz (2008), Tarı and Bıdırdı (2009), Koyuncu (2010), Lebe and Ersungur (2012), and Aydemir and Genç (2015). However, some Turkish studies, such as Emir et al. (2013) and Çiftçi & Yıldız (2015), report neutral results, indicating no significant impact of openness on FDI. Overall, while both international and domestic literature generally support a positive relationship between openness and FDI, nuances exist in some domestic findings regarding neutrality.

4 Conclusion

Regardless of their level of development, achieving strong and sustainable growth in all country economies depends on expanding their production capacities. To expand their production capacities, countries initially need to have sufficient levels of domestic savings and require the machinery, equipment, and technology necessary to convert these savings into investments. These requirements are particularly met by foreign capital investments in underdeveloped or developing countries where domestic savings are insufficient and technological development is limited, which are the primary sources of financing for investment and production. Therefore, many developing country economies implement various regulatory policy measures to attract foreign direct investments. In this regard, accurately identifying the investment criteria of multinational companies is of great importance for a country like Türkiye, whose development and progress depend on successfully integrating into the global economy. For this purpose, this study provides a comparative analysis of domestic and international research findings, highlighting differences through literature review.

In comparing the determinants of FDI between domestic and international literature, our analysis reveals several key insights. Firstly, both sets of literature consistently identify economic growth, market size and trade openness as significant contributors to FDI, with similar findings across both contexts. These determinants are universally acknowledged for their impact on increasing FDI. Secondly, the effect of labor cost on FDI presents contrasting results. International literature predominantly finds mixed results, indicating varying correlations between labor cost and FDI decisions. Conversely, domestic studies often highlight a negative correlation, indicating that labor cost in the national context is associated with firms operating in the Turkish economy being more labor-intensive rather than capital-intensive. Finally, the variable of FDI highlights a notable distinction between domestic and international perspectives. Domestic literature suggests that international investors are particularly sensitive to exchange rate volatility and prioritize economic stability, leading to less environmental degradation. This contrasts with international findings that show a more neutral impact of FDI on environmental outcomes. Overall, the comparative analysis underscores that while some determinants of FDI are consistently recognized across both domestic and international literature, there are key areas where perspectives diverge, reflecting different priorities and sensitivities in varying contexts.

References

Açıkalın, S., Gül, E., & Yaşar, E. (2006). Ücretler ve büyüme ile doğrudan yabancı yatırımlar arasındaki ilişkinin ekonometrik analizi. *Dumlupınar Üniversitesi Sosyal Bilimler Dergisi, 16*, 271–282.

Agarwal, J. P. (1980). Determinants of foreign direct investment: A survey.

Aliber, R. Z. (1970). A theory of direct foreign investment. The international corporation, 12–36.

Ang, J. B. (2008). Determinants of foreign direct investment in Malaysia. *Journal of Policy Modeling, 30*(1),

Aw, Y. T., & Tang, T. C. (2010). The determinants of inward foreign direct investment: The case of Malaysia. *International Journal of Business and Society, 11*(1), 59–76.

Aydemir, O., & Genç, E. (2015). Uluslararası sermaye hareketlerinin belirleyicileri, doğrudan yabancı yatırımlar kapsamında bir analiz, Türkiye örneği. *Optimum ekonomi ve Yönetim bilimleri Dergisi, 2*(2), 17–41.

Aytekin, G. K. (2019). Doğrudan Yabancı Yatırımlar DYY, Belirleyicileri ve Ekonomik Etkileri. *Ufuk Üniversitesi Sosyal Bilimler Enstitüsü Dergisi, 8*(15), 255–278.

Batmaz, N., & Tunca, H. (2007). Türkiye'de Doğrudan Yabancı Sermaye Yatırımlarının Bölgesel Belirleyicileri Üzerine Bir Eş Bütünleşme Analizi (1992–2003). *Sosyal Bilimler Araştırmaları Dergisi, 2*(1), 199–224.

Bénassy-Quéré, A., Fontagné, L., ve Lahrèche-Révil, A. (2005) How does FDI react to corporate taxation? *International Tax and Public Finance*, 12(5), 583–603.

Bhavan, T., Xu, C., & Zhong, C. (2011). The relationship between foreign aid and FDI in South Asian economies. *International Journal of Economics and Finance, 3*(2), 143–149.

Bilgili, F., Tülüce, N. S. H., ve Doğan, İ. (2012). The determinants of FDI in Turkey: A Markov regime-switching approach. *Economic Modelling, 29*(4), 1161–1169.

Blonigen, B. A. (1995) Explaining Japanese foreign direct investment in the United States (Doctoral dissertation, University of California, Davis).

Blonigen, B., & Feenstra, R. C. (1996). Protectionist threats and foreign direct investment.

Blonigen, B. A. (1997). Firm-specific assets and the link between exchange rates and foreign direct investment. The American Economic Review, 447–465.

Blonigen, B. A., & Piger, J. (2014). Determinants of foreign direct investment. Canadian Journal of Economics, 47(3), 775–812.

Borensztein, E., De Gregorio, J., & Lee, J. W. (1998). How does foreign direct investment affect economic growth?. *Journal of International Economics, 45*(1), 115–135.

Buch, C. M., & Kleinert, J. (2008). Exchange rates and FDI: Goods versus capital market frictions. *World Economy, 31*(9), 1185–1207.

Caves, R. E. (1974). Multinational firms, competition, and productivity in host-country markets. *Economica, 41*(162), 176–193.

Chakrabarti, A. (2001). The determinants of foreign direct investments: Sensitivity analyses of cross-country regressions. *Kyklos, 54*(1), 89–114.

Çiftci, F., & Yıldız, R. (2015). Doğrudan Yabancı Yatırımların Ekonomik Belirleyicileri: Türkiye Ekonomisi Üzerine Bir Zaman Serisi Analizi. *Business & Economics Research Journal, 6*(4).

Culem, C. G. (1988). The locational determinants of direct investments among industrialized countries. *European Economic Review, 32*(4), 885–904.

De Angelo, C. F., Eunni, R. V., & Fouto, N. M. M. D. (2010). Determinants of FDI in emerging markets: evidence from Brazil. *International Journal of Commerce and Management, 20*(3), 203–216.

De Castro, P. G., Fernandes, E. A., & Campos, A. C. (2013). The determinants of foreign direct investment in Brazil and Mexico: an empirical analysis. *Procedia Economics and Finance, 5*, 231–240.

Demirhan, E., & Masca, M. (2008). Fulltext search in archive. *Prague economic papers, 17*(4), 356–369.

Deichmann, J., Karidis, S., & Sayek, S. (2003). Foreign direct investment in Turkey: Regional determinants. *Applied Economics, 35*(16), 1767–1778.

Demirhan, E., ve Masca, M. (2008). Determinants of foreign direct investment flows to developing countries: A cross-sectional analysis. *Prague Economic Papers, 4*(4), 356–369.

Donghui, Z., Yasin, G., Zaman, S., & Imran, M. (2018). Trade openness and FDI inflows: A comparative study of Asian countries. *European Online Journal of Natural and Social Sciences, 7*(2), pp. 386.

Dunning, J. H. (1980). Toward an eclectic theory of international production: Some empirical tests. In The Eclectic Paradigm: A Framework for Synthesizing and Comparing Theories of International Business from Different Disciplines or Perspectives (pp. 23–49). London: Palgrave Macmillan UK.

Emir, M., Uysal, M., & Doğru, B. (2013). Ülkenin risklilik durumu ile ülkeye gelen doğrudan yabancı yatırım arasındaki ilişki: Türkiye örneği.

Erdal, F., & Tatoglu, E. (2002). Locational determinants of foreign direct investment in an emerging market economy: Evidence from Tukey. *Multinational Business Review, 10*, 21–27.

Faeth, I. (2009). Determinants of foreign direct investment–a tale of nine theoretical models. *Journal of Economic Surveys, 23*(1), 165–196.

Felisoni de Angelo, C., Eunni, R. V., & Manoel Martins Dias Fouto, N. (2010). Determinants of FDI in emerging markets: Evidence from Brazil. *International Journal of Commerce and Management, 20*(3), 203–216.

Froot, K. A., & Stein, J. C. (1991). Exchange rates and foreign direct investment: An imperfect capital markets approach. *Quarterly Journal of Economics, 106*, 1191–1217.

Galego, A., Vieira, C., ve Vieira, I. (2004) The CEEC as FDI attractors: A menace to the EU periphery? *Emerging Markets Finance and Trade, 40*(5), 74–91.

Harris, R. S., & Ravenscraft, D. (1991). The role of acquisitions in foreign direct investment: Evidence from the US stock market. *The Journal of Finance, 46*(3), 825–844.

Healy, P. M., ve Palepu, K. G. (1993). The effect of firms' financial disclosure strategies on stock prices. Accounting horizons, 7(1), 1.

Hymer, S. (1976) The International Operations of National Firms: A Study of Foreign Direct Investment. MIT Press, Cambridge, MA.

Jadhav, P. (2012). Determinants of foreign direct investment in BRICS economies: Analysis of economic, institutional and political factor. *Procedia-Social and Behavioral Sciences.*

Johnson, A. (2006). The effects of FDI inflows on host country economic growth. *The Royal Institute of technology. Centre of Excellence for studies in Science and Innovation* http://www.infra.kth.se/cesis/research/publications/working papers, 5.

Kar, M. ve Tatlisöz, F. (2008). Türkiye'de Doğrudan Yabanci Sermaye Hareketlerini Belirleyen Faktörlerin Ekonometrik Analizi. Karamanoğlu Mehmet Bey Üniversitesi Sosyal Ve Ekonomik Araştırmalar Dergisi, 2008(1), 436–458.

Kaya, V., & Yılmaz, Ö. (2003). Türkiye'de Doğrudan Yatirimlarin Belirleyicileri: 1970–2000. Atatürk Üniversitesi İktisadi ve İdari Bilimler Dergisi, 17(3–4).

Kiliçarslan, Z. (2018). The relationship between exchange rate volatility & foreign direct investment in Turkey: Toda & Yamamoto Causality Analysis. *International Journal of Economics & Financial Issues, 8*(4), 61–67.

Kinoshita, Y., & Campos, N. F. (2006). Determinants and effects of foreign direct investment in transition economies. In *Managing development* (pp. 261–285). Routledge.

Kok, R., & Acikgoz Ersoy, B. (2009). Analyses of FDI determinants in developing countries. *International Journal of Social Economics, 36*(1/2), 105–123.

Koyuncu, F. T. (2010). Türkiye'de Seçilmiş Makroekonomik Değişkenlerin Doğrudan Yabanci Sermaye Yatirimlari Üzerindeki Etkisinin Yapisal Var Analizi: 1990–2009 Dönemi. *Ekonomi Bilimleri Dergisi, 2*(1), 55–62.

Kumari, R., & Sharma, A. K. (2017). Determinants of foreign direct investment in developing countries: A panel data study. *International Journal of Emerging Markets, 12*(4), 658–682.

Kurtaran, A. (2007). Doğrudan Yabanci Yatirim Kararlari ve Belirleyicileri. *Atatürk Üniversitesi Sosyal Bilimler Enstitüsü Dergisi, 10*(2), 367–382.

Lall, S. (1978). Transnationals, domestic enterprises, and industrial structure in host LDCs: a survey. *Oxford Economic Papers, 30*(2), 217–248.

Lansbury, M., Pain, N., & Smidkova, K. (1996). Foreign direct investment in Central Europe since 1990: An econometric study. *National Institute Economic Review, 156*, 104–114.

Lebe, F., & Ersungur, Ş. M. (2011). Türkiye'de Doğrudan Yabancı Sermaye Yatırımını Etkileyen Ekonomik Faktörlerin Ampirik Analizi. *Atatürk Üniversitesi İİBF Dergisi*, 10, 321–339.

Lebe, F., & Ersungur, Ş. (2012). Türkiye'de Doğrudan Yabanci Sermaye Yatirimini Etkileyen Ekonomik Faktörlerin Ampirik Analizi. Atatürk Üniversitesi İktisadi ve İdari Bilimler Dergisi, 25.

Leitão, N. C. (2010) Foreign direct investment: The Canadian experience. International Journal of Economics and Finance, 2(4), 82–88.

Leitão, N. C., ve Faustino, H. C. (2010). Determinants of foreign direct investment in Portugal. *Journal of Applied Business and Economics, 11*(3), 19–26.

Lim, D. (1983) Fiscal incentives and direct foreign investment in less developed countries. The Journal of Development Studies, 19(2), 207–212.

Lim, E. G. (2001) Determinants of, and the relation between, foreign direct investment and growth a summary of the recent literature.

Lunn, J. (1980). Determinants of US direct investment in the EEC: Further evidence. *European Economic Review, 13*(1), 93–101.

Mucuk, M. (2011). Doğrudan Yabancı Sermaye Yatırımları: Teori ve Türkiye Uygulaması. Konya: Çizgi Kitabevi. North, DC (2002). Kurumlar, Kurumsal Değişim ve Ekonomik Performans, Çev: GÇ Güven, İstanbul: Sabancı Üniversitesi Yayınlan.

Mudiyanselage, M. M. R., Epuran, G., & Tescaşiu, B. (2021). Causal Links between Trade Openness and Foreign Direct Investment in Romania. Journal of Risk and Financial Management, 14(3), 90. https://doi.org/10.3390/jrfm14030090.

Nigh, D. (1985). The effect of political events on United States direct foreign investment: A pooled time-series cross-sectional analysis. *Journal of International Business Studies.*

Polat, B. (2015) Determinants of FDI into Central and Eastern European Countries: Pull or Push Effect? Eurasian Journal of Economics and Finance, 3(4), 39–47.

Ranjan, V., & Agrawal, G. (2011). FDI inflow determinants in BRIC countries: A panel data analysis. *International Business Research*, *4*(4), 255.

Saini, N., & Singhania, M. (2018). Determinants of FDI in developed and developing countries: A quantitative analysis using GMM. *Journal of Economic Studies*, *45*(2), 348–382.

Schneider, F., & Frey, B. S. (1985). Economic and political determinants of foreign direct investment. *World Development*, *13*(2), 161–175.

Stevens, G. V. (1998). Exchange rates and foreign direct investment: A note. *Journal of Policy Modeling*, *20*(3), 393–401.

Tarı, R., & Bıdırlı, H., (2009). Türkiyede Doğrudan Yabancı Sermaye Yatırımlarının Temel Belirleyicileri: 1990-2006 Dönemine İlişkin Ekonometrik Analiz. *Dumlupınar Üniversitesi Sosyal Bilimler Dergisi*, (24). 1–15.

UNCTAD. (1998) World Investment Report: Trends and Determinants. United Nations Conference on Trade and Development, New York and Geneva.

United Nations Conference on Trade and Development. (2019). *World investment report 2019: Special economic zones*. United Nations.

Vergil, H., & Çeştepe, H. (2006). Döviz Kuru Değişkenliği Ve Yabanci Doğrudan Yatirim Akimlari: Türkiye Örneği. *İstanbul Üniversitesi İktisat Fakültesi Mecmuası*, *55*(1), 975–986.

Vijayakumar, N., Sridharan, P., & Rao, K. C. S. (2010). Determinants of FDI in BRICS Countries: A panel analysis. *International Journal of Business Science & Applied Management (IJBSAM)*, *5*(3), 1–13.

Walch, N., ve Wörz, J. (2012) The impact of country risk ratings and of the status of EU integration on FDI inflows in CESEE countries. *Focus on European Economic Integration*, *3*, 8–26.

Wang, Z. Q., & Swain, N. J. (1995). The determinants of foreign direct investment in transforming economies: Empirical evidence from Hungary and China. *Weltwirtschaftliches Archiv*, *131*(2), 359–382.

Wheeler, D., & Mody, A. (1992). International investment location decisions: The case of US firms. *Journal of International Economics*, *33*(1–2), 57–76.

Yapraklı, S. (2006). Türkiye'de Diş Ticaret Fiyatlari ile Reel Döviz Kuru Arasindaki İlişki: Ekonometrik Bir Analiz. *Hacettepe Üniversitesi İktisadi ve İdari Bilimler Fakültesi Dergisi*, *24*(1), 69–87.

Havva Nesrin Tiryaki[1]

An Analysis of Global and Macroeconomic Factors Affecting Sovereign CDS in Türkiye

1 Introduction

Countries' credit risks have become crucial determinants for investors when making direct and portfolio investment decisions in a particular country. The credit default swaps (CDS) market has emerged as a critical instrument for hedging and trading credit risk, serving as a prominent gauge of the likelihood of default for both corporate and sovereign institutions.

CDS, flexible financial tools, are vital to economic stability. They hedge against potential losses on debts associated with foreign exchange-based bonds, empowering financial institutions to lend to those with the highest creditworthiness. Their practical applications include streamlining the assumption of credit risks and providing insights into credit risk through pricing (Stulz, 2010).

A sovereign CDS is created to compensate overseas investors in case of a government default. According to Yu (2016), a debtor country may fail due to either a lack of economic resources to repay its debts or a deliberate refusal to fulfill its responsibility. Sovereign CDS spreads offer insights into the level of credit risk associated with a particular sovereign entity. Therefore, CDS can be a valuable tool to mitigate the risk associated with sovereign credit and enhance overall financial stability.

Understanding the reasons that impact the CDS premium is crucial to the research process. It is essential to comprehend how and to what degree Türkiye's macroeconomic and external variables influence this indicator, which assesses the country's riskiness about foreign markets, to mitigate the risk they cause.

For this purpose, the study examines the effects of stock return, inflation, global economic policy uncertainty index (GEPU), and volatility index (VIX) on Türkiye's CDS premiums using the ARDL and Toda Yamamoto (T-Y) causality analysis methods. The following sections of the study include a literature review, an introduction of the data used, the methodology, and the findings obtained through analysis.

---

1 Ph.D., Asst. Prof., Bilecik Şeyh Edebali University, Faculty of Economics and Administrative Sciences Department of Economics, nesrin.ozkan@bilecik.edu.tr, ORCID: https://orcid.org/0000-0002-0083-0827.

2 Literature Review

To understand how sovereign CDS premiums affect nations' credibility, evaluating their causes is important. The literature suggests that numerous factors influence CDS spread. Kartal et al. (2022) investigated the determinants of credit default swap (CDS) premiums across three distinct categories. The initial category encompasses global factors. Global factors encompass market volatility, measured by the market volatility (VIX index) and oil price indices. In addition, Naifar (2020) recommends various global uncertainty indices. The second category utilizes macroeconomic data from countries, including exchange rates, inflation, non-financial corporation foreign currency debt, and reserves. The third category concentrates on the correlation with market factors, including interest rates, non-performing loans, and stock index/price/return.

This section primarily examines research analyzing the impact of country-specific macroeconomic and financial variables on CDS data. Afterward, it studies research on global variables.

From studies focusing on country-specific macroeconomic variables, Hilschier and Nosbusch (2010) studied how much macroeconomic fundamentals can explain changes in CDS spreads. They analyzed the influence of country-specific factors and global determinants on government debt prices using data from 31 emerging market economies between 1994 and 2007. They discover that disparities in country fundamentals account for a significant portion of the fluctuation in sovereign debt prices in the market.

From 1997 to 2008, Baldacci et al. (2011) investigated country risk premiums in 46 developing market economies. Researchers found that countries with outstanding debts and deficits are likelier not to pay their debts. Spreads go down when political risk goes down. This is especially true during financial turmoil, when markets are more likely to be affected by institutional uncertainty. So, they concluded that global financial conditions and investors' willingness to take risks considerably affect domestic fundamentals.

Liu and Morley (2012) conducted a study to examine the correlation between credit default swap (CDS) premiums and domestic macroeconomic factors, precisely the interest rate and exchange rate, in France and the US. The study utilized the Granger causality test and focused on the period from 2005 to 2010 for France and from 2008 to 2010 for the US. The researchers discovered that the exchange rate mainly influences the country's CDS spreads, whereas domestic interest rates have little impact.

From 2005 to 2010, Aizenman et al. (2013) evaluated fifty countries' fiscal space and other economic variables to determine the price of sovereign risk.

According to dynamic panel estimates, macroeconomic factors such as fiscal space are significant indicators of sovereign risk. They also noted that inflation and the ratio of external debt to GDP are important indicators, as is the government debt to tax ratio.

Alternatively, numerous authors highlight the significance of global issues in determining sovereign bond spreads. Due to the contagion effect, Dailami et al. (2008) found that US interest rate policy affects emerging market bond spreads. However, US interest rates affect secondary market spreads and vary depending on the national debt.

Özatay et al. (2009) examine how global financial conditions, US macroeconomic news and national factors affected the developing emerging market bond index (EMBI) spread in 18 developing nations from 1997 to 2006. The results show that global liquidity, willingness to take risks, crisis spread, and US macroeconomic news on developing markets affect EMBI spread over time.

Furthermore, Qian and Luo (2016), Drago et al. (2017), and Galil et al. (2014) have concluded that changes in stock return or price have a substantial impact on the premiums of country and firm credit default swaps (CDS). The authors assert that a robust stock market is correlated to lower CDS premiums.

Longstaff et al. (2011) investigated CDS spreads in 26 countries and argued that global factors affect CDS data more. The analysis reveals that global financial market characteristics and global risk premiums, specifically US stock market risk premiums, volatility, and bond market, influence country credit risk more than local economic fundamentals.

The VIX index is another significant global variable utilized in the studies. Pan and Singleton (2008), Hibbert and Pavlova (2017), Park et al. (2019), Akçelik and Fendoğlu (2019), Kartal (2020), and Varlık and Obekcan (2023) have established noteworthy correlations between the VIX index and CDS premiums.

A recent, detailed study by Naifar (2020) examined the global implications of four Gulf Cooperation Council (GCC) countries. Naifar (2020) used a regression model to include more global variables and consider market situations like the VIX index, Merrill Lynch Option Volatility Estimate (MOVE) index, crude oil volatility index (OVX), GVZ index, US economic policy uncertainty index (EPU) for global risk factor, 10-year U.S. treasury interest rate as world interest rate, and TED spread. His analysis showed that global financial uncertainty affected the VIX and MOVE indices the most.

3 Data and Methodology

3.1 Explanation of Variables

The study used hypothetical variables influenced by literature studies on Türkiye's CDS prime. The data cover the period from January 2009 to April 2024. The variables used, the expected coefficient markers, and the data sources obtained for the variables are described below.

CDS: The exchange rate premium for Türkiye's five-year US dollar bonds. Daily data from Investing.com[2] is converted by the website to monthly. LCDS represents the series logarithm in the study and is a dependent variable in the ARDL model.

BIST 100 Index: The BIST 100 Index uses the highest 100 shares, Borsa Istanbul's market value, and transaction volume to depict country-specific market dynamics in stock return statistics. According to the literature research, rising equities returns have lowered CDS premiums. The variable's coefficient signal is predicted to be negative. The data came from Investing.com.[3] The study uses LSR to depict the series logarithm.

Inflation: Türkiye's macroeconomic influence on the CDS premium is calculated using the data from the Republic of Türkiye Central Bank's EVDS[4] System as a % change in CPI. The positive predicted ratio of the INF variable.

GEPU: The Global Economic Policy Uncertainty Index (GEPU), inspired by Naifar (2020), measures global uncertainty. Baker et al. (2016) found that about 2,000 US newspapers used terms like "economy," "politics," and "uncertainty" in their articles. The data sources from Policyuncertainty.com[5]. The LGEPU variable represents the logarithm and is expected to be positive.

VIX: The VIX, or volatility index, is a short-term market indicator of flexibility or unpredictability used in research to indicate worldwide financial uncertainty. The Chicago Stock Exchange (CBOE) devised it in 1993 to assess US stock exchange 30-day predicted volatility. Its association with the CDS should be beneficial in this investigation. Daily data comes from Investing.com[6], converted to monthly. The study uses LVIX for the series logarithm.

---

2 https://www.investing.com/rates-bonds/turkey-cds-5-year-usd-historical-data.

3 https://www.investing.com/indices/ise-100.

4 https://evds2.tcmb.gov.tr/index.php?/evds/serieMarket.

5 https://www.policyuncertainty.com/global_monthly.html.

6 https://tr.investing.com/indices/cboe-vix-volatility-historical-data.

As indicated in equation (1), the empirical analysis will consider four factors of the CDS, which have been extensively studied in the literature. These factors encompass macroeconomic, market, and global variables in Türkiye.

$$LCDS_t = \alpha_0 + \alpha_1 LSR_t + \alpha_2 INF_t + \alpha_3 LGEPU_t + \alpha_4 LVIX_t + \varepsilon_t \tag{1}$$

The ARDL methodology (Pesaran & Shin, 1995; Pesaran et al., 2001) was chosen for its efficiency in limited and finite sample sizes. Linear transformation with the ARDL methodology can also be used to develop a dynamic error correction model (ECM) using variables with varied stationarity (Banerjee et al., 1993). The following model was developed for the ARDL test, which tests cointegration to determine long-term causality. The ARDL form of equation (1) can be shown as in equation (2):

$$\begin{aligned} \Delta LCDS_t = {} & \alpha_0 + \Sigma_{i=1}^{n_1} \sigma_{1i} \, \Delta LCDS_{t-1} + \Sigma_{i=0}^{n_2} \beta_{2i} \, \Delta LSR_{i,t-i} + \Sigma_{i=0}^{n_3} \partial_{3i} \, \Delta INF_{i,t-i} \\ & + \Sigma_{i=0}^{n_4} \pi_{4i} \Delta LGEPU_{i,t-i} + \Sigma_{i=0}^{n_5} \varnothing_{5i} \, \Delta LVIX_{i,t-i} + \delta_{1i} LCDS_{t-1} \\ & + \delta_{2i} LSR_{t-1} + \delta_{3i} INF_{t-1} + \delta_{4i} LGEPU_{t-1} + \delta_{5i} LVIX_{t-1} \, \varepsilon_{it} \end{aligned} \tag{2}$$

The model in equation 2, Δ represents the first differential operator, α_0 represents the fixed term, σ_1, β_2, ∂_3, π_4, and $\varnothing_5$ represent the short-term coefficients, and δ_1, δ_2, δ_3, δ_4 and δ_5 represent the long-term coefficients. The ε represents the error term. Before testing a long-term relationship between variables in Equation 2, it is necessary to use the F test. The hypothesis of the F test suggests that there is no significant long-term relationship between the variables, as indicated by the equation $H_0=\delta_1=\delta_2=\delta_3=\delta_4=\delta_5=0$. The alternative hypothesis, denoted as $H1 = \delta_1 \neq \delta_2 \neq \delta_3 \neq \delta_4 \neq \delta_5 \neq 0$. Suppose the F test results exceed the critical values specified in Pesaran et al. (2001), indicating support for the long-term relationship. In that case, the second phase assumes the inclusion of the error correction term in equation 3 of the ARDL model.

$$\begin{aligned} \Delta LCDS_t = {} & \alpha_0 + \Sigma_{i=1}^{n_1} \sigma_{1i} \, \Delta LCDS_{t-1} + \Sigma_{i=0}^{n_2} \beta_{2i} \, \Delta LSR_{i,t-i} + \Sigma_{i=0}^{n_3} \partial_{3i} \, \Delta INF_{i,t-i} \\ & + \Sigma_{i=0}^{n_4} \pi_{4i} \Delta LGEPU_{i,t-i} + \Sigma_{i=0}^{n_5} \varnothing_{5i} \Delta LVIX_{i,t-i} + \delta_1 LCDS_{t-1} \\ & + \delta_2 LSR_{t-1} + \delta_3 INF_{t-1} + \delta_4 LGEPU_{t-1} + \delta_5 LVIX_{t-1} + \delta_6 ECT_{t-1} \varepsilon_{it} \end{aligned} \tag{3}$$

In equation 3, the ECT_{t-1} term represents the error correction term, whereas δ6 denotes the error correction coefficient. If δ6 is statistically significant and has a negative value, it indicates the rate at which a temporary shock returns to stability in the long run.

This study also examines the causal relationship between explanatory variables and CDS in Türkiye using the Granger causality test conducted by Toda and Yamamoto (1995). The first step in modeling using the Toda-Yamamoto approach is determining the VAR system's appropriate lag length (k) and the variables' maximum order of integration (dmax). Later, the appropriate lag length k determined within the VAR model is artificially modified by adding it to the maximum order of integration dmax (p = k + dmax). However, the added terms' coefficients are not considered in the model. In the causality procedure, the dmax value must not exceed the k value (Allou et al., 2020).

3.2 Empirical Findings

The study initially establishes the sequence of integration for the series. Thus, Table 1 displays the outcomes of the ADF and PP unit root tests conducted on the variables utilized.

Table 1: Unit Root Test Results

	Level		First Difference			
Variables	**ADF**	**PP**	**ADF**	**Decision**	**PP**	**Decision**
LCDS	-3.8223* (-4.0089)	-3.8830* (-4.0089)	-13.3175* (-4.0092)	I (1)	-13.3212* (-4.0092)	I (1)
LSR	0.3251* (-4.0089)	0.2855* (-4.0089)	-12.2348* (-4.0092)	I (1)	-12.2113 (-4.0092)	I (1)
INF	-2.8599* (-3.4661)	-6.1242* (-3.4661)	-10.9967* (-3.4669)	I (1)	-	I (0)
LGEPU	-2.5104* (-3.4663)	-2.7967* (-3.4663)	-11.3381* (-3.4667)	I (1)	-20.5881* (-3.4663)	I (1)
LVIX	-4.4799* (-3.4663)	-6.8374* (-3.4663)	-	I (0)	-	I (0)
Note: * indicates that the unit root will be rejected at the 1 % significance level.						

The findings in Table 1 demonstrate that the INF and LVIX series exhibit stationary behavior at level I (0), while the remaining series display stationary behavior at the first difference. Given the varying degrees of stagnation in the series, with the highest stagnance level being I (1), utilizing the ARDL test as outlined in

Pesaran et al. (2001) is suitable. Furthermore, in the T-Y causality test, the maximum order of integration, denoted as dmax, is equivalent to 1.

After determining the series' stationary levels, the presence of a long-term cointegration relationship among them was investigated using the ARDL bounds test. The dependent variable used in the model is LCDS, as shown in Table 2, and the critical values presented by Pesaran et al. (2001).

Table 2: Cointegration Bound Test Results

MODEL	Specification	F-Statistic	%1 Anlamlılık Düzeyinde Kritik Değerler	
LCDS= f (LSR, INF, LGEPU, LVIX)	ARDL (2, 2, 6, 0, 0)	4.4874	I (0)	I (1)
			3.29	4.37

The calculated F statistic value for the model (4.4874) is greater than the upper value of 4.37 at the 1 % significance level. Therefore, the null hypothesis H_0, which assumes that the long-term coefficients in the model in Equation 2 are not related, is rejected, and the presence of cointegration is accepted.

Table 3 displays the coefficients of the calculated ARDL (2,2,6,0,0) model for both the short-run and long-run.

Table 3: Short Run and Long Run Results of ARDL

Short-run coefficients	**Coefficient**	**Prob.**
D (LCDS (-1))	-0.1752	0.0182
D (LSR)	-1.0951	0.0000
D (LSR (-1))	-0.4868	0.0008
D (INF)	0.0097	0.1174
D (INF (-1))	-0.0252	0.0006
D (INF (-2))	-0.0104	0.1479
D (INF (-3))	-00258	0.0004
D (INF (-4))	-0.0098	0.1198
D (INF (-5))	-0.0151	0.0137
CointEq (-1) *	-0.0766	0.0000
Long run coefficients	**Coefficient**	**Prob.**
LSR	-0.1593	0.5929
INF	0.3853	0.0483
LGEPU	0.5592	0.1823
LVIX	1.5756	0.1733
C	-3.4531	0.4591

(continued on next page)

Table 3: Continued

Diagnostics tests	
Heteroskedasticity	$\chi^2(14) = 1.1078$ prob. (0.3545)
Serial correlation	$\chi^2(22) = 0.8336$ prob. (0.6157)
Functional form misspecification	F (0.0105) = 1.38 prob. (0.9181)
Normality	Jarque-Bera = 84.42 prob. (0.00)

According to Table 3, the short-term effects of LGEPU and LVIX variables on LCDS have not been calculated. Besides, as expected, the LSR variable negatively affects the LCDS in the short term, both when it occurs and in the following period. The INF, however, has a significant positive impact not in the current period but in the first, third, and fifth periods, as expected. The CointEq (-1) variable is between 0 and -1 at the significance level of %1. The estimated ratio of -0.07 indicates that 7 % of the imbalance that occurred in the previous month was corrected in the current month. This means that, after a shock, it takes about 14 months for LCDS, LBIST, and INF to re-establish their long-term balance.

According to the long-term results in Table 3, only the INF variable has a positive impact of 5 % significance, which aligns with expectations. An inflation increase of 1 % increases the CDS premium by 0.3853.

In addition, the significance tests in Table 3 indicate that the model has the correct configuration and does not have variable variance and auto-correlation problems. Also, in Figure 1, CUSUM and CUSUMSQ stability plots are used, whereas CUSUM stays within the critical 5 % significant level, while CUSUMSQ does not.

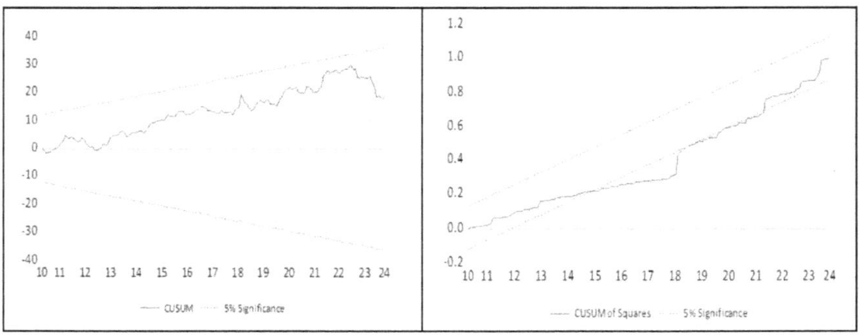

Figure 1: CUSUM and CUSUMSQ Stability Test Results

The study also utilized T-Y causality analysis to determine the causal relationship between variables. A VAR model has been constructed to ascertain the optimum length of lag. The identification of the optimum number of lags is determined by various criteria, including the Akaike Information Criteria (AIC) with the lowest value, the Final Prediction Error (FPE), the Schwarz Information Criterion (SC), the Hannan-Quinn Criterion (HQ), and the likelihood ratio (LR). Table 2 displays the arranged order of VAR lags based on five criteria, ultimately determining that the most favorable lag order, k, is 2. Next, we must create a VAR model adjusted to the maximum integration of variables (dmax) for prediction. Since the higher level of the used variables is I(1), dmax is regarded as 1. Therefore, a Toda-Yamamoto test was employed, with three lags (k+dmax). Table 4 shows the VAR lag order selection criteria.

Table 4: VAR Lag Order Selection Criteria

Lag	LogL	LR	FPE	AIC	SC	HQ
0	-480.4095	NA	0.000171	5.516016	5.606087	5.552549
1	242.1109	1395.778	6.18e-08	-2.410352	-1.869928*	-2.191158*
2	281.5941	74.03100*	5.24e-08*	-2.574933*	-1.584157	-2.173079
3	290.4945	16.18251	6.31e-08	-2.391983	-0.950854	-1.807468
4	306.0240	27.35305	7.05e-08	-2.284364	-0.392882	-1.517188
5	315.2766	15.77146	8.47e-08	-2.105416	0.236419	-1.155579
6	329.5413	23.50437	9.64e-08	-1.983424	0.808764	-0.850927
7	347.6388	28.79152	1.05e-07	-1.904987	1.337554	-0.589828
8	368.3163	31.72111	1.12e-07	-1.855867	1.837026	-0.358048

Table 5: The results of the Toda-Yamamoto Granger Causality Test

Null Hypothesis	**Chi-sq**	**(k+dmax)**	**Prob.**	**Direction of Causality**
LSR does not cause LCDS	9.0455	3	0.0287	**LSR→LCDS**
INF does not cause LCDS	2.1929	3	0.5333	INF # LCDS
LGEPU does not cause LCDS	2.7951	3	0.4243	LGEPU # LCDS
LVIX does not cause LCDS	5.6986	3	0.1272	LVIX # LCDS
LCDS does not cause LSR	16.2215	3	0.0010	**LCDS →LSR**
INF does not cause LSR	2.0473	3	0.5626	INF # LSR
LGEPU does not cause LSR	7.9465	3	0.0471	**LGEPU →LSR**
LVIX does not cause LSR	6.0802	3	0.1078	LVIX # LSR

(continued on next page)

Table 5: Continued

Null Hypothesis	Chi-sq	(k+dmax)	Prob.	Direction of Causality
LCDS does not cause INF	22.4445	3	0.0001	**LCDS →INF**
LSR does not cause INF	48.1316	3	0.0000	**LSR →INF**
LGEPU does not cause INF	6.0418	3	0.1096	LGEPU# INF
LVIX does not cause INF	3.4080	3	0.3329	LVIX # INF
LCDS does not cause LGEPU	9.1712	3	0.0271	**LCDS → LGEPU**
LSR does not cause LGEPU	2.5899	3	0.4592	LSR # LGEPU
INF does not cause LGEPU	1.9529	3	0.5822	INF # LGEPU
LVIX does not cause LGEPU	4.6459	3	0.1996	LVIX # LGEPU
LCDS does not cause LVIX	3.7152	3	0.2939	LCDS # LVIX
LSR does not cause LVIX	1.9407	3	0.5848	LSR # LVIX
INF does not cause LVIX	0.0564	3	0.9965	INF # LVIX
LGEPU does not cause LVIX	1.2089	3	0.7509	LGEPU# LVIX

Table 5 displays the chi-square and probability values computed using the modified Wald test. along with the presence and directions of causation between the variables. Therefore. a bidirectional relationship was seen between LCDS and LSR. with a statistically significant association at the 5 % significance level. In addition. LCDS and LSR induce fluctuations in the INF. A causal relationship exists between LCDS and LGEPU and between LGEPU and LSR.

4 Conclusion

The CDS premium. which provides real-time information about countries' ability to repay their debts. is influenced by financial and macroeconomic data. The country's CDS premiums rise as the risk of my credit failure increases.

Studies in the literature have been followed to explain the reasons for the fluctuations in Türkiye's CDS premiums using macroeconomic. global. and market data. According to ARDL and T-Y analysis findings. a negative relationship was found between domestic stock market returns in the short term. as expected. in line with Qian and Luo (2016). In addition. a bidirectional causality between stock market returns and CDS has been determined.

While a change in the inflation rate causes the CDS to rise after a period. it also causes the CDS to rise at a 5 % significance level in the long run. In this analysis for Türkiye. no significant effects of the GEPU and VIX variables used to represent global effects on CDS were found. This shows that the factors affecting the CDS premium in Türkiye are primarily domestic dynamics.

According to the T-Y analysis. the change in CDS premiums significantly affects domestic variables such as inflation and stock return.

References

Aizenman, J., Hutchison, M., & Jinjarak, Y. (2013). What is the risk of European sovereign debt defaults? Fiscal space, CDS spreads and market pricing of risk. *Journal of International Money and Finance, 34*, 37–59.

Akcelik, F., & Fendoglu, S. (2019). *Country risk premium and domestic macroeconomic fundamentals when global risk appetite slides* (No. 1904). Research and Monetary Policy Department. Central Bank of the Republic of Turkey.

Allou, E. J., Adeleye, B. N., Cheng, J., & Abdul, R. (2020). Is there a nexus between China outward foreign direct investment and welfare in Côte d'Ivoire? Empirical evidence from the Toda–Yamamoto procedure. *African Development Review, 32*(3), 499–510.

Baker, S. R., Bloom, N., & Davis, S. J. (2016). Measuring economic policy uncertainty. *The quarterly journal of economics, 131*(4), 1593–1636.

Baldacci, E., Gupta, S., & Mati, A. (2011). Political and fiscal risk determinants of sovereign spreads in emerging markets. *Review of Development Economics, 15*(2), 251–263.

Baldacci, E., Gupta, S., & Mati, A. (2011). Political and fiscal risk determinants of sovereign spreads in emerging markets. *Review of Development Economics, 15*(2), 251–263.

Banerjee, A., Dolado, J. J., Galbraith, J. W., & Hendry, D. (1993). *Co-integration, error correction and the econometric analysis of non-stationary data*. Oxford University Press.

Dailami, M., Masson, P. R., & Padou, J. J. (2008). Global monetary conditions versus country-specific factors in the determination of emerging market debt spreads. *Journal of International Money and Finance, 27*(8), 1325–1336.

Drago, D., Di Tommaso, C., & Thornton, J. (2017). What determines bank CDS spreads? Evidence from European and US banks. *Finance Research Letters, 22*, 140–145.

Galil, K., Shapir, O. M., Amiram, D., & Ben-Zion, U. (2014). The determinants of CDS spreads. *Journal of Banking & Finance, 41*, 271–282.

Hibbert, A. M., & Pavlova, I. (2017). The drivers of sovereign CDS spread changes: Local versus global factors. *Financial Review, 52*(3), 435–457.

Hilscher, J., & Nosbusch, Y. (2010). Determinants of sovereign risk: Macroeconomic fundamentals and the pricing of sovereign debt. *Review of Finance, 14*(2), 235–262.

Kartal, M. T. (2020). The behavior of Sovereign Credit Default Swaps (CDS) spread: Evidence from Turkey with the effect of Covid-19 pandemic. *Available at SSRN 3642652.*

Kartal, M. T., Ertuğrul, H. M., & Ayhan, F. (2022). Determinants of sovereign credit default swap (CDS) spreads in emerging countries: Evidence from Turkey. *Hacettepe Üniversitesi İktisadi ve İdari Bilimler Fakültesi Dergisi, 40*(4), 742–761.

Liu, Y., & Morley, B. (2012). Sovereign credit default swaps and the macroeconomy. *Applied Economics Letters, 19*(2), 129–132.

Longstaff, F. A., Pan, J., Pedersen, L. H., & Singleton, K. J. (2011). How sovereign is sovereign credit risk? *American Economic Journal: Macroeconomics, 3*(2), 75–103.

Naifar, N. (2020). What explains the sovereign credit default swap spreads changes in the GCC region?. *Journal of Risk and Financial Management, 13*(10), 245.

Özatay, F., Özmen, E., & Şahinbeyoğlu, G. (2009). Emerging market sovereign spreads. Global financial conditions and US macroeconomic news. *Economic Modelling, 26*(2), 526–531.

Pan, J., & Singleton, K. J. (2008). Default and recovery implicit in the term structure of sovereign CDS spreads. *The Journal of Finance, 63*(5), 2345–2384.

Park, Y. J., Kutan, A. M., & Ryu, D. (2019). The impacts of overseas market shocks on the CDS-option basis. *The North American Journal of Economics and Finance, 47*, 622–636.

Pesaran, M. H., & Shin, Y. (1995). *An autoregressive distributed lag modelling approach to cointegration analysis* (Vol. 9514. pp. 371–413). Cambridge. UK: Department of Applied Economics. University of Cambridge.

Pesaran, M. H., Shin, Y., & Smith, R. J. (2001). Bounds testing approaches to the analysis of level relationships. *Journal of Applied Econometrics, 16*(3), 289–326.

Qian, Z., & Luo, Q. (2016). Regime-dependent determinants of China's sovereign credit default swap spread. *Emerging Markets Finance and Trade, 52*(1), 10–21.

Stulz, R. M. (2010). Credit default swaps and the credit crisis. *Journal of Economic Perspectives, 24*(1), 73–92.

Toda, H. Y., & Yamamoto, T. (1995). Statistical inference in vector autoregressions with possibly integrated processes. *Journal of Econometrics*, *66*(1–2), 225–250.

Varlık, S., & Obekcan, M. (2023). Ülke Risk Priminin Belirleyicisi Olarak Merkez Bankası Kredibilitesi: Türkiye'den Kanıtlar. *Bulletin of Economic Theory and Analysis*, *8*(2), 128–155.

Yu, S. (2016). The effect of political factors on sovereign default. *Review of Political Economy*, *28*(3), 397–416.

Nuran H. Belet[1]

Global Shocks and Turkish Economy: A Dummy Variable Approach

1 Introduction

Throughout history, economic shocks have had a profound impact on global economies, with events such as the Great Depression of 1929, the Financial Crisis of 2008, and most recently, the unprecedented economic downturn brought about by the COVID-19 pandemic. Turkey, like many other countries, felt the effects of financial crises in 1994, 2001, 2008–2009, and during the COVID-19 pandemic. The country experienced extended lockdowns, business closures, high unemployment rates, and reduced economic growth as a result of the pandemic. Despite these challenges, Turkey demonstrated a remarkable level of resilience in response to the COVID-19 shock when compared to other nations. As an emerging economy, Turkey effectively managed the COVID-19 crisis and successfully navigated the economic impacts it brought about, ensuring the stability of its economy. Turkey, a transcontinental country situated at the intersection of Europe and Asia with a substantial youthful demographic, stood out as one of the rare nations globally to steer clear of an economic downturn during the initial phase of the pandemic. After impressive 1.8 % expansion in GDP in the preceding year and promising indicators in the first six months of 2021 (notable 7.2 % and 21.7 % year-on-year growth in the first and second quarters, respectively) (Chan, 2021; Çakmaklı et al., 2021).

Turkey's economy has struggled with macroeconomic instability for a considerable period of time. Historically, the causes of this instability were clear and easily identified. Factors such as irresponsible monetary policies, unsustainable fiscal spending, inadequate financial regulation, and inconsistent exchange-rate policies have all played a role in this. However, Turkey should be commended for successfully addressing and rectifying these traditional sources of vulnerability as it emerged from the crisis of 2001, and 2008–2009 (Rodrik, 2012; Türel, 2010).

1 Ph.D., Asst. Prof. Nuran H. Belet, Ankara Hacı Bayram Veli University, Faculty of Economics and Administrative Sciences, Department of Economics, nuran.belet@hbv.edu.tr. Orcid: 0000-0003-0394-6339.

In the present day, monetary policy is executed under an inflation-targeting structure and overseen by an autonomous Central Bank.

The impact of global shocks on economy encompasses a complexity, thus, it is crucial to comprehend the implications of global shocks on the Turkish economy in order to formulate efficient economic policies. The most recent four crises in the Turkish economy were not merely the consequence of misguided policies enacted by Turkey alone; they were also significantly shaped by the broader systemic characteristics of the global economy. The objective of this study is to conduct a comparative analysis of the four economic crises that have impacted the Turkish economy over the past three decades. The aim is to investigate the interaction between these crises and macroeconomic variables within a system that is also affected by and integrated with the international economic system.

This study explores the impact of 1994 economic crisis, 2001 crisis, 2008 financial crisis, and 2020 COVID-19 pandemic crisis across three macroeconomic indicators gross savings, economic growth, Debt to GDP ratio, and current account deficit for the period of 1991–2023. Utilizing the time series Dummy variable (T-DV) method enables a comprehensive and resilient examination of these repercussions, providing valuable guidance for decision-makers. With the ongoing expansion of global economic interconnectedness, research of this nature becomes progressively essential in preserving domestic economic integrity.

2 Literature Review

Extensive research has been conducted on the correlation between financial market volatility, global shocks and economic cycles in Emerging Market Economies (EMEs). According to (Claessens et al., 2010; Claessens et al., 2011), a thorough examination of recessions and recoveries in 45 countries indicates that EMEs experience financial market disruptions more frequently compared to developed economies. This conclusion is supported by the findings of (Aizenman & Powell, 2003), who posit that the prevalence of weak legal systems and substantial information verification costs intensify the impact of volatility on production, employment, and welfare in these markets. (Koray, 2006) conducted an analysis on the impact of inflation variability on the Turkish economy during the period of 1981 to 1989. By employing a vector autoregressive model, the study revealed that shocks in inflation variability had a negative effect on output, money supply, and the real exchange rate, while the trade balance and interest rate experienced a positive response. This indicates a significant influence of inflation variability on economic activity in Turkey.

In a separate study, (Haque, 2017) examined the effects of fiscal shocks on the Bangladeshi economy using structural vector autoregression methodology. Concentrating on the time frame of 2003 to 2015, the research shed light on the macroeconomic dynamics influenced by fiscal shocks, offering insights into how fiscal policy decisions can impact economic activity in Bangladesh.

Furthermore, (Altunöz, 2023)delved into the validation of the Financial Kuznets Curve hypothesis and the classical Kuznets curve as they relate to income distribution and financial development in the Turkish economy. Through the employment of the ARDL bound test approach, the study scrutinized the correlation between these factors, contributing to the comprehension of economic development trends in Turkey based on income distribution and financial development.

(Aghion et al., 2004) delved deeper into this connection by utilizing a dynamic open economy model that takes into account financial frictions. Their findings suggest that economies with moderate levels of financial development are more susceptible to instability, indicating that Emerging Market Economies (EMEs) tend to endure greater volatility when compared to highly developed or severely underdeveloped economies. This instability is frequently exacerbated by credit limitations and foreign-currency-denominated debt, as emphasized by (Felipe et al., 2004) and (Neumeyer & Perri, 2005).

The examination of how uncertainty influences investment is a pivotal research topic. (Cespedes, 2013) explore the impact of global uncertainty shocks on investment and consumption in Emerging Market Economies (EMEs) in comparison to developed nations. Their results suggest that EMEs undergo more profound and enduring decreases in investment after uncertainty shocks, largely due to heightened credit limitations. These findings align with previous studies by (Pindyck, 1991) and (Dixit, 1992), which demonstrate that uncertainty can result in investment hysteresis, causing firms to postpone investment choices until uncertainty is resolved.

Additional analysis is presented by (Ramey & Ramey, 1995) showcasing a significant inverse correlation between the rate of economic expansion and its fluctuations. This adverse impact is accredited to the unpredictability surrounding advancements in GDP. The interplay between financial constraints and overseas indebtedness in foreign denominations is notably harmful, as the repercussions of currency devaluations on balance sheets can trigger pronounced constraints on credit availability and exacerbate economic downturns.

The correlation between trade openness and economic growth has been a topic of extensive discussion within academic literature. Traditional and modern economic theories both suggest that trade openness can stimulate economic growth through improving market efficiency and resource allocation. Various research, such as that conducted by (Dollar, 1992), (Edwards, 1993), and (Sachs & Warner, 1995), presents compelling evidence in favor of the beneficial effects of trade openness on economic growth. These studies highlight that decreasing trade restrictions results in increased economic integration, ultimately leading to enhanced productivity and expansion.

The consensus on the beneficial connection between trade openness and economic growth is not undisputed. (Levine & Renelt, 1992) as well as (Rodríguez & Rodrik, 2000) offer critiques, contending that numerous studies overlook crucial factors influencing growth and employ inadequate metrics of openness. Their analysis indicates that the relationship is nuanced and contingent on specific circumstances.

The primary cause behind the economic crisis in Turkey in 1994 was due to fiscal imbalances and an overvalued exchange rate (Cömert & Yeldan, 2018; Öniş & Güven, 2011). In the period leading up to the crisis, Turkey had implemented expansionary fiscal policies that resulted in substantial budget deficits (Koch et al., 2002; Ismihan et al., 2003). These deficits were funded by borrowing from domestic markets, which consequently caused high interest rates. Moreover, Turkey adhered to a fixed exchange rate system that overestimated the value of the Turkish Lira, resulting in a decline in competitiveness in global markets (Öniş & Güven, 2011). The simultaneous occurrence of high interest rates and an overvalued currency led to a speculative attack on the Turkish Lira, resulting in a significant devaluation and a notable economic downturn (Koray, 2006; Cömert & Yeldan, 2018).

Prior to the onset of the crisis, the government had introduced economic policies with the goal of promoting liberalization, such as deregulating the financial sector and reducing trade barriers (Utkulu & Özdemir, 2004). Unfortunately, the absence of fiscal restraint in conjunction with these measures exacerbated the economic vulnerabilities (Utkulu & Özdemir, 2004; Türel, 2010).

The immediate aftermath of the crisis was marked by a severe recession, resulting in a 5 % contraction of the GDP and a staggering inflation rate of over 100 %. In response to this economic turmoil, the government-initiated austerity measures that entailed reductions in public spending, tax increases, and allowing the Lira to float freely (Er et al., 2014; Utkulu & Özdemir, 2004).

While these stringent policies did contribute to stabilizing the economy, they came at a substantial cost in terms of both social and economic repercussions.

The 2001 turmoil in Turkey had its origins in inherent weaknesses within the banking sector and unsustainable fiscal policies. Before the crisis unfolded, Turkey had initiated a disinflation strategy supported by the International

Monetary Fund (IMF), which comprised of stringent monetary measures and a crawling peg exchange rate mechanism (Er et al., 2014). Nevertheless, political unrest and a lack of fundamental reforms eroded trust in the economic agenda. A combination of a substantial public sector shortfall, soaring inflation rates, and a delicate banking industry resulted in a decline in investor trust (Uygur, 2010). In February 2001, a political upheaval sparked a massive capital flight, ultimately resulting in the devaluation of the Turkish Lira and a banking sector meltdown (Türel, 2010; Uygur, 2010).

Following the crisis, the Turkish government collaborated with the IMF to initiate a thorough reform agenda. The program encompassed various significant actions such as restructuring the banking industry, enhancing fiscal responsibility, and introducing structural changes to boost efficiency and competitiveness (Er et al., 2014; Rodrik, 2009). The banking sector underwent consolidation, with the closure or merger of vulnerable institutions, and regulatory supervision was reinforced. These initiatives played a crucial role in rebuilding trust in the economy, resulting in a swift turnaround. By 2002, Turkey had achieved a return to economic growth, and inflation was successfully managed (Orhangazi & Yeldan, 2021).

Responding to the crisis, the Turkish government implemented expansionary monetary and fiscal measures to revitalize the economy. Interest rates were decreased, while fiscal initiatives such as tax reductions and increased public spending were enacted. These strategies effectively mitigated the crisis's impact and spurred a swift recovery (Cömert & Yeldan, 2018). By 2010, the economy demonstrated robust growth, registering a 9 % expansion. Nevertheless, the crisis emphasized Turkey's urgent need to address fundamental vulnerabilities, such as the substantial current account deficit and reliance on short-term capital inflows (Orhangazi & Yeldan, 2021; Türel, 2010).

Prior to the COVID-19 outbreak, Turkey was already contending with elevated inflation rates and a devalued currency as a result of ongoing geopolitical strife and economic disparities (Tuysuz et al., 2022). The pandemic further exacerbated these challenges, resulting in a significant downturn in economic performance due to the implementation of lockdowns and restrictions. The tourism industry, a crucial source of revenue for the country, was especially severely impacted. In an attempt to counteract the adverse effects, the Turkish government put in place a number of measures aimed at lessening the blow (Orhangazi & Yeldan, 2021; Çakmaklı et al., 2021). These initiatives consisted of financial stimulus efforts such as direct payments to vulnerable households, postponement of tax obligations, and financial aids to businesses. The Central Bank of Turkey likewise made the decision to decrease interest rates in order to stimulate economic growth. Despite these interventions, the crisis underscored the fundamental vulnerabilities present in the economy, particularly pertaining to the management of inflation and the stability of the currency (Chan, 2021; Altunöz, 2023).

The comparison of the impact of four crises on various macroeconomic indicators of Turkey is presented in Table 1 below.

Table 1: Comparison of four crisis

Year	**1994**	**2001**	**2008-09**	**COVID-19**
Type of Crisis	Twin	Twin	Monetary	Health and Economic
Phase of the mid-term cycle in the world economy	Ascending	Decline Transition	Decline	Decline
Status of capital flows to developing economies	Increase	Rapid Increase	Decrease	Decrease
Phase of the mid-term national cycle	Ascending	Decline	Decline	Decline
Exchange rate regime before the crisis	Managed Float	Crawling Peg	More or Less Free Float	Managed Float
Magnitude of external shock from capital flows (billion $)	8.8	16.7	43.2	Significant (not data specified)
Current account deficit / GDP %	3.6	3.7	5.9	4.7
Public sector deficit / GDP %	8.2	8.6	1.5	10
Domestic public debt / GDP %	18.0	30.9	30.3	40
External debt / GDP %	66.6	54.1	35.8	50.7
Public external debt / Total external debt %	64.6	53.0	36.1	49.5
Annual inflation rate %	67.8	49.2	11.5	14.3
Crisis Timeline (start / end)	February 1994 / March 1995	February 2001 / February 2002	August 2008 / November 2009?	March 2020/July 2021
Duration of Crisis (months)	14	13	16?	16
Output Loss Caused by the Crisis (as % of GDP)	6.7	5.9	7.9	10 (approx.)

Source(Türel, 2010: 57),various economic reports and the author's own calculations.

Unlike previous studies, our research delves into the correlation between global shocks and external debt, savings rates, and the current account deficit, presenting a distinct viewpoint from earlier studies. Through the analysis of these factors, our goal is to offer a thorough comprehension of the impact of global disruptions on economic results, particularly with regards to external debt, savings patterns, and the current account deficit.

Furthermore, we take a comprehensive view of how global disturbances influence the economy and different macroeconomic factors. By employing a robust econometric model, our research offers detailed understanding of the complex interactions between global shocks and economic metrics, thereby adding significant insights to the realm of macroeconomics.

3 Data and Methodology

The study uses time series dummy variable approach to explore the relationship between independent variables, global shocks of 1994, 2001, 2008, and 2020, and dependent variables GDP per capita growth, gross savings, current account deficit, and external debt service for the period of 1991 to 2023.

3.1 Variables, Description, and Source

Table 1 below shows the variables, source and description of dependent and independent variables.

Table 2: Description of Variables

Variables	Description	Unit	Source
GDP	GDP per capita	Annual Percentage	WDI
Unemp	Unemployment rate	Annual Percentage	WDI
DS	External Debt Service as percentage of GDP	Annual Percentage	WDI
CAD	Current Account Deficit	Annual Percentage	WDI
Gr Sav	Gross Savings rate	Annual Percentage	WDI
FDI	Foreign Direct Investment	Annual Percentage	WDI
inf	Inflation	Annual Percentage	WDI
Open I	Openness Index	Annual Percentage	Calculated Proxy
DV (1995, 2001, 2008, 2020)	Dummy Variables	0 and 1	N/A

The openness index for Turkey is calculated using the following formula by (Ozturk & Radouai, 2020; Keho, 2017).

$$\text{Openness Index (TOI)} = \frac{\text{Imports (M)} + \text{Exports (X)}}{\text{GDP (PPP)}}$$

The dummy variables for each shock will be 1 for next four years from the year the shock occurred and 0 otherwise.

3.2 Model Specification

The study uses time series dummy variable model to determine the relationship between dependent and independent variables. Using time series dummy variable approach is efficient and robust for shock analysis. It captures the abrupt fluctuations by integration dummy variables into model, enabling efficient quantification and repercussion of the shocks. In-addition, time series dummy variables is particularly effective to isolate the effect of economic shocks. With dummy variable for each shock i.e., 1994, 2001, 2008, and 2020, it effectively isolates the effect of each shock on the dependent variables.

The mathematical form is represented below.

$$y_t = \propto_t + \sum_{t=1}^{n} \beta_t X_t + \sum_{t=1}^{m} \gamma_t DV_t + \varepsilon_t \quad (1)$$

Where, y_t represents the dependent variables, X_t represents independents variables, DV_t represents the dummy variables, $\propto_t$ represents the constant, β_t and γ_t represents the parameters of independent variables and dummy variables, and ε_t represents the residuals respectively.

3.3 Estimation and Results

Before estimating time-series model, it is crucial to verify the stationarity of variables. The Dickey-Fuller test was utilized for this purpose. The findings can be found in Table 2 provided.

Table 3: Dicky-Fuller Unit Root Test

Variables	T-Statistics	P-Value	Order of Integration
gdp	-5.724	0.0000	*I(0)*
unemp	-4.634	0.0001	*I(1)*
debt	-7.484	0.0000	*I(1)*
cad	-3.882	0.0022	*I(0)*
grsav	-6.977	0.0000	*I(1)*
fdi	-5.224	0.0000	*I(1)*
inf	-5.774	0.0000	*I(1)*
openI	-5.216	0.0000	*I(1)*

The estimation of model is given in Table 3 below.

Table 4: Estimation Results

	(1)	(2)	(3)	(4)
	gdp	grsavd1	cad	debtd1
unempd1	-1.738**	-0.0438	1.433***	0.00142
	(-3.79)	(-0.13)	(4.34)	(1.69)
debtd1	-3.733**	-1.911*	-7.717	
	(-3.42)	(-2.65)	(-0.11)	
grsavd1	0.111			-0.00126*
	(0.40)			(-2.65)
cad	-0.330			0.00412**
	(-1.01)			(3.67)
fdid1	0.171	0.694	-0.00849	0.00262
	(0.21)	(1.13)	(-0.01)	(1.72)
infd1	-0.0494	0.0451	0.0268	0.000168*
	(-1.12)	(1.40)	(0.88)	(2.14)
opend1	2.67*	5.881	-1.160	0.0259

(continued on next page)

Table 4: Continued

	(1)	(2)	(3)	(4)
	(2.28)	(0.76)	(-0.16)	(1.33)
dv1994	-4.294***	0.113	3.372***	0.00148
	(-4.16)	(0.14)	(4.54)	(0.75)
dv2001	-0.0110	-0.0401	0.166	-0.000567
	(-0.01)	(-0.04)	(0.17)	(-0.22)
dv2008	-2.884	-0.252	-1.293	0.00189
	(-1.79)	(-0.21)	(-1.12)	(0.61)
dv2020	-1.004*	-0.839	1.435	0.000572
	(-2.66)	(-0.74)	(1.33)	(0.19)
_cons	4.001***	-0.0153	-3.685***	-0.00209
	(5.94)	(-0.03)	(-7.60)	(-1.69)
N	32	32	32	32

t statistics in parentheses
* $p < 0.05$, ** $p < 0.01$, *** $p < 0.001$

The results of the regression analysis suggest notable connections between different economic indicators. Specifically, the unemployment rate (unempd1) is found to have a considerable adverse impact on GDP, as evidenced by a coefficient of -1.738 and a significance level of p < 0.01, indicating that higher levels of unemployment are associated with a decrease in GDP. In contrast, unemployment is shown to have a significant positive effect on the current account deficit (CAD), with a coefficient of 1.433 and a significance level of p < 0.001, suggesting that rising unemployment worsens the CAD. Additionally, debt (debtd1) is found to significantly negatively affect both GDP (coefficient: -3.733, p < 0.01) and gross savings (coefficient: -1.911, p < 0.05), underscoring the harmful consequences of elevated debt levels on economic performance and savings. The openness index (opend1) demonstrates a positive and statistically significant coefficient (coefficient: 2.67, p < 0.1), indicating a favorable impact of trade openness (free trade) on GDP.

Gross savings (grsavd1) display a notable inverse correlation with the current account deficit (CAD), with a coefficient of -0.00126 and a significance level below $p < 0.05$, indicating that increased savings contribute to a reduction in the deficit. It is important to highlight that the dummy variable representing the year 1994 (dv1994) has a significant adverse effect on GDP (coefficient: -4.294, $p < 0.001$) and a significant positive impact on CAD (coefficient: 3.372, $p < 0.001$), reflecting potential economic upheavals or policy changes during that specific period. Initially, prior to the 1994 crises, there were record government deficits accompanied by a significant interest burden and elevated levels of inflation (Koch et al., 2002; Cömert & Yeldan, 2018). Moreover, the increasing imports and minimal exports also led to high amount of deficit (Kriwoluzky & Rieth, 2018; Koch, et al., 2002). In-addition, during the periods of crisis in 1994 and 2001, the Turkish government enacted rigorous monetary and fiscal measures, which likely exacerbated the impact the 1994 crisis, on the contrary, the policies adopted during the latest crisis i.e., 2008, and COVID-19 leaned towards being more expansive in nature, as a result the impact of crisis were averted (Cömert & Yeldan, 2018; Türel, 2010).

The dummy variable COVID-19 (dv2020) has a significant and negative impact on GDP, underscoring the adverse impact of COVID-19 pandemic on Turkish economy, although the impact is weak in nature, due to rapid policy responses in the form expansionary fiscal and monetary policies which led to increase in the demand during the COVID-19 period, keeping the economy afloat (Tuysuz et al., 2022; Çakmaklı et al., 2021).

Additional factors, such as foreign direct investment (fdid1), inflation rate (infd1), and level of openness (opend1), usually do not demonstrate substantial influences on the variables being measured, except for a slight favorable impact of fdid1 on debt (coefficient: 0.00262, $p < 0.05$). The constant values have significance in relation to GDP and CAD, featuring both positive and negative coefficients correspondingly, suggesting foundational effects in a scenario where other factors remain consistent.

While the shocks of 1994 and 2020 COVID-19 has a significant and negative impact on GDP, and 1994 shocks has a positive impact on CAD, on the other hand, shock of 2001 and 2008 shows no significant impact on GDP and other independent variables i.e., gross savings, external debt, and current account deficit. This indicates the shocks these shocks have no impact on these variables. To analyze the relationship in-depth, and robustly, a policy-oriented study is recommended.

4 Conclusion

This research employs a time series dummy variable model to evaluate the influence of four significant shocks – specifically, the crises of 1994, 2001, 2008, and 2020, including the COVID-19 pandemic – as well as other independent variables on four key economic indicators: GDP growth, current account deficit, gross savings, and external debt for the period spanning from 1991 to 2023 in Turkey. The analysis indicates that factors such as unemployment, external debt, the 1994 crisis, and the 2020 COVID-19 crisis have a detrimental and statistically significant effect on GDP growth, emphasizing their critical role in ensuring sustainable economic development in Turkey. The substantial and beneficial impact of the openness index on GDP suggests that the level of openness in the Turkish economy is a key factor in driving economic growth. In-addition, the adverse effect of high gross savings on external debt illustrates that savings play a significant role in lowering external borrowing for Turkey. Moreover, Gross savings display a notable inverse correlation with the current account deficit (CAD), indicating that increased savings contribute to a reduction in the deficit. The economic crises experienced by Turkey in 2001 and 2008 did not have a substantial effect on key indicators such as gross savings, external debt, and current account deficit. Therefore, a comprehensive analysis is necessary to fully understand the correlation.

The findings of our study lead to the following policy recommendations:

- Modifying the monetary policy framework focused on inflation targeting, which has positioned the CBRT as a surrogate for financial interests, to incorporate additional economic goals.
- Shifting the focus from an excessive emphasis on 'exporting at all costs' to a more unified and equitable industrial strategy.
- Enhancing productivity and job creation in the agricultural industry, which has been adversely affected by the experimental nature of World Bank-driven 'reform' measures;
- Exploring practical remedies for the challenge of 'jobless growth' may have been the key domains where these potentials and prospects came to fruition.
- Shifting focus towards utilizing domestic avenues such as domestic borrowing and increasing domestic savings as opposed to relying on external borrowing to fund the economy.

References

Çakmaklı, C., Demiralp, S., Yeşiltaş, S., & Yıldırım, M. A. (2021). An evaluation of the Turkish economy during COVID-19. *SWP Working Papers*.

Öniş, Z., & Güven, A. B. (2011). Global crisis, national responses: The political economy of Turkish exceptionalism. *New Political Economy, 16*(5), 585–608.

Aghion, P., Bacchetta, P., & Banerjee, A. V. (2004). Financial development and the instability of open economies. *Journal of Monetary Economics, 51*(6), 1077–1106.

Aizenman, J., & Powell, A. (2003). Volatility and financial intermediation. *Journal of International Money and Finance, 22*(5), 657–679.

Altunöz, U. (2023). Analysis of Turkey's classic and financial kuznets curves in regards to the level of development and increased distribution. *Journal of Ekonomi, 10*, 60–69.

Cömert, H., & Yeldan, E. (2018). A tale of three crises in Turkey: 1994, 2001 and 2008–09. *ERC Working Papers in Economics 18/09*.

Cespedes, L. (2013). The impact of uncertainty shocks in emerging economies. *Journal of International Economics, 90*(2).

Chan, L. (2021). *Post-Covid prospects: Turkey*. Hong Kong: HKTDC Research.

Claessens, n., Kose, A., & Terrones, M. E. (2010). Recessions and financial disruptions in emerging markets: A Bird's Eye View. *SSRN Electronic Journal*, 55–84.

Claessens, S., Kose, M. A., & Terrones, M. E. (2011). How do business and financial cycles interact? . *MF Working Paper WP/11/88*, 1–54.

Dixit, A. (1992). Investment and hysteresis. *The Journal of Economic Perspectives, 6*(1), 107–132.

Dollar, D. (1992). Outward-oriented developing economies really do grow more rapidly: Evidence from 95 LDCs, 1976–1985. *Economic Development and Cultural Change, 40*(3), 523–544.

Edwards, S. (1993). Openness, trade liberalization, and growth in developing countries. *Journal of Economic Literature, 31*(3), 1358–1393.

Er, P. H., Karacar, Z., & Ozturk, R. H. (2014). Analysis of economic growth of Turkey: Past and pre 2001 financial crisis. *Journal of Economics Finance and Accounting, 1*(1), 26–37.

Felipe, L., Chang, C. R., & Velasco, A. (2004). Balance sheets and exchange rate policy. *American Economic Review, 94*(4), 1183–1193.

Haque, S. (2017). How do fiscal shocks affect the macroeconomic dynamics in the Bangladesh Economy. 24.

Ismihan, M., Metin-Ozcan, K., & Tansel, A. (2003). Macroeconomic instability, capital accumulation and growth: The case of Turkey 1963–1999. *Bilkent University Working Paper No. 02-05*, 29.

Keho, Y. (2017). The impact of trade openness on economic growth: The case of Cote d'Ivoire. *Taylor & Francis Journals, 5*(1).

Koch, L., Chaudhary, M. A., & Bilquees, F. (2002). February 2001 crisis in Turkey: Causes and consequences. *The Pakistan Development Review, 40*(4), 467–486.

Koray, F. (2006). Inflation variability and the Turkish economy. *Applied Economics*, 787–793.

Kriwoluzky, A., & Rieth, M. (2018). *Policy responses to Turkey's crisis: Independent central bank and international credit.* Berlin: Deutsches Institut für Wirtschaftsforschung (DIW).

Levine, R., & Renelt, D. (1992). A sensitivity analysis of cross-country growth regressions. *The American Economic Review, 82*(4), 942–963.

Neumeyer, P. A., & Perri, F. (2005). Business cycles in emerging economies: The role of interest rates. *Journal of Monetary Economics, 52*(2), 345–380.

Orhangazi, Ö., & Yeldan, A. E. (2021). The re-making of the Turkish crisis. *Development and Change*, 460–503.

Ozturk, O., & Radouai, N. (2020). Does trade openness contribute to economic growth and development of Morocco? *The Third Economics, Business and Organization Research (EBOR) Conference* (pp. 443–453). Business and Organization Research (EBOR).

Pindyck, R. S. (1991). Irreversibility, uncertainty, and investment. *Journal of Economic Literature, XXIX*, 1110–1148.

Ramey, G., & Ramey, V. A. (1995). Cross-country evidence on the link between volatility and growth. *American Economic Association*, 1138–1151.

Rodríguez, F., & Rodrik, D. (2000). Trade policy and economic growth: A Skeptic's guide to the cross-national evidence. *NBER Macroeconomics Annual*, 261–325.

Rodrik, D. (2009). The Turkish economy after the crisis. *Discussion Paper No. 2009/9.*

Rodrik, D. (2012). The Turkish economy after the Global Financial Crisis. *Ekonomi-tek Volume, 1*(1), 41–61.

Sachs, J. D., & Warner, A. (1995). Economic reform and the process of global integration. *Brookings Paper-s on Economic Activity, 1.*

Türel, O. (2010). Türkiye'de 1994, 2001 ve 2008-9 Ekonomiik Krizlerinin Karşilaştirmali Analizi. *Ekonomik Yaklaşım, 21*(75), 27–75.

Tuysuz, S., Baycan, T., & Altuğ, F. (2022). Economic impact of the COVID-19 outbreak in Turkey: Analysis of vulnerability and resilience of regions and diversely affected economic sectors. *Asia-Pacific Journal of Regional Science, 6*, 1133–1158.

Utkulu, U., & Özdemir, D. (2004). Does trade liberalization cause a long run economic growth in Turkey. *Economics of Planning, 37*, 245–266.

Uygur, E. (2010). The global crisis and the Turkish economy. *Discussion Paper, No. 2010/3.*

Meral Çabaş[1] & Bengü Açdoyuran[2] & Nazife Özge Beşer[3]

Are the Shocks to the Load Capacity Factor Permanent or Temporary in Türkiye?

1 Introduction

Climate change is an important global problem affecting the world and its people, and its effects have become more pronounced in recent years. Changes in climatic conditions adversely affect water resources and biodiversity, leading to deterioration of water quality, food security and the regularity of seasons, which threatens human health (Guloglu et al., 2023). The environmental impacts of human activities on natural resources and production areas have increased interest in sustainability management and policies on a global scale (Rugani et al., 2014). Increasing pressures on the ecosystem lead to serious imbalances in the geographical distribution of resources, negatively affecting biodiversity and human well-being. In order to protect the welfare of future generations, current resource consumption and waste production should not exceed the capacity of natural resources to renew themselves and eliminate waste (Yılancı & Pata, 2020). In this context, the United Nations Framework Convention on Climate Change, Kyoto Protocol and Paris Agreement encourage countries to develop and implement climate policies (IPCC, 2023).

Today, the environment is seen as an economic and limited resource, emphasizing the importance of problems such as global warming, climate change and environmental pollution. For this reason, many scientists have conducted studies focusing on environmental problems. Environmental economists and policy makers have focused on climate change and the increasing threats brought by this change (Başoğlu & Akyazı, 2015). Increasing attention to climate change has led

---

1 Ph.D., Independent Researcher, dogacabas@hotmail.com, ORCID: 0000-0002-3335-3297.

2 Ph.D., Asst. Prof., Agri Ibrahim Cecen University, Faculty of Economics and Administrative Sciences, Department of Economics, bcetin@agri.edu.tr, ORCID: 0000-0003-1513-1110.

3 Ph.D., Assoc. Prof., Agri Ibrahim Cecen University, Faculty of Economics and Administrative Sciences, Department of Economics, nokilic@agri.edu.tr, ORCID: 0000-0002-6738-3638.

economists and policymakers to re-evaluate their impact on the environment, highlighting the paradigm of sustainable economic growth (Alam et al., 2016). Considering the recent studies conducted by researchers on the environment, it is seen that they prefer ecological footprint instead of CO_2 emissions to evaluate environmental degradation (Al-Mulali et al., 2015; Wang et al., 2011; Galli et al., 2012). However, while CO_2 and ecological footprint reflect the damage caused by human activities to the environment, they ignore how nature responds to these damages. The ecological footprint is an indicator that measures the sustainability of a population's consumption in global hectares. However, the ecological footprint only measures the pressure that humans have on the environment and can ignore the sustainability of the resources provided by nature. In addition to producing harmful atmospheric pollutants, human activities also negatively affect biodiversity, causing serious ecological problems. This situation has become so critical that it can no longer be ignored by policy makers (Sinha & Schneider, 2024). On the other hand, although ecological footprint is a widely used tool by ecological and political communities, it has received serious criticism (Van den Bergh & Verbruggen, 1999; Moffatt, 2000; Rapport, 2000; Lenzen et al. (2007). At this point, Siche et al. (2010) proposed the load capacity factor (LCF) calculated based on the biological capacity (BC)/EF ratio, in order to simultaneously evaluate the pressures of human activities on air, water and soil and the capacity of natural resources to cope with these pressures. There are many studies on the determinants and macroeconomic effects of biocapacity. These studies generally focused on the effects of environmental degradation and changes in natural capital on output. The purpose of this study is to examine whether the load capacity factor in Türkiye is stationary for the period 1961–2022. Ecological footprint per person and biocapacity per person were used to calculate the load capacity factor, and the load capacity variable was obtained by dividing the biocapacity by the ecological footprint. In this study, 3 different categories of unit root tests were used to measure stationarity. First, 6 different traditional unit root tests were applied; second, structural break unit root tests allowing for 1 and 2 structural breaks were applied. Finally, since both groups of unit root tests do not take into account slow structural changes in economies, the FKPSS unit root test based on the Fourier approach, which is capable of capturing smooth structural changes, was applied and deep and comprehensive information about the LCF series in Türkiye was obtained.

2 Literature Survey

Ongoing industrial activities in many countries around the world have brought environmental problems that affect the biocapacity and ecological footprint of these countries. The negative effects of industrialization and rapid technological developments pose an increasing environmental problem. In this context, scientists who conduct solution-oriented studies on environmental problems are developing various methods to evaluate human-induced environmental pressure. Ecological footprint, one of these indicators, is recognized as an indicator of sustainability and sustainable development. The study of ecological footprint is of great importance in the implementation areas of global and local policies that combat global warming, climate change and environmental degradation Yılancı et al. (2022). In recent years, many studies have attempted to assess the determinants and responses of biocapacity relative to a range of socioeconomic indicators in various settings. In many studies, the effect on economic growth and subjective well-being has been examined using biocapacity and natural capital data (Zhang et al., 2017; Chen et al., 2019; Hassan et al., 2019; Ali et al., 2021). Some of the results obtained concluded that biocapacity and subjective well-being are positively related in high-income countries. In addition, it was concluded that biocapacity and financial development interact significantly; other studies have not found a significant relationship between economic growth and biocapacity.

In recent years, the effects of factors affecting the environment on economic growth and development, as well as the permanence of shocks on these factors, have become one of the attention-grabbing issues. Therefore, studies have been conducted on the stability of environmental factors. In many of the studies conducted in this field in the existing literature, especially CO_2 emissions have been used as data as an environmental factor (Acar & Lindmark 2017; Ye et al. 2020; Lee et al. 2023; Borowiec & Paiez, (2023). The results obtained from the studies vary in terms of stationarity. However, these studies do not provide comprehensive information in terms of environmental factors. Because CO_2 emissions alone are not a variable that provides sufficient information about the changes occurring in resources such as forestry, mining, water and soil. Therefore, researchers have begun to examine the ecological footprint and the stability of other components in their analyzes in order to obtain more comprehensive information about environmental problems. Considering that CO_2 emissions seem insufficient as an indicator of environmental degradation, Solarin and Bello (2018), revealing the increasing importance of the ecological footprint as a collective environmental indicator, examined the stability of the ecological footprint of 128 countries for the period 1961–2013. As a result of the analysis, it was concluded that the ecological footprint data for 96 countries were not stationary. Ulucak and Lin (2017), who investigated the stability of the ecological footprint components of the USA, concluded that the ecological footprint is not stationary. In their study covering the period 1961–2016, Yilanci and Pata (2020) examined the stability of ecological balance for a sample of E-7 and G-7 countries with unit root tests with structural breaks. The analysis showed that the ecological balance was stable in Germany, Italy, Japan, the USA, China, India, Indonesia, Mexico and Türkiye, while it was not stable in the remaining five countries. Caglar et al. (2021) examined the resilience of the ecological footprint and its subcomponents to shocks (economic, political, pandemics, etc.) in EU-5 countries and found that there is a unit root for all countries except the built-up land footprint for Spain and the grassland and forest footprint for the UK. Most of the studies have concluded that the ecological footprint is non-stationary. In addition, there are studies that conclude that the ecological footprint is stationary. Applying Panel KPSS stationarity test for G20 countries, Bilgili and Ulucak (2018) find that the ecological footprint is stationary for these countries. Moreover, Bilgili et al. (2019) applied stationarity test for 60 countries selected from Asia, Africa, America and Europe. As a result of the analysis, they found that there is stationarity in countries in other regions except Asian countries.

3 Data and Methodology

3.1 Data

This study focuses on understanding whether environmental shocks have a lasting impact on environmental balance in Türkiye. The load capacity factor (LCF) is used to represent the environmental balance and is derived from the ratio of biocapacity data to ecological footprint. Biocapacity and ecological footprint data are compiled from the Global Footprint Network (2024). Econometric analyses were carried out with the help of annual data from 1961 to 2022. In the analyses, the stationarity of the environmental balance is comprehensively investigated with traditional unit root tests such as ADF, PP, KPSS, DF-GLS, ERS-PO and NG-Perron, unit root tests with 1 and 2 structural breaks such as ZA, DF and LP, and FKPSS nonlinear unit root test based on the fourier approach. The analyses are performed with Eviews and Gauss econometrics packages.

3.2 Econometric Methodology

In this study, the stationarity of the LCF series is analyzed with 3 different categories of unit root tests. Firstly, ADF (Augmented Dickey-Fuller, 1981), PP (Philips Perron, 1988), KPSS (Kwiatkowsky-Philips-Schmidt-Shin, 1992), DF-GLS (Dickey-Fuller Generalized Least Squares, 1996), ERS-PO (ERS Point Optimal, 1996) and NG-Perron (2001) tests were used. Despite the assumption that ADF, PP, and KPSS unit root tests are weak when the sample is small, Elliot et al. (1996) developed ADF-GLS and Ng and Perron (2001) developed NG-Perron tests. However, all these tests do not take into account structural changes in economies. Secondly, unit root tests with structural breaks were used in the study. Perron (1989) included structural changes (1929 Great Depression, 1973 Oil Crisis) as dummy variables for the first time in his study. Ziwot-Andrews (1992) criticized Perron's inclusion of exogenous structural changes in the model and found it to be unobjective. Arguing that this may vary from person to person, they developed the unit root test (UFT) in which the structural change is calculated endogenously. However, this test was criticized for allowing only one structural change and Lumsdaine-Papell (1997) improved the test and developed the LP unit root test that allows two structural changes. It has been argued that unit root tests that allow more than two structural changes are not more robust (Lee & Strazicich, 2003). Another weakness of these tests using dummy variables is that they ignore slow structural changes in economies. To overcome this problem, Fourier functions constructed with trigonometric terms have been introduced. Fourier functions are unaffected by the number, shape and location of structural changes and have robust results. Accordingly, the stationarity of LCF is finally tested with the Fourier KPSS (FKPSS) unit root test developed by Becker et al. (2006).

4 Econometric Findings and Discussion

This section presents the empirical evidence from the econometric analysis and a discussion of these findings. Traditional unit root tests are applied and the results are reported in Table 1.

Table 1: Traditional Unit Root Test Findings

Unit Root Tests	Intercept		Trend and Intercept	
	Level	First Difference	Level	First Difference
ADF test	-3.16639	-10.66109***	-2.161886	-11.95446***
PP test	-2.425925	-10.27317***	-1.878493	-11.97225***
KPSS test	0.954599	0.496379***	0.218989	0.022735***
DF-GLS test	0.820439	-2.001582**	-1.112781	-2.661091***
ERS-PO test	606.3956	1.817413***	29.15336	5.409475**
NG-Perron test	0.37895	-14.47777***	-4.43569	-26.08769***

Note: *, **, *** indicate the significance level of 10 %, 5 % and 1 %, respectively.

When we look at the results of the traditional unit root tests, the test statistics of the ADF and PP tests, which test the null hypothesis "The series contains a unit root", are smaller in absolute value than the critical values for both models (intercept, intercept & trend). This indicates that the null hypothesis cannot be rejected and the series is unit rooted. Unlike ADF and PP, the null hypothesis of the KPSS test is "The series is stationary". Since the statistic obtained from this test is greater than the table value, the null hypothesis is rejected and the LFC series is unit rooted. The difference of LCF series is stationary at I(1) degree in all three tests. The null hypothesis of the DF-GLS, ERS-PO and NG-Perron (MZa, MZt) unit root tests is "The series is unit rooted". The fact that DF-GLS and NG-Perron (MZa) test statistics are less than the critical value in absolute value and the ELS-PO test statistic Pt is larger than the critical value shows that LCF has unit root. In the second part of the analysis, LCF series unit root tests with structural breaks were tested and the findings are reported in Table 2.

Table 2: Unit Root Test Findings with Structural Breaks

Unit R. Tests		Intercept				Trend			
	Conditions	Level	Break Date	First Difference		Level	Break Date	First Difference	
ZA test		-3.014530	2009	-7.8206***	1978	-3.0353	2008	-7.3974***	1981
DF test	Innovative outlier	-3.831532	2021	-12.4908***	1978	-2.6750	2008	-12.5095***	1978
	Additive outlier	-2.502452	2000	-12.6688***	1977	-2.7707	2008	-12.7635***	1978
		Test Statistic		**Break Date**		**Critical Values**			
LP test	Model AA	-4.8300		1973–2011		-6.94(%1)	-6.24(%5)	-5.96(%10)	
	Model BB	-5.1486		1973–2012		-7.34	-6.82	-6.49	

Note: *, **, *** indicate the significance level of 10 %, 5 % and 1 %, respectively.

Unit root test results with structural breaks show that for all three tests and all models, the LCF series contains unit root at level and becomes stationary at the first difference. When we examine the dates of structural breaks; It refers to the first major "Oil Crisis" in 1973 and the "Global Financial Crisis" that broke out in the USA in 2008. The year 2000 is also observed as the date of a structural break for the Turkish economy, in which the macroeconomic balance that paved the way for the 2001 crisis was deviated. While 2011 and 2012 again point to an economic crisis, 2021 appears as a structural change in which the effects of the

Covid-19 pandemic that affected the whole world are seen. In the last part of the analysis, the FKPSS unit root test based on the Fourier approach was applied and the results are reported in Table 3.

Table 3: Fourier KPSS Unit Root Test Findings

Variable	Model	MinSSR	FKPSS	Frequency (k)	F Test Stat.
LCF	Constant	3.420048	0.423401 (6)	1	36.71342
LCF	Constant & trend	0.196486	0.068814 (5)	1	51.69630

Note: The inside () refers to the bandwidth automatically found by the Newey-West method.

The fact that the F test statistics for both models of the Fourier KPSS test, which can capture smooth transition structural changes, are greater than the table values indicates that the trigonometric terms are significant and the FKPSS test can be used. The fact that the FKPSS test statistics are larger than the table values in Becker et al. (2006) indicates that the LCF contains a unit root for both models. Guneysu (2023) also found that the LCF series for Türkiye has a unit root.

5 Conclusion and Recommendation

This study empirically examines the stationarity of the load capacity factor in Türkiye using annual data for the period 1961–2022. A chronological unit root methodology is followed in the analysis to check the robustness of the stationarity of the LCF series and robust findings are obtained. All test results show that the load capacity factor is non-stationary in Türkiye. This confirms that environmental shocks have persistent effects in Türkiye. Increasing the load capacity factor is possible with an increase in ecological footprint or a decrease in biocapacity. In this regard, it would be beneficial for policy makers to take these results into consideration when regulating environmental policies. Accelerating the green energy transition, increasing and expanding the use of green technologies, efficient use of resources, more funding to tackle the climate crisis and many other green measures can reduce the ecological footprint. Waste management is also critical to maintain biocapacity. Pollution of agricultural lands, water resources and forests with waste must be prevented. It is the responsibility of the whole world to preserve the ecological balance and leave a habitable planet for the future.

References

Acar, S., & Lindmark, M. (2017). "Convergence of CO_2 emissions and economic growth in the OECD countries: Did the type of fuel matter?". *Energy Sources B Energy Economics Planning, 12*, 618–627. https://doi.org/10.1080/15567249.2016.1249807.

Al-Mulali, U., & Ozturk, I. (2015). "The effect of energy consumption, urbanization, trade openness, industrial output, and the political stability on the environmental degradation in the MENA (Middle East and North African) region". *Energy, 84*, 382–389.

Alam, M. M., Murad, M. W., Noman, A. H. M., & Ozturk, I. (2016). "Relationships among carbon emissions, economic growth, energy consumption and population growth: Testing Environmental Kuznets Curve hypothesis for Brazil, China, India and Indonesia". *Ecological Indicators, 70*: 466–479.

Ali, A., Usman, M., Usman, O., & Sarkodie, S. A. (2021). "Modeling the effects of agricultural innovation and biocapacity on carbon dioxide emissions in an agrarian-based economy: Evidence from the dynamic ARDL simulations". *Frontiers Energy Research, 8*: 381. https://doi.org/10.3389/fenrg.2020.592061.

Başoğlu, A., & Akyazı, H. (2015). "The IS-LM model adapted to environment in macroeconomic analysis of environmental issues". *Marmara Üniversitesi İ.İ.B. Dergisi, 37*(2), 129–152.

Becker, R., Enders, W., & Lee, J. (2006). "A stationary test in the presence of an unknown number of smooth breaks". *Journal of Time Series Analysis, 27*(3), 381–409.

Bilgili, F., & Ulucak, R. (2018). "Is there deterministic, stochastic, and/or club convergence in ecological footprint indicator among G20 countries?". *Environmental Science and Pollution Research, 25*(35), 35404–35419.

Bilgili, F., Ulucak, R., & Koçak, E. (2019). Implications of environmental convergence: Continental evidence based on ecological footprint. In *Energy and environmental strategies in the era of globalization*. Berlin: Springer.

Borowiec, J., & Papież, M. (2023). "CO_2 emissions convergence in countries at different levels of development. Do environmental policy and globalisation matter?". Available at SSRN 4380388.

Caglar, A. E., Balsalobre-Lorente, D., & Akin, C. S. (2021). "Analysing the ecological footprint in EU-5 countries under a scenario of carbon neutrality: Evidence from newly developed sharp and smooth structural breaks in unit root testing". *Journal of Environmental Management, 295*, 113155.

Chen, S., Saud, S., Saleem, N., & Bari, M. W. (2019). "Nexus between financial development, energy consumption, income level, and ecological footprint in CEE countries: Do human capital and biocapacity matter?". *Environmental*

Science and Pollution Research, 26(31), 31856–31872. https://doi.org/10.1007/s11356-019-06343-z.

Elliott, G., Rothenberg, T., & Stock, J. H. (1996). "Efficient tests for an autoregressive unit root". *Econometrica, 64*, 813–836. http://dx.doi.org/10.2307/2171846.

Galli, A., Kitzes, J., Niccolucci, V., Wackernagel, M., Wada, Y., & Marchettini, N. (2012). "Assessing the global environmental consequences of economic growth through the Ecological Footprint. A focus on China and India". *Ecological Indicators, 17*, 99–107.

Global Footprint Network (2024). Biocapacity and ecological footprint Data, https://www.footprintnetwork.org.

Guloglu, B., Caglar, A. E., & Pata, U. K. (2023). "Analyzing the determinants of the load capacity factor in OECD countries: Evidence from advanced quantile panel data methods". *Gondwana Research, 118*, 92–104.

Guneysu, Y. (2023). "Türkiye'de finansal gelişme, küreselleşme ve sanayileşmenin yük kapasite faktörü üzerindeki etkisi". *Gümüşhane Üniversitesi Sosyal Bilimler Dergisi, 14*(3), 934–946.

Hassan, S. T., Baloch, M. A., Mahmood, N., & Zhang, J. (2019). "Linking economic growth and ecological footprint through human capital and biocapacity". *Sustainable Cities and Society, 47*, 101516. https://doi.org/10.1016/j.scs.2019.101516.

IPCC. (2023). Synthesis report of the IPCC sixth assessment report (AR6), https://www.ipcc.ch/report/ar6/syr/downloads/report/IPCC_AR6_SYR_LongerReport.pdf.

Kwiatkowski, D., Phillips, P.C., Schmidt, P., Shin, Y., (1992). "Testing the null hypothesis of stationarity against the alternative of a unit root: How sure are we that economic time series have a unit root? J". Econom. 54 (1–3), 159–178.

Lee, J., & Strazicich, M. C. (2003). "Minimum Lagrange Multiplier Unit Root Test with TwoStructural Breaks". *The Review of Economics and Statistics, 85* (4), 1082–1089.

Lee, J., Yücel, A. G., & Islam, M. T. (2023). "Convergence of CO_2 emissions in OECD countries". *Sustainable Technology and Entrepreneurship, 2*(1), 100029.

Lenzen, M., Hansson, C. B., & Bond, S. (2007). "On the bioproductivity and land-disturbance metrics of the Ecological Footprint". *Ecological Economics, 61*(1), 6–10.

Lumsdaine, R. L., & Papell, D. H. (1997). "Multiple trend breaks and the unit root hypothesis". *The Review of Economics and Statistics, 79*(2), 212–218.

Moffatt, I. (2000). "Ecological footprints and sustainable development". *Ecological Economics, 32*(3), 359–362.

Ng, S., & Peron, P. (2001). "Lag length selection and the construction of Unit Root Tests with good size and power". *Econometrica*, *69*(6), 1519–1554. http://dx.doi.org/10.1111/1468-0262.00256

Perron, P. (1989). "The great crash, the oil price shock, and the unit root hypothesis". *Econometrica*, *57*(6), 1361–1401.

Rapport, D. J. (2000). "Ecological footprints and ecosystem health: Complementary approaches to a sustainable future". *Ecological Economics*, *32*(3), 367–370.

Rugani, B., Roviani, D., Hild, P., Schmitt, B., & Benetto, E. (2014). "Ecological deficit and use of natural capital in Luxembourg from 1995 to 2009". *Science of the Total Environment*, *468*, 292e301. https://doi.org/10.1016/j.scitotenv.2013.07.122.

Siche, R., Pereira, L., Agostinho, F., & Ortega, E. (2010). "Convergence of ecological footprint and emergy analysis as a sustainability indicator of countries: Peru as case study". *Communications in Nonlinear Science and Numerical Simulation*, *15*(10), 3182–3192.

Sinha, A., & Schneider, N. (2024). "Stationarity and convergence of LCF series". *Gondwana Research*, *129*, 425–440.

Solarin, S. A., & Bello, M. O. (2018). "Persistence of policy shocks to an environmental degradation index: The case of ecological footprint in 128 developed and developing countries". *Ecological Indicators*, *89*, 35–44. doi: 10.1016/j.ecolind.2018.01.064.

Ulucak, R., & Lin, D. (2017). "Persistence of policy shocks to ecological footprint of the USA". *Ecological Indicators*, *80*, 337–343. https://doi.org/10.1016/j.ecolind.2017.05.020

Van den Bergh, J. C., & Verbruggen, H. (1999)."Spatial sustainability, trade and indicators: An evaluation of the ecological footprint". *Ecological Economics*, 29(1), 61–72.

Wang, S. S., Zhou, D. Q., Zhou, P., & Wang, Q. W. (2011). "CO_2 emissions, energy consumption and economic growth in China: A panel data analyses". *Energy Policy*, *39*(9), 4870–4875.

Ye, C., Chen, Y., Inglesi-Lotz, R., & Chang, T. (2020). "CO_2 emissions converge in China and G7 countries? Further evidence from Fourier quantile unit root test". *Energy Environment*, *31*(2), 348–363. http://dx.doi.org/10.1177/0958305X19867082.

Yılancı, V., & Pata, U. K. (2020). "Are shocks to ecological balance permanent or temporary? Evidence from LM unit root tests". *Journal of Cleaner Production*, *276*, 124294.

Yılancı, V., Pata, U. K., & Cutcu, I. (2022). "Testing the persistence of shocks on ecological footprint and sub-accounts: Evidence from the big ten emerging markets". *International Journal of Environmental* Research, *16*(1), 10. https://doi.org/10.1007/s41742-021-00391-5.

Zhang, S., Shi, Q., & Cheng, M. (2017). "Renewable natural capital, the biocapacity, and subjective well-being". *Journal of Cleaner Production*, *150*, 277–286. https://doi.org/10.1016/j.jclepro.2017.03.021.

Zivot, E., & Andrews, D. (1992). "Further evidence on the great crash, the oil price shock, and the Unit Root Hypothesis". *Journal of Business & Economic Statistics*, *10*(3): 251–270.

Gülsüm Akarsu[1]

Impacts of Uncertainty, Economic Complexity and Environmental Policy Stringency on Foreign Direct Investments: Panel Data Quantile Regression Analysis

1 Introduction

Countries have made various arrangements and provide incentives to attract foreign direct investments into their economies which may accelerate growth through technology and knowledge spillover, job creation and capital transfers. The foreign investors on the other hand expect to increase their profits, reduce their production costs, access to local markets and resources of host countries. In the world, there are differences in the foreign direct investment inflows (FDI) among countries. Therefore, it is important to investigate the factors affecting FDI.

The aim of this study is to analyze the determinants of FDI by focusing on uncertainty, economic complexity, and environmental policy stringency by employing data on 32 countries[2] over 2002–2020 and quantile regression analysis. In the literature, various studies examine the impact of different factors on FDI, but up to our knowledge there is not any study focusing on these three factors together by using quantile regression analysis. Due to high sunk costs associated with FDI and its adverse effects on economic activities such as production and investment, uncertainty may decrease FDI (Bommadevara & Sakharkar, 2021).

Although Islam and Beloucif (2024) mention the market seeking as the most significant characteristic of FDI, Sadeghi et al. (2020) put forward efficiency-seeking behavior of FDI as the recent trend, thus consider the economic

---

1 Ph.D., Assoc. Prof., Ondokuz Mayis University, Faculty of Economic and Administrative Sciences Department of Political Science and Public Administration, akarsu.gulsum@gmail.com, ORCID: 0000-0002-4877-1969.

2 Australia, Austria, Belgium, Brazil, Canada, Chile, China, Czechia, Denmark, Finland, France, Germany, Greece, Hungary, Ireland, Israel, Italy, Japan, South Korea, Mexico, Netherlands, New Zealand, Norway, Poland, Portugal, Russian Federation, Slovak Republic, South Africa, Spain, Sweden, Switzerland, and United Kingdom.

complexity indicating country's capabilities, social capital, institutional quality, productive knowledge, sophistication of economic activities, and easier creation of upstream and downstream linkages among the determinants of FDI.

For environmental policy strength, the studies consider different theoretical views: pollution haven hypothesis, pollution halo effect, race to the bottom, environmental Kuznets curve (EKC) and Porter's hypothesis (Bae et al., 2022). According to pollution haven hypothesis, firms prefer the countries with flax environmental policy to reduce costs, but in contrast, Porter's hypothesis suggests that stringent environmental policies may lead to innovation which reduces costs, improves firm performance, and provide competitive advantage, therefore, may attract FDI. Race to the bottom assumes that countries may relax environmental regulations to accelerate FDI. However, FDI may also reduce pollution through green and environmentally friendly technology transfer which is known as pollution halo effect. Lastly, EKC hypothesizes an inverse U-shaped relationship between economic development and pollution. Through its favorable impact on economic development, FDI may also reduce pollution.

2 Literature Review

In the literature, huge number of studies have analyzed the factors affecting the foreign direct investment flows. There are different theories such as eclectic ownership, location, and internationalization advantage theory, product life cycle theory, institutional theory, and resource-based view (Zhang et al., 2023). According to Islam and Beloucif (2024), the main determinants of FDI inflows are domestic investment, natural resource endowment, political stability, level of crime and corruption, labor productivity, governance indicators, corporate tax, market size, trade openness, exchange rate, inflation rate, interest rate, investment and business climate, unemployment rate, growth, labor cost, infrastructure quality and human capital. They concluded that the most important determinant is the market size. However, in the literature various other factors have also been examined. As mentioned by Sadeghi et al. (2020), throughout history, based on the changes in sectoral composition of FDI, the dominant factor has changed from natural resource endowment, market size, institutions, trade agreements, technology, skills to high technology and labor with high skills. This study focuses on the studies analyzing the effects of uncertainty, economic complexity, and environmental policy stringency on FDI.

Findings of Bommadevara and Sakharkar (2021) indicate adverse impact of economic policy uncertainty on FDI based on dynamic panel data estimation for 22 countries between 2000 and 2019. But they showed the importance of economic growth, human capital, exchange rates, trade openness, financial openness, exchange rate stability, financial development, environmental and institutional quality for FDI attraction. For 19 high income countries over 2000–2021 using panel ARDL model, Haque et al. (2022) found adverse effects of economic policy uncertainty, inflation and exchange rate, but the favorable impacts of growth and trade openness which may ensure investor confidence. The adverse impact of uncertainty was also shown by Marasanti and Verico (2023) for 10 ASEAN countries between 2015 and 2019 employing system panel GMM estimation method in addition to favorable effects of currency depreciation on FDI. Using panel data on 48 Asian countries over 2000–2020 and system GMM estimation technique, Zhang et al. (2023) found adverse impacts of global and domestic economic policy uncertainty on FDI, but domestic policy uncertainty's negative effect declines with financial development.

Sadeghi et al. (2020) showed the positive effect of economic complexity on FDI for a panel of 79 countries over 1980–2014 using dynamic panel data GMM-system estimation method along with adverse effects of real wage and geographical distance and favorable effects of market size, institutional quality, trade openness, domestic investment, human capital, common language, and financial development. In contrast, Uttama and Arwatchanakarn (2023) reported a negative effect of economic complexity but positive effect of productive capacity on FDI of 17 Asian countries between 1995 and 2020. In addition to the significant effect of economy's size, Gomez-Zaldıvar et al. (2021) found favorable and spillover impacts of economic complexity on 32 Mexican states' FDI and 85 Mexican manufacturing sector industrial groups' FDI over 1999–2018 based on spatial panel data analysis.

Sarmidi, Nor, and Ridzuan (2015) found the evidence of pollution haven effect and favorable impact of corruption on FDI employing dynamic panel data model and GMM estimation for 110 developed and developing countries over 2005–2012. But this adverse effect of environmental policy stringency declines with low corruption level. Other studies showing the negative effect of stringent environmental policies are Bialek and Weichenrieder (2021) and Kalamova and Johnstone (2011).

3 Methodology

In this study, author performs analysis based on panel data quantile regression method introduced by Powell (2022). The advantage of the method is that explanatory variables' impacts can be found on the distribution of foreign direct investment inflows $\left(fdi_{it}\right)$, therefore, the method provide outlier robust estimates, heterogenous impacts of explanatory variables, and better performance against nonnormality. Equation (1) shows the model employed in which explanatory variables' effects can be obtained at different quantiles of foreign direct investment inflows' conditional distribution. In this equation, fdi_{it} is explained as a function of lagged fdi_{it} (fdi_{it-1}) and following explanatory variables (X_{it}): logarithm of per capita real GDP $(lpcrgdp_{it})$, unemployment rate $(unemployment_{it})$, gross fixed capital formation rate $(capital_{it})$, industrialization $(industry_{it})$, inflation rate $(inflation_{it})$, logarithm of real effective exchange rate index $(lreexcr_{it})$, trade openness $(trade_{it})$, control of corruption $(corruption_{it})$, government effectiveness $(governmenteff_{it})$, regulatory quality $(regulatoryq_{it})$, political stability $(politicalstab_{it})$, economic complexity index (eci_{it}), uncertainty index $(uncertainty_{it})$ and environmental policy stringency $(environment_{it})$.

$$Q_{fdi_{it}}\left(q|X_{it}, fdi_{it-1}\right) = fdi_{it-1}\alpha_0\left(q\right) + X_{it}{'}\alpha_1\left(q\right) \quad (1)$$

where, $q \in (0,1)$ for country i and year t under the assumption of $P(fdi_{it} \leq fdi_{it-1}\alpha_0(q) + X_{it}{'}\alpha_1(q)|Instruments_i) = q_i$ and non-separable disturbance term. Fixing d given by $d_{it} = \left(fdi_{it-1}, X_{it}^{'}\right)$, one can obtain $fdi_d = d_{it}{'}\alpha\left(u^*\right)$ for randomly selected $u^* \sim U(0,1)$. The model is estimated by Generalized Methods of Moment estimation method producing consistent and asymptotically normal estimates shown by Powell (2022). The method does not specify and give estimates of nonadditive fixed effects. One may expect favorable impacts of $lpcrgdp_{it}$, $capital_{it}$, $industry_{it}$, $trade_{it}$, $corruption_{it}$, $governmenteff_{it}$, $regulatoryq_{it}$, $politicalstab_{it}$, and eci_{it} and adverse effects of $unemployment_{it}$, $inflation_{it}$, $lreexcr_{it}$, $uncertainty_{it}$, and $environment_{it}$.

4 Data

The dataset covers panel data series on 32 countries over 2002–2020 based on data availability and includes foreign direct investment inflows (% of GDP) [BX.KLT.DINV.WD.GD.ZS], real GDP per capita (constant 2021 international $, PPP) [NY.GDP.PCAP.PP.KD], unemployment rate [SL.UEM.TOTL.ZS], gross fixed capital formation (% of GDP) [NE.GDI.FTOT.ZS], industry share (% of GDP) [NV.IND.TOTL.ZS], inflation (annual %) [FP.CPI.TOTL.ZG], real effective exchange rate index (2010 = 100) [PX.REX.REER], trade openness [NE.TRD.GNFS.ZS], corruption control [CC.EST], government effectiveness [GE.EST], regulatory quality [RQ.EST], political stability [PV.EST], economic complexity index, uncertainty index and environmental policy stringency index. All the data are taken from The World Bank (2024), World Development Indicators Database except economic complexity index, uncertainty index and environmental policy stringency index. Economic complexity index, uncertainty index and environmental policy stringency index are obtained from The Growth Lab at Harvard (2019), Ahir et al. (2022) and Organisation for Economic Co-operation and Development [OECD] (2024).

5 Results

Table 1 shows the estimation results for pooled dynamic panel data model, fixed effects panel dynamic model and panel data quantile regression model. The estimation results are provided for different quantiles of foreign direct investment inflows. Results indicate the statistically significant positive effect of per capita real GDP and adverse effect of economic complexity based on the estimation of dynamic panel data model with fixed effects.

The statistically significant dynamic effect indicates the evidence of persistency in FDI and this effect increases at higher quantiles. Real GDP per capita was found to decrease FDI at almost all quantiles, therefore, foreign investors may not be market seeker as opposite to the findings of the most studies. The findings show the adverse effects of unemployment. FDI becomes much more responsive to unemployment rate at the higher levels of FDI. Moreover, findings show lower FDI associated with an increase in country's *capital* , *inflation* , *trade* and control of corruption at lower quantiles, and FDI increases because of increase in *capital* , *inflation* , *trade* and control of corruption at an increasing rate for higher quantiles. The FDI's responsiveness to industrialization and $regulatoryq_{it}$ in absolute values was found to exhibit a U shape as FDI increase. At lower quantiles of FDI, industrialization and $regulatoryq_{it}$ increases FDI, but at higher quantiles, FDI become much more responsive and adverse effects of industrialization and $regulatoryq_{it}$ were shown. Results indicate favorable impact of currency appreciation at higher quantiles of FDI, whereas adverse effects at lower quantiles. $governmenteff_{it}$ was shown to have adverse effects on FDI at almost all quantiles. Results indicate that adverse effects of $politicalstab_{it}$ at lower quantiles become favorable for higher quantiles of FDI.

Figure 1 demonstrates quantile treatment effect estimates for eci_{it}, $uncertainty_{it}$ and $environment_{it}$. Results indicate that as foreign direct investment inflows increase, the negative impact of economic complexity decreases, and its effect becomes positive at the 90th quantile. Therefore, at the highest ratio of foreign direct investment inflows in GDP, economic complexity increases ratio of foreign direct investment inflows. Sadeghi et al. (2020) and Gomez-Zaldıvar et al. (2021) also show favorable impact of economic complexity, but adverse effect is in line with the findings of Uttama and Arwatchanakarn (2023). Responsiveness of FDI to environmental policy stringency increases at higher quantiles of FDI. More stringent environmental policies cause much more declines in FDI at higher FDI ratios. Sarmidi et al. (2015), Bialek and Weichenrieder (2021), and Kalamova and Johnstone (2011) found the evidence of pollution haven effect, similarly.

Table 1: Results

Variable	Pooled	FE	q1	q2	q3
fdi_{it-1}	0.45***	0.33*	0.02***	0.26***	0.35***
	(1.17×10^{-1})	(1.25×10^{-1})	(4.10×10^{-4})	(2.72×10^{-5})	(2.83×10^{-6})
$lpcrgdp_{it}$	-0.68	15.92**	-1.20***	-0.78***	-0.17***
	(8.96×10^{-1})	(4.84)	(3.35×10^{-2})	(1.02×10^{-3})	(1.25×10^{-4})
$unemployment_{it}$	-0.09	0.21	-0.02***	0.02***	-1.54×10^{-3}***
	(8.72×10^{-2})	(1.48×10^{-1})	(1.14×10^{-3})	(1.80×10^{-5})	(6.16×10^{-6})
$capital_{it}$	-0.07	-0.38	-0.08***	-0.04***	-0.02***
	(5.94×10^{-2})	(2.49×10^{-1})	(1.18×10^{-3})	(4.01×10^{-5})	(5.71×10^{-6})
$industry_{it}$	-0.04	0.46	0.13***	0.10***	0.07***
	(1.17×10^{-1})	(3.82×10^{-1})	(4.46×10^{-4})	(7.24×10^{-6})	(5.86×10^{-6})
$inflation_{it}$	0.05	0.18	-0.05***	-0.04***	-0.05***
	(1.95×10^{-1})	(1.88×10^{-1})	(1.38×10^{-3})	(5.47×10^{-5})	(1.11×10^{-5})
$lreexcr_{it}$	-1.01	-5.78	-1.06***	-0.30***	-0.49***
	(3.51)	(5.62)	(3.07×10^{-2})	(1.30×10^{-3})	(1.11×10^{-4})
$trade_{it}$	0.06**	-0.03	-0.04***	-0.02***	-8.43×10^{-4}***
	(1.71×10^{-2})	(7.75×10^{-2})	(2.68×10^{-4})	(4.59×10^{-6})	(6.81×10^{-7})
$corruption_{it}$	1.53	1.97	-1.02***	-1.09***	0.09***
	(1.61)	(2.71)	(3.35×10^{-2})	(1.48×10^{-3})	(9.84×10^{-5})

Table 1: Continued

Variable	Pooled	FE	q1	q2	q3	
$governmenteff_{it}$	-0.56	-0.63	-0.12***	0.52***	-0.78***	
	(1.67)	(1.80)	(2.20×10^{-2})	(1.04×10^{-3})	(1.79×10^{-4})	
$regulatoryq_{it}$	-1.21	-1.91	2.85***	1.96***	1.31***	
	(1.18)	(5.26)	(4.98×10^{-2})	(1.64×10^{-3})	(8.16×10^{-5})	
$politicalstab_{it}$	-0.41	0.01	-0.88***	-0.50***	-0.58***	
	(1.11)	(2.49)	(7.73×10^{-3})	(4.46×10^{-4})	(1.22×10^{-4})	
eci_{it}	-0.40	-7.85***	-0.39***	-0.23***	-0.40***	
	(4.19×10^{-1})	(2.12)	(3.89×10^{-3})	(2.13×10^{-4})	(3.28×10^{-5})	
$uncertainty_{it}$	-4.18	-4.91	-0.48***	-0.60***	-0.84***	
	(3.03)	(4.29)	(8.41×10^{-3})	(1.09×10^{-3})	(9.24×10^{-5})	
$environment_{it}$	-0.47	-2.30	-0.17***	0.03***	-0.07***	
	(3.55×10^{-1})	(1.25)	(4.06×10^{-3})	(2.86×10^{-4})	(3.68×10^{-5})	
Variable	**q4**	**q5**	**q6**	**q7**	**q8**	**q9**
fdi_{it-1}	0.39***	0.38***	0.50***	0.57***	0.64***	0.62***
	(9.00×10^{-6})	(3.94×10^{-3})	(7.88×10^{-7})	(1.04×10^{-2})	(1.82×10^{-4})	(2.88×10^{-5})
$lpcrgdp_{it}$	-0.41***	-0.44***	-0.24***	0.15	-0.07***	-0.87***
	(2.48×10^{-4})	(4.74×10^{-2})	(1.04×10^{-4})	(1.31×10^{-1})	(9.77×10^{-3})	(5.44×10^{-4})
$unemployment_{it}$	-0.01***	1.19×10^{-3}	-0.04***	-0.05***	-0.10***	-0.21***
	(2.47×10^{-5})	(9.00×10^{-3})	(8.07×10^{-6})	(8.64×10^{-3})	(4.37×10^{-4})	(6.24×10^{-5})
$capital_{it}$	-4.25×10^{-3}***	0.01**	0.04***	0.08***	0.09***	0.10***
	(2.32×10^{-5})	(6.40×10^{-3})	(2.76×10^{-6})	(6.23×10^{-3})	(4.29×10^{-4})	(4.59×10^{-5})
$industry_{it}$	0.03***	0.01*	-0.05***	-0.09***	-0.17***	-0.31***
	(9.65×10^{-6})	(7.67×10^{-3})	(3.69×10^{-6})	(3.90×10^{-3})	(8.47×10^{-4})	(2.90×10^{-5})
$inflation_{it}$	-0.03***	0.01	0.04***	0.06***	0.09***	0.13***
	(2.61×10^{-5})	(1.51×10^{-2})	(7.64×10^{-6})	(1.20×10^{-2})	(1.64×10^{-3})	(1.09×10^{-4})
$lreexcr_{it}$	-0.09***	-0.70***	-0.87***	0.72***	-1.03***	0.80***
	(1.04×10^{-3})	(1.90×10^{-1})	(1.13×10^{-4})	(1.58×10^{-1})	(8.62×10^{-3})	(2.11×10^{-3})
$trade_{it}$	0.01***	0.02***	0.03***	0.04***	0.09***	0.15***
	(3.09×10^{-6})	(3.41×10^{-4})	(8.92×10^{-7})	(9.27×10^{-4})	(9.57×10^{-5})	(9.58×10^{-6})

(continued on next page)

Table 1: Continued

Variable	q4	q5	q6	q7	q8	q9
$corruption_{it}$	0.30***	0.37***	0.99***	0.35**	2.03***	2.80***
	(5.53×10^{-4})	(6.81×10^{-2})	(1.22×10^{-4})	(1.41×10^{-1})	(8.83×10^{-3})	(4.44×10^{-4})
$governmenteff_{it}$	-0.18***	4.42×10^{-3}	-0.68***	-0.12	-0.38***	-0.02***
	(4.30×10^{-4})	(6.23×10^{-2})	(1.60×10^{-4})	(1.55×10^{-1})	(1.33×10^{-2})	(5.68×10^{-4})
$regulatoryq_{it}$	0.49***	0.40***	-0.35***	-0.54***	-2.12***	-3.46***
	(2.46×10^{-4})	(8.16×10^{-2})	(9.07×10^{-5})	(1.44×10^{-1})	(4.22×10^{-3})	(5.17×10^{-4})
$politicalstab_{it}$	-0.50***	-0.24***	-0.07***	0.45***	0.22***	0.51***
	(3.29×10^{-4})	(7.79×10^{-2})	(3.85×10^{-5})	(1.06×10^{-1})	(5.07×10^{-3})	(6.08×10^{-4})
eci_{it}	-0.37***	-0.41***	-0.18***	-0.16***	-0.08***	0.12***
	(6.65×10^{-5})	(3.54×10^{-2})	(2.76×10^{-5})	(2.89×10^{-2})	(2.83×10^{-3})	(4.29×10^{-4})
$uncertainty_{it}$	-0.85***	-0.82***	-1.16***	-0.09	-1.90***	0.23***
	(5.33×10^{-4})	(6.98×10^{-2})	(2.49×10^{-4})	(1.18×10^{-1})	(1.11×10^{-2})	(1.05×10^{-3})
$environment_{it}$	-0.13***	-0.34***	-0.27***	-0.46***	-0.50***	-0.55***
	(1.44×10^{-4})	(5.90×10^{-2})	(1.50×10^{-5})	(2.70×10^{-2})	(3.49×10^{-3})	(3.59×10^{-4})

10 %, 5 %, and 1 % levels of statistical significance are shown by *, **, ***, respectively. The values in parentheses are Driscoll-Kraay robust standard errors for pooled (Pooled) and fixed effects (FE) dynamic panel data models and standard errors. Quantile regression model was estimated employing Markov Chain Monte Carlo numerical optimization method in Stata using qregpd command.
Source: (Author's calculation)

Moreover, increases in uncertainty decreases FDI at an increasing rate up to 70th quantile, and afterwards this adverse effect declines and become favorable at the highest level of FDI ratio in GDP. Bommadevara and Sakharkar (2021), Haque et al. (2022), Marasanti and Verico (2023), and Zhang et al. (2023) are the studies showing negative effects of uncertainty.

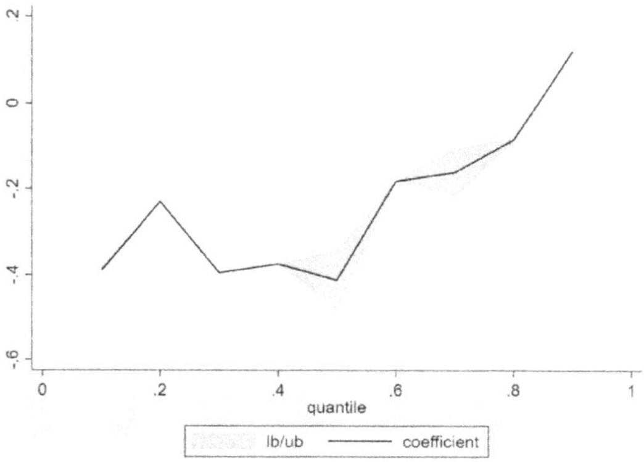

a. eci_{it}

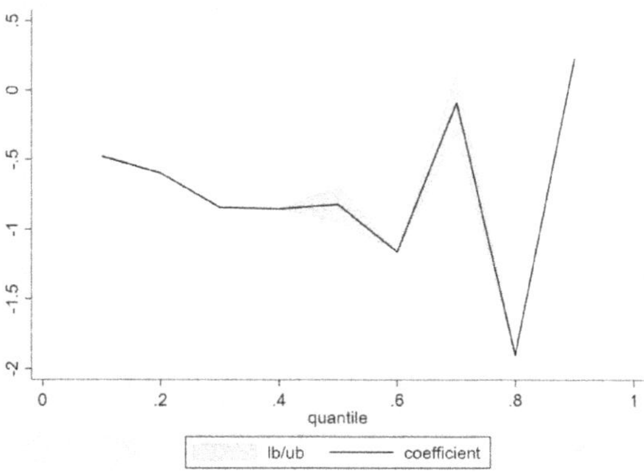

b. $uncertainty_{it}$

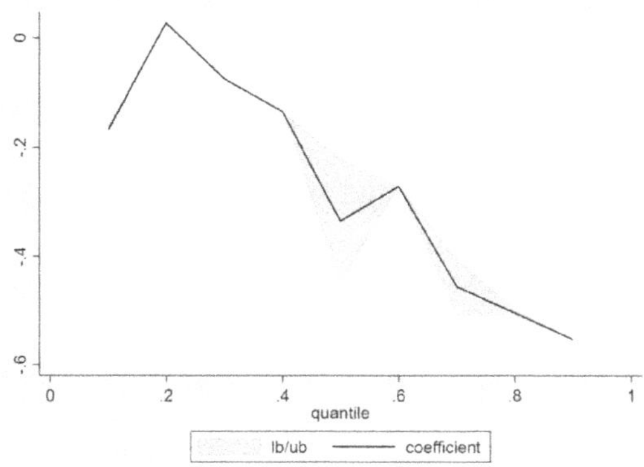

c. $environment_{it}$

Figure 1: *Quantile Treatment Effect Estimates for* eci_{it}, $uncertainty_{it}$ *and* $environment_{it}$

Source: (Author's calculation)

6 Conclusion

This study analyzes the effects of uncertainty, economic complexity, and environmental policy stringency on FDI using panel data quantile regression analysis and data on 32 countries over 2002–2020. Results show different effects of these factors at different quantiles. Negative effects of economic complexity and uncertainty were found to decline as FDI increase. At the highest FDI, economic complexity and uncertainty were shown to affect FDI favorably. However, adverse effect of stringent environmental policies increase at higher FDI.

Therefore, countries should decrease uncertainty to attract much more FDI and implement various policies related to governance, financial and economic stabilization, labor market, trade, infrastructural development, economic sophistication, and industrialization to attract FDI which may transfer environmentally friendly and green technologies for their sustainable development, transition to low carbon economy and increasing their competitiveness.

References

Ahir, H., Bloom, N., & Furceri, D. (2022). World Uncertainty Index (Working Paper). Retrieved from www.nber.org, www.policyuncertainty.com. (29763). Retrieved 14.05.2024, from National Bureau of Economic Research www.nber.org, www.policyuncertainty.com

Bae, M. G., Wang, Y. C., & Liu, N. (2022). Revisiting the relationship between the strength of environmental regulation and Foreign Direct Investment. *Frontiers in Psychology, 13*, 899918. doi: 10.3389/fpsyg.2022.899918

Bialek, S., & Weichenrieder, A. J. (2021). Do Stringent Environmental Policies Deter FDI? M&A versus Greenfield. *Environmental & Resource Economics, 80*(3), 603–636. doi: 10.1007/s10640-021-00600-x

Bommadevara, R., & Sakharkar, A. (2021). Do good institutions and economic uncertainty matter to Foreign Direct Investment? *Asian Economic and Financial Review, 11*(6), 471–487. doi: 10.18488/journal.aefr.2021.116.471.487

Gomez-Zaldıvar, M., Llamosas-Rosas, I., & Gomez-Zaldıvar, F. (2021). The relationship between economic complexity and the pattern of Foreign Direct Investment flows among Mexican States. *The Review of Regional Studies, 51*, 64–88.

Haque, M. A., Biqiong, Z., Arshad, M. U., & Yasmin, N. (2022). Role of uncertainty for FDI inflow: Panel econometric analysis of selected high-income nations. *Cogent Economics & Finance, 10*(1), 2156677. doi: 10.1080/23322039.2022.2156677

Islam, M. S., & Beloucif, A. (2024). Determinants of Foreign Direct Investment: A systematic review of the empirical studies. *Foreign Trade Review, 59*(2), 309–337. doi: 10.1177/00157325231158846

Kalamova, M., & Johnstone, N. (2011). *Environmental Policy Stringency and Foreign Direct Investment.* Retrieved from Paris: https://www.oecd.org/en/publications/environmental-policy-stringency-and-foreign-direct-investment_5kg8ghvf85d5-en.html

Marasanti, A. T. P., & Verico, K. (2023). Role of socioeconomic uncertainty on Foreign Direct Investment (FDI) inflows and economic growth in ASEAN. *Journal of ASEAN Studies, 11*(2), 367–389. doi: 10.21512/jas.v11i2.8766

Organisation for Economic Co-operation and Development [OECD]. (2024). OECD.Stat Environmental Policy Stringency Index. Retrieved from stats.oecd.org. Retrieved 09.06.2024 stats.oecd.org

Powell, D. (2022). Quantile regression with nonadditive fixed effects. *Empirical Economics*, 63(5), 2675–2691

Sadeghi, P., Shahrestani, H., Kiani, K. H., & Torabi, T. (2020). Economic complexity, human capital, and FDI attraction: A cross country analysis. *International Economics, 164*, 168–182. doi:https://doi.org/10.1016/j.inteco.2020.08.005

Sarmidi, T., Nor, A. H. S. M., & Ridzuan, S. (2015). Environmental Stringency, Corruption and Foreign Direct Investment (FDI): Lessons From global evidence. *Asian Academy of Management Journal of Accounting and Finance, 11*(1), 85–96.

The Growth Lab at Harvard, U. (2019). Growth Projections and Complexity Rankings (Publication no. doi: 10.7910/DVN/XTAQMC). Retrieved 09.06.2024, from Harvard Dataverse https://doi.org/10.7910/DVN/XTAQMC

The World Bank. (2024). World Development Indicators. Retrieved from databank.worldbank.org. Retrieved 09.06.2024 databank.worldbank.org

Uttama, N., & Arwatchanakarn, P. (2023). How do economic complexity and productive capacities foster foreign direct investment flows? Evidence from the Asian economies. *Economics Bulletin, 43*(1), 629–643. Retrieved from https://www.scopus.com/inward/record.uri?eid=2-s2.0-85192227281&partnerID=40&md5=5e4a8e3eb81f669b7296c1c52400dab1

Zhang, B., Ma, J., Khan, M. A., Repnikova, V., Shidlovskaya, K., Barykin, S., & Ahmad, M. S. (2023). The effect of economic policy uncertainty on Foreign Direct Investment in the era of global value chain: Evidence from the Asian Countries. *Sustainability, 15*, 6131. doi: 10.3390/su15076131

Murat Ergül[1] & Ali Rauf Karataş[2]

Analysis of the Presence of Income Convergence for Selected European Countries

1 Introduction

Following World War II, which is widely regarded as a global turning point across numerous socio-economic dimensions, notably in Europe, significant disparities in development began to emerge. Consequently, economic analyses, which previously emphasized the dimension of time, found themselves compelled to also consider spatial concepts. From micro to macro perspectives, it is understood that income disparities between places, cities, regions, countries, and even continents lead to social and economic consequences such as migration, social unrest, overcrowded cities, inadequate education and healthcare services, pollution, and scarcity. At this juncture, the convergence hypothesis proposed by the neoclassical growth model discusses scenarios that may arise following growth disparities. The hypothesis posited by the neoclassical perspective has spawned a substantial literature by testing the existence of economic growth and convergence across many countries. Fundamentally, the Solow (1956)-Swan (1956) economic growth models imply income convergence, suggesting that relatively poorer countries will experience higher growth rates than richer countries and will eventually reach the same income level over the long term. Thus, poorer countries converge towards the income levels of richer countries. This hypothesis has been tested in studies such as those by Barro (1991), Barro and Sala-i Martin (1992), and Sala-i Martin (1996), which argue for the possible existence of a negative relationship between current income levels and growth rates. Neoclassical growth models, assuming the same level of technology, population growth rate, and savings rate, externally attribute the faster growth of lower-income countries to attracting foreign investments with high returns on capital. Therefore,

1 Ph.D., Asst. Prof., Karabük University, Faculty of Economics and Administrative Sciences Department of Economics, muratergul@karabuk.edu.tr, ORCID: 0000-0003-2117-7561.
2 Ph.D., Asst. Prof., Karabük University, Faculty of Economics and Administrative Sciences Department of Economics, aliraufkaratas@karabuk.edu.tr, ORCID: 0000-0003-1031-6722.

a relatively lower-income country with high growth rates converges in the long term towards a higher-income country. In this perspective, it is accepted that every economy can reach a common long-term equilibrium level, and "absolute convergence" can be discussed. However, it should be noted that over time, this convergence evolves into "conditional convergence". As emphasized earlier, for convergence to occur, economic units (countries and/or regions) being tested for convergence must have similar levels of technology and other economic parameters. If deep structural differences exist among these units, different long-term equilibriums can emerge, referred to in the literature as "club convergence". The types of convergence used to test whether countries are converging economically under different economic parameters are classified as "β convergence" and "σ convergence". Under the general assumptions of traditional neoclassical views, β convergence is named when relatively poorer countries show higher growth rates compared to richer countries, indicating a tendency to catch up in terms of per capita income and production levels. This type of convergence, tested for numerous countries and groups of countries in the literature, is defined as capital being relatively scarce in an economy, thus having relatively high marginal productivity of capital. The regression equation testing unconditional β convergence can be expressed as follows:

$$log\left(\frac{\vartheta i,t+T}{\vartheta i,t}\right)=\alpha+\beta ln\left(\vartheta_{i,t}\right)+\varepsilon_i \quad (1)$$

In this regression equation, "$\vartheta_{i,t}$" represents the per capita income of country "i" in period "t", "$\vartheta_{i,t+T}$" represents the per capita income of country "i" in period "t+T", a and β represent the estimation parameters and "ε_i" represents the error term. If β>0 is determined, it is concluded that the estimated series converge to each other.

"σ" convergence deals with whether inequalities in income distributions are increasing or decreasing. Therefore, if it is concluded that inequalities in income distribution are decreasing as a result of a test, it means that there is "σ" convergence. We can mathematically express the regression equation of σ convergence as follows;

$$\sigma_t=\sqrt{\frac{1}{N}\sum_{i=1}^{N}.\left[\ln\left(\vartheta_{i,t}\right)-\bar{ln}(\vartheta_t)\right]^2} \quad (2)$$

In this equation, " σ_t ", represents the standard deviation of income distribution in period "t" and "N" is the total number of countries. In addition, " $\vartheta_{i,t}$ " represents the per capita income level of country "i" in period t. $ln(\vartheta_t)$ represents the average of the logarithm of per capita incomes in all countries subject to the test in the same period "t".

In this paper, we use the new unit root test used by Payne et all. (2022), which is basically a blend of the PANIC procedure in Bai and Ng (2004) and the two-break test of Lee and Strazicich (2003), to examine both the convergence trends over the selected period and the possible presence of structural breaks.

2 Literature Review

Especially after the 1980s, on the basis of neo-classical doctrines, a very intensive literature has been created on whether the economies of less developed countries have converged to the economies of developed countries. In the following years, the possible existence of convergence has been analyzed in terms of per capita income and growth rates as well as other variables such as technology level, CO_2 emissions, agricultural production, government revenues, and energy demand. At the same time, the income and development disparities that emerged not only between countries but also between regions within countries, especially after World War II, created a literature trying to determine whether the regions within the same country are converging or not. In this sense, Carlino and Mills (1993) can be shown as one of the studies testing the existence of regional convergence. In the study, the time period covering the years 1929–1990 is analyzed for 8 different regions in the United States. The study, which tests whether the per capita income levels between the regions in question have converged or not with cross-sectional and time series techniques, detects both stochastic convergence and β-convergence among the U.S. regions. Puente (2017) examines whether there is a convergence between Spanish regions in terms of per capita incomes and the factors that may affect the convergence process. Using data from the 1980 to 2015 period, the study identifies the existence of interregional convergence. Comparing income differences in Spanish regions with those in other European countries, the study emphasizes that regional differences are not very different from other European regions, but the rate of per capita income convergence in Spanish regions is quite low. Soukiazis (2003), on the other hand, investigates the potential for per capita income convergence among seven different regions in Portugal. Conducted for the years 1981–1996, the study explores absolute and conditional β-convergence as well as σ-convergence, concluding that conditional convergence is slower than absolute convergence among Portuguese regions using nonlinear least squares methods. Numerous similar studies have explored income convergence at the regional level, including Tian et al. (2016) for Chinese regions, Kholodilin et al. (2012) for Russian regions, Zeren and Yilanci (2011) for Turkish regions, and Kharisma and Saleh (2013) for Indonesian regions. Many studies have also examined national-level income convergence, attempting to elucidate the presence, speed, and structure of per capita income convergence between countries.

Table 1: Literature Review

The Author(s) Of The Study	Time Interval	Countries And/Or Country Groups	Metadology	The Obtained Result
Asongu, S. (2014)	1981–2009	38 African Countries	Cross-Section And Time Series Analisis	Both Endogenous And Neoclassical Growth Theories Are Supported
Kant, C. (2019)	1951–2013	46 Different Countries	Computing Catch-Up Index	Convergence Has Not Been Detected
Czasonis, M., & Quinn, M. (2012)	1971–2007	29 European Countries	Regression and Unit Root Test	Mixed Results Can Be Obtained Depending On The Applied Method And The Independent Variable Used
Dawson, J. W., & Strazicich, M. C. (2010)	1900–2001	20 OECD Countries	LM Unit Root Test	23 Out Of 29 Countries' Incomes Are Converging. And WWII Is Identified As The Time Period Of Structural Breaks.
Djennas, M., & Ferouani, B. (2014)	1980–2011	52 African Countries	Generalised Gini Coefficient Framework	Nearly No Evidence Of Convergence Is Found For The Sample Countries.

Table 1: Continued

The Author(s) Of The Study	Time Interval	Countries And/Or Country Groups	Metadology	The Obtained Result
Jena, D., & Barua, A. (2020)	1995–2017	28 European Countries	OLS,FE and RE	The Presence Of Sigma Convergence Is Detected. Additionally, It Is Concluded That Government Expenditures And Trade Reduce Income Disparities, Especially In Low-İncome Countriese
Huffman, N., & Huffman, W. (2021)	1990–2014	35 SSA Countries	OLS	Evidence For Counditional Convergence Is Found.
Borsi, M. T., & Metiu, N. (2015)	1970–2010	21 European Countries	Panel Unit Root Test	There Is No Convergence In All Of The Sample Countries
Das, R. C., Das, U., & Das, A. (2021).	2006Q1-2017Q2	BRICS Nations	Panel Unit Root Test	There Is Strong Evidence Of Sigma Convergence For The Sample Period.
Chong, T. T. L.,et all. (2008).	1950–200	15 OECD Countries	Nonlinear Unit Root Test	Mix Results Are Found.

3 Data Description and Summary Statistics

In the study, per capita Gross Domestic Product (GDP) data of Austria, Belgium, Finland, France, Greece, Spain, Luxembourg, Italy, the Netherlands, and Portugal, which are European Union countries, are utilized to assess the potential existence of convergence. The constructed datasets consist of annual data spanning the period from 1960 to 2022. These data are sourced from the World Bank database and are presented in US dollars for the year 2015.

Table 2: Descriptive Statistics of Per Capita Income Series (US Dollars)

	AUSTRIA	BELGIUM	FINLAND	FRANCE	GREECE	ITALY	LUXEMBOURG	NETHER-LANDS	PORTUGAL	SPAIN
Mean	31238.49	29202.87	30351.07	27386.53	15541.34	24885.19	68649.31	32529.29	13882.19	18725.71
Median	32097.24	29797.41	29278.36	28664.13	15690.88	27851.12	70779.97	31614.65	15088.41	19187.18
Max.	46697.86	44184.69	46792.75	38832.02	24072.79	34081.09	112417.9	50546.79	22125.94	28087.90
Min.	12051.15	11734.31	11425.29	11169.65	5030.744	9687.647	25520.18	14131.96	3934.945	6215.036
Std. Dev.	11061.91	9792.199	11507.95	8361.155	4703.069	7592.815	31128.70	10790.39	5527.098	6576.654
Skewness	-0.184876	-0.172002	-0.040686	-0.368873	-0.463369	-0.561594	0.053561	-0.019012	-0.284004	-0.188223
Kurtosis	1.705064	1.756554	1.636246	1.897303	2.795241	1.925568	1.387397	1.704711	1.720346	1.738882
Jarque-Bera	4.760640	4.369305	4.899421	4.620557	2.364521	6.341887	6.856402	4.407951	5.145390	4.546842
Probability	0.092521	0.112517	0.086319	0.099234	0.306585	0.041964	0.032445	0.110364	0.076330	0.102959
Observations	63	63	63	63	63	63	63	63	63	63

Notes:Std.Dev. is Standart Deviation, Min.is the Minimum Value and Max. is the Maximum Value

Tablo 1 presents the descriptive statistics of the selected European countries included in the analysis. Across these series for the selected countries, it is observed that per capita income ranges from a minimum of \$5,030 for Greece to a maximum of \$112,417 for Luxembourg. Additionally, the close proximity of the mean and median values indicates that the data have a symmetric distribution. Moreover, the descriptive statistics presented in the table highlight that Luxembourg has the highest income level among the sample countries when considering per capita GDP values. Alongside numerous summary statistics in Table 1, the graphical representation of these series is provided in Graph 1.

In this graph, it is evident that Luxembourg stands out in terms of per capita income level compared to the other countries.

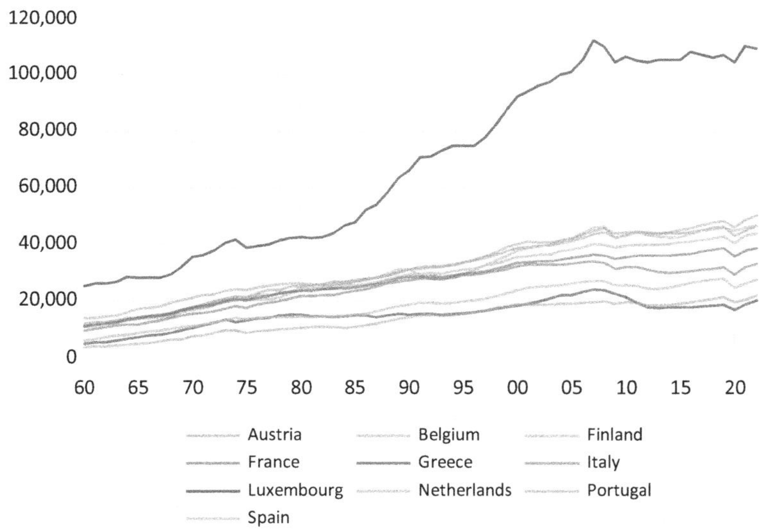

Graph 1: Graphical Representation of the Series

4 Emprical Findings

In this study, the potential existence of income convergence among the 10 selected European countries is tested using a new unit root test that diverges from other unit root tests by simultaneously considering structural breaks and cross-correlations in conjunction with factor structures, as utilized by Payne et al. (2022). The proposed 'new test' integrates the Two-Break Test and PANIC procedure, allowing for cross-correlations in the factor structure. To estimate the breakpoints and factor terms, the Langrange Multiplier test is employed. The findings are presented in Tables 3 and 4, where Table 3 includes the dates of structural breaks at the level, and Table 4 determines the dates of breakpoints including the Trend variable.

Table 3: Two Level Breaks

Countries	LM	p-val	Brk1	Brk2
Austria	-4,67[a]	**0**	1978	1980
Belgium	-1,8	0,56	1999	2002
Finland	-12,07[a]	**0**	1987	1989
France	-3,34[a]	**0,03**	1979	1981
Greece	-4,6	0	1999	2001
Italy	-4,5	0	1978	1986
Luxembourg	-1,12	0,96	1985	2003
Netherlands	-1,39	0,84	1978	1981
Portugal	-3,12[a]	**0,04**	1982	1985
Spain	-2,38	0,22	1994	2003

Note: Significance level is denoted as x^a (10 %) with respect to rejection of the null hypothesis.

Based on the findings obtained using the Two-Level Breaks model, it is observed that the null hypothesis is rejected at the 10 % significance level for Austria, Finland, France, and Portugal. Therefore, the p-values reported in the table are found to be less than 0.1, indicating rejection of the null hypothesis. The rejection of the null hypothesis suggests the presence of income convergence in these countries. Additionally, structural breaks are identified in the economies of these countries during specific years: 1978–1980 for Austria, 1987–1989 for Finland, 1979–1982 for France, and 1982–1985 for Portugal.

Table 4: Two Trend Breaks

Countries	LM	p-val	Brk1	Brk2
Austria	-8,63[a]	**0**	1982	2001
Belgium	-3,26	0,43	1984	2002
Finland	-2,8	0,73	1990	2002
France	-2,67	0,81	1981	2002
Greece	-5,22[a]	**0**	1978	1983
Italy	-3,98	0,12	1990	1995

Table 4: Continued

Countries	LM	p-val	Brk1	Brk2
Luxembourg	-4,74[a]	**0,02**	1982	2003
Netherlands	-4,55[a]	**0,03**	2000	2003
Portugal	-4,04[a]	**0,1**	1980	1990
Spain	-4,19[a]	**0,07**	1979	1982

Note: Significance level is denoted as x[a] (10 %) with respect to rejection of the null hypothesis.

According to Table 4, which presents the results of the model including the Trend variable, it differs from Table 3 in terms of identifying countries showing convergence tendencies and their breakpoint dates. Based on the findings of this model, the countries where per capita income convergence is observed are listed as Austria, Greece, Luxembourg, Netherlands, Portugal, and Spain. Therefore, Table 4 indicates that more countries exhibit a convergence tendency compared to the Two-Level Breaks model.

Furthermore, examining the structural breakpoint dates according to this model reveals the following periods:

- For Austria: 1982–2001
- For Greece: 1978–1983
- For Luxembourg: 1982–2003
- For Netherlands: 2000–2003
- For Portugal: 1980–1990
- For Spain: 1979–1982

Both models indicate that Belgium and Italy are countries where convergence is not observed, as shown in the tables.

5 Conclusion

Numerous national and international economic authorities are known to seek remedies to mitigate disparities between regions. Particularly since World War II, both inter-country and intra-country regional disparities have persisted despite extensive efforts aimed at reducing them. Economic policymakers are deeply concerned about these disparities, prompting rigorous studies in this area. Neo-classical predictions, which posit that low-income countries can catch up with high-income countries in the long term through their experience of high growth rates, have been tested across numerous countries and groups. However, this theory has yet to reach a definitive conclusion.

In this study, conducted by Payne et al. (2022), the potential existence of income convergence among 10 European countries is examined using a new unit root test spanning the period from 1960 to 2022. The findings present countries where income convergence is valid and highlight the structural break dates within these national economies. According to the results, based on a model including trends, it is concluded that 6 out of the 10 countries exhibit a tendency towards income convergence.

References

Asongu, S. (2014). AFrican development: Beyond income convergence. *South African Journal of Economics*, *82*(3), 334–353.

Bai, J., & Ng, S. (2004). A PANIC Attack on Unit Roots and Cointegration. *Econometrica*, *72*(4), 1127–1177.

Barro, R. J. (1991). Economic growth in a cross section of countries. *The Quarterly Journal of Economics*, *106*(2), 407–443.

Barro, R. J., & Sala-İ-Martin, X. (1992). Convergence. *Journal of Political Economy*, *100*(2), 223–251.

Borsi, M. T., & Metiu, N. (2015). The evolution of Economic Convergence in the European Union. *Empirical Economics*, *48*, 657–681.

Carlino, G. A., & Mills, L. O. (1993). Are US regional incomes converging?: A time series analysis. *Journal of Monetary Economics*, *32*(2), 335–346.

Chong, T. T. L., Hinich, M. J., Liew, V. K. S., & Lim, K. P. (2008). Time Series Test of nonlinear convergence and transitional dynamics. *Economics Letters*, *100*(3), 337–339.

Czasonis, M., & Quinn, M. (2012). Income convergence in Europe: Catching up or falling behind?. *Acta Oeconomica*, *62*(2), 183–204.

Das, R. C., Das, U., & Das, A. (2021). BRICS Nations and income convergence: An insight from the quarterly data for 2006Q1–2017Q2. *Global Business Review*, *22*(4), 1054–1069.

Dawson, J. W., & Strazicich, M. C. (2010). Time-Series Tests of income convergence with two structural breaks: Evidence from 29 countries. *Applied Economics Letters*, *17*(9), 909–912.

Djennas, M., & Ferouani, B. (2014). Growth and income convergence in Africa. *Journal of Economics and Development Studies*, *2*(4), 63–76.

Huffman, N., & Huffman, W. (2021). Convergence Theory and conditional income convergence among Sub-Saharan African countries. *Agricultural Economics*, *52*(6), 915–925.

Jena, D., & Barua, A. (2020). Trade, governance and income convergence in the European Union: Evidence on the "Theory of relative backwardness". *Research in Globalization*, *2*, 100013.

Kant, C. (2019). Income convergence and the catch-up index. *The North American Journal of Economics and Finance*, *48*, 613–627.

Kharisma, B., & Saleh, S. (2013). Convergence of income among provinces in Indonesia, 1984–2008: A panel data approach. *Journal of Indonesian Economy and Business*, *28*(2), 167–187.

Kholodilin, K. A., Oshchepkov, A., & Siliverstovs, B. (2012). The Russian regional convergence process: Where is it leading?. *Eastern European Economics*, *50*(3), 5–26.

Lee, J., & Strazicich, M. C. (2003). Minimum Lagrange Multiplier Unit Root Test with Two Structural Breaks. *Review of Economics and Statistics*, *85*(4), 1082–1089.

Payne, J. E., Lee, J., Islam, M. T., & Nazlioglu, S. (2022). Stochastic convergence of per capita greenhouse gas emissions: New Unit Root Tests with breaks and a factor structure. *Energy Economics*, *113*, 106201.

Puente, S. (2017). Regional convergence in Spain: 1980–2015. *Banco De Espana Article*, *21*, 17.

Sala-İ-Martin, X. X. (1996). The classical approach to convergence analysis. *The Economic Journal*, *106*(437), 1019–1036.

Solow, R. M. (1956). A contribution to the theory of economic growth. *The Quarterly Journal of Economics*, *70*(1), 65–94.

Soukiazis, E. (2003). Regional convergence in Portugal. *Documento De Trabalho/ Discussion Paper*, *14*.

Swan, T. (1956). Economic growth and capital aggregation. *Economic Record*, *32*(2), 343–361.

Tian, X., Zhang, X., Zhou, Y., & Yu, X. (2016). Regional income inequality in China revisited: A perspective from Club Convergence. *Economic Modelling*, *56*, 50–58.

Zeren, F., & Yilanci, V. (2011). Türkiye'de Bölgeler Arasi Gelir Yakinsamasi: Rassal Katsayili Panel Veri Analizi Uygulamasi/Convergence among the regions of Turkey: Evidence From random coefficient models. *Business and Economics Research Journal*, *2*(1), 143.

Burhan Durgun[1]

Analysis of the Pollution Haven Hypothesis in Türkiye

1 Introduction

Environmental degradation, exacerbated by climate change and man's destructive impact on the environment, has become a major issue on the national and international agenda. In order to prioritize environmental protection, it is necessary to encourage the production, consumption, and trade relations to become more environmentally friendly. Precautions are taken to help nature renew itself and protect biodiversity.

As environmental protection policies become increasingly stringent, industries that are unable to adapt seek new production centers. Consequently, capital flows are directed towards countries where existing production technologies can be utilized. This influx of foreign direct investment (FDI) has given rise to the pollution haven hypothesis (PHH) in the academic literature. This hypothesis, initially proposed by Pethig (1976), posits that the increased costs associated with more stringent environmental regulations are a primary factor driving the reallocation of FDI from developed to less developed or developing countries (Yilanci vd., 2023). The phenomenon of pollution havens, which can attract higher levels of FDI in the context of lower environmental standards, represents a significant contributing factor to the global increase in environmental pressure (Balsalobre-Lorente vd., 2019). Developing countries have a comparative cost advantage due to low pollution controls. In this case, developing countries specialize in dirty goods while developed countries specialize in more environmentally friendly products (Gill vd., 2018).

As Dinda (2006) posits, when multinational companies seeking to circumvent stringent regulations relocate to developing countries, developed countries reap environmental benefits from trade, while developing countries experience a net loss. Moreover, Copeland and Taylor (1994) predicted that such a flow of

1 Ph.D. Res. Asst., Dicle University, Faculty of Economics and Administrative Sciences, Department of Economics, burhan.durgun@dicle.edu.tr ORCID: 0000-0001-7742-6059.

FDI from the US to Mexico would be an environmental disaster for Mexico, and an employment disaster for the US.

While discussing the polluting impact of FDI, Porter & van der Linde (1995) argue that in the presence of well-designed environmental regulations, innovation can be encouraged and resource utilization efficiency can be achieved, thus reducing firm-source pollution. Birdsall and Wheeler (1993) also found that the pollution haven hypothesis is not valid in their analysis of Latin American countries. In other words, FDI can enhance the environmental quality of the host country. This phenomenon, known as the pollution halo (PH/Porter Hypothesis) hypothesis, refers to the fact that multinational companies in developed countries contribute to the reduction of environmental pollution in the host country by transferring efficient and environmentally friendly technologies to developing countries (Yilanci et al., 2023).

FDI has been demonstrated to have a number of beneficial effects, including the expansion of the capital stock, the reduction of deficits in the balance of payments, the generation of employment opportunities and increased competition, and the introduction of dynamism to the host country's market (Aydemir et al., 2012). Nevertheless, FDI is unlikely to be effective in the presence of significant technological gaps, disparate trade regimes, and political risks (De Mello Jr., 1999). Gray (2002) lists conditions for the effectiveness of FDI such as international trade in goods and services, the presence of subsidiaries of multinational companies, the international mobility of physical, financial, human and knowledge capital provided through FDI, and institutional quality. In this respect, it is more appropriate to consider trade openness and FDI together in the analysis.

Since renewable energies are clean energy sources, they are likely to reduce emissions. The increase in the level of urbanization is related to the increase in economic activities that increase the amount of emissions (Sadorsky, 2014). In light of the theoretical framework presented, this study examines the impact of FDI, trade openness, renewable energy, and urbanization on environmental degradation. This study, which tests the PHH hypothesis, differs from other studies with asymmetric linkages, a wide range of observations, and an extended model. The following sections of the paper include a literature summary, data, methodology, analysis results, and finally, a conclusion.

2 Literature Review

The PHH hypothesis is one of the most frequently studied topics in the literature due to the phenomenon of globalization. Studies testing different countries and country groups with different analysis techniques have deepened the literature. A summary of recent studies on the FDI-environment relationship is provided below.

Ur Rahman et al. (2019) tested the trade openness, FDI, agriculture and carbon dioxide (CO2) emissions linkages in Pakistan between 1975 and 2016 using ARDL and NARDL methods. The findings of the study provide evidence for the pollution haven and Environmental Kuznets Curve (EKC) hypotheses. Rana and Sharma (2019) evaluated the relationship between FDI, growth, trade and CO2 emissions in India for the period 1982–2013 through causality analysis. EKC and PHH hypotheses were confirmed in the study. Balsalobre-Lorente et al. (2019) examined the link between FDI and ecological footprint in MINT countries in the period 1990–2013. The study confirmed the EKC hypothesis and found an inverted U-shaped relationship between pollution and FDI. Mert and Caglar (2020) analyzed the relationship between FDI and carbon emissions in Türkiye for the period 1974–2018 using asymmetric cointegration and causality tests and confirmed the PH hypothesis. Gyamfi et al. (2021) investigated the relationship between energy consumption, FDI, real GDP and CO2 emissions in 15 oil-producing and 28 non-oil-producing countries for the period 1990–2016 using panel cointegration and panel quantile regression methods. The study proved the PHH hypothesis for both country groups and emphasized the environmental protection effect of renewable energy consumption. Bulus and Koc (2021) investigated the effects of FDI, GDP per capita, public expenditures, renewable energy, exports and imports on CO2 emissions in Korea for the period 1970–2018 using ARDL method. The study found an N-shaped relationship between GDP and emissions and a positive relationship between FDI and emissions. Abid and Sekrafi (2021) examined the relationship between trade openness, FDI, growth and CO2 emissions in 31 developed and 100 developing countries using the GMM method. Their findings implied the PHH hypothesis in developing countries and both the EKC and PH hypotheses in developed countries. In their 2022 study, Temurlenk & Lögün employed the NARDL method to investigate the influence of FDI, trade openness, economic growth, and the manufacturing sector on CO2 emissions in Türkiye over the 1974–2017 period. Their findings supported the PH hypothesis in the short run but revealed that both positive and negative shocks can contribute to pollution in the long run. Balsalobre-Lorente et al. (2022) examined the interconnections between economic complexity, FDI, renewable energy consumption, and urbanization in PIIGS countries for the period 1990–2019, employing a panel cointegration method. The study confirmed the EKC and PHH hypotheses and demonstrated that renewable energy mitigates environmental pressure. Abbasi et al. (2023) employed the PMG ARDL method to test the EKC, pollution haven, and PH hypotheses in 13 Asian countries over the period 1985–2020. Their findings indicate that GDP per capita, energy consumption, FDI, and urbanization are positively correlated with CO2 emissions. The quadratic terms validated the EKC, PH, and PHH hypotheses in these countries. In a study published in 2023, Yilanci and colleagues examined the impact of economic growth, FDI, and trade openness in the fisheries sector on the fishing footprint in Indonesia between 1976 and 2018. The authors employed the autoregressive distributed lag (ARDL) method to analyze the data. The study confirmed the existence of both the EKC hypothesis and the PH hypothesis. Kisswani and Zaitouni (2023) tested the PHH hypothesis for Malaysia, Philippines, Singapore and Thailand for the period 1971–2014 using ARDL method. According to the findings of the study, the PHH hypothesis is confirmed for the Philippines, while the PH hypothesis is confirmed for Malaysia and Singapore. Kim and Seok (2023) analyzed the linkages between FDI, income and CO2 in Korea for the period 1971–2020 using the ARDL method. The findings of the study show that the PHH hypothesis is valid at low income level and the PH hypothesis is valid at high income level. Zheng et al. (2024) conducted an investigation into the impact of energy intensity, economic growth, population, and FDI on CO2 emissions in China for the period 1980–2019, utilising the NARDL method. The findings of the study provide evidence for the PHH hypothesis in both the short and long run.

Upon a review of the extant literature on the relationship between FDI and environmental degradation, it becomes evident that the PHH hypothesis is corroborated in certain studies, whereas the PH hypothesis is validated in others. Mixed results were found in studies using multi-country or non-linear methods. In studies specific to Türkiye, Temurlenk and Lögün (2022) argue that the PHH hypothesis is valid, while Mert and Caglar (2020) argue that the PH hypothesis is valid. This lack of consensus necessitates a re-evaluation of the situation from different perspectives.

3 Data and Methodology

This study investigates the impact of FDI on environmental degradation in Türkiye and employs time series methods. The units and sources of carbon dioxide emissions (*lco2*), foreign direct investment (*fdi*), trade openness (*ltro*), renewable energy consumption (including hydroelectricity) (*lrnw*) and urbanization rate (*lurb*) used in the study are given in the Table 1. The analysis period is limited to the period 1970–2022 according to data availability. In order to harmonize the data, variables other than *fdi* are modeled in their natural logarithms. In the model of the research, *lco2* is the dependent variable, *fdi* is the asymmetric independent variable, *ltro* and *lrnw* are symmetric independent variables and *lurb* is the control variable.

Table 1: Data Description and Units

Abbreviation	Variable	Unit	Source
lco2	Carbon dioxide emissions	Per capita, tonnes	Energy Institute
fdi	Foreign Direct Investments	Net Inflows, % of GDP	Worldbank, WDI
ltro	Trade Openness	Trade, % of GDP	Worldbank, WDI
lrnw	Renewable Energy Consumption (Inc. hydro)	Per capita, MegaJoule	Energy Institute
lurb	Urbanization Rate	Urban/Total Population	Worldbank, WDI

The model is subjected to rigorous testing and the resulting findings are subjected to rigorous evaluation. First, information on stationarity and cointegration analysis from time series methods is provided.

The ADF test, developed by Dickey and Fuller (1981) as an improvement on the DF (1979) test, is employed in the presence of high-order autoregressive processes. However, the test is unable to identify structural breaks. Structural breaks may affect the stationarity of time series, Enders and Lee (2012) maintained that one or more structural breaks can be captured by low frequency Fourier functions. The test they developed is a Dickey-Fuller (1979) type regression model in which the deterministic term is expressed as a Fourier function. Their proposed approach can be considered a complement to the Fourier LM test and the DF-GLS unit root tests (Enders & Lee, 2012: 196). The regression model for the test is designated as $\Delta y_t = \rho y_{t-1} + c_1 + c_2 t + c_3 \sin\left(\frac{2\pi kt}{T}\right) + c_4 \cos\left(\frac{2\pi kt}{T}\right) + e_t$. In order to ascertain the appropriateness of the test, it is necessary to ascertain whether the sine and cosine terms are statistically significant with respect to the specified model. The null hypothesis employed for this test are $H_0 : c_3 = c_4 = 0$ (classic ADF should be used). The critical values for the estimated $F\left(\breve{k}\right)$ test statistic are provided in Enders and Lee (2012). If the estimated test statistic exceeds the relevant critical table value, the null hypothesis is rejected. The hypotheses of the flexible Fourier unit root test are $H_0 : \rho = 0$ and $H_1 : \rho < 0$. The critical values for the test statistic are provided in Enders and Lee (2012). If the absolute value of the calculated test statistic exceeds the relevant critical table value, the null hypothesis is rejected.

3.1 Nonlinear ARDL Approach

The asymmetric method developed by Shin et al. (2014) considers both the positive and negative partial sum decompositions of the explanatory variables, thus enabling the asymmetric effect of the independent variable on the dependent variable to be quantified. The test can also demonstrate both long-run and short-run asymmetric effects. As in the classical ARDL test, variables may exhibit stationarity at different levels, including I(0) and/or I(1). The long-run asymmetric model of the study is designated as $lco2_t = \beta^+ fdi_t^+ + \beta^- fdi_t^- + u_t$. In accordance with the variables under consideration in the present study, the nonlinear unconstrained error correction model proposed by Shin et al. (2014) is designated as

$$\begin{aligned}\Delta lco2_t = {} & \alpha_0 + \sum_{i=1}^{p-1}\alpha_{1i}\Delta lco2_{t-i} + \sum_{i=0}^{q-1}\left(\omega_{1i}^{+}\Delta fdi_{t-i}^{+} + \omega_{1i}^{-}\Delta fdi_{t-i}^{-}\right) + \sum_{i=0}^{q-1}\omega_{2i}\Delta ltro_{t-i} \\ & + \sum_{i=0}^{q-1}\omega_{3i}\Delta lrnw_{t-i} + \sum_{i=0}^{q-1}\omega_{4i}\Delta lurb_{t-i} + \alpha_3 lco2_{tt-1} + \alpha_4 fdi_{t-1}^{+} + \alpha_5 fdi_{t-1}^{-} \\ & + \alpha_6 ltro_{t-1} + \alpha_7 lrnw_{t-1} + \alpha_8 lurb_{t-1} + \varepsilon_i .\end{aligned}$$

In the test, the null hypothesis for the long run is $H_0 : \frac{-\alpha_4}{\alpha_3} = \frac{-\alpha_5}{\alpha_3}$ while the null hypothesis for the short run is $H_0 : \omega_{1i}^{+} = \omega_{1i}^{-}$. The null hypothesis for the short and long run is $H_0 : \omega_{1i}^{+} = \omega_{1i}^{-}$ for each i and $H_0 : \frac{-\alpha_4}{\alpha_3} = \frac{-\alpha_5}{\alpha_3}$. In order to proceed to the cointegration stage, it is necessary to detect an asymmetric effect in at least one of the periods. The null hypothesis of the cointegration test is $H_0 : \alpha_3 = \alpha_4 = \alpha_5 = 0$. The test statistic employed in this test is identical to that proposed by Pesaran et al. (2001) in their F test statistic. If the estimated value exceeds the upper bound value, the null hypothesis is rejected.

4 Results

Table 2 presents descriptive statistics and a correlation matrix for the series. All correlations between variables are positive and statistically significant. While the correlations of *fdi* are relatively low, other correlations are quite high.

Table 2: Descriptive Statistics and Correlation Matrix

	lco2	*fdi*	*ltro*	*lrnw*	*lurb*
lco2	1				
fdi	0.7003	1			
ltro	0.9115	0.6024	1		
lrnw	0.9327	0.5280	0.8783	1	
lurb	0.9767	0.6684	0.9507	0.9317	1
Mean	0.9886	0.8325	3.5386	8.4248	4.0710
Median	1.0508	0.4654	3.7316	8.5810	4.1376
Maximum	1.6159	3.6234	4.3965	9.6448	4.3441
Minimum	0.1000	0.0194	2.2082	6.6699	3.6437
Std. Dev.	0.4181	0.8417	0.5389	0.7526	0.2206

In order to ascertain the potential long-run relationship between the variables, it is necessary to determine their respective stationarity levels. The Table 2 presents the results of the ADF unit root test. According to the ADF test results, all variables except lurb exhibit unit roots at the level and are stationary in the first difference. lurb, however, is stationary at the level.

It is possible that structural changes occurring during the analysis period may affect the stationarity of the variables. Therefore, the stationarity of the variables is also tested using the Fourier ADF unit root test, which determines the breaks gradually. According to the results in the table, the level values of *lco2*, *ltro* and *lrnw* have unit roots, while the level values of *fdi* and *lurb* are stationary. According to the results of the difference series, all variables except *ltro* are stationary. However, in order to use FADF statistics, the F test must also be significant. According to the results, FADF test statistics are reliable only for level and difference values of *lco2* and level values of *lurb*, while classical ADF test statistics are reliable for other variables. In this case, *lurb* is subject to I(0) process while other variables are subject to I(1) process.

Table 3: ADF and FADF Unit Root Tests

Test	ADF			FADF		Decision
Variable	t-İst.	Prob.	Freq.	F Test Stat.	FADF Test Stat.	
lco2	-1.9379	0.3129	5	7.7980**	-1.6863	I1 (FADF)
Δlco2	-7.6174***	0.0000	5	7.1402*	-6.0245***	
fdi	-2.1245	0.2362	1	6.0544	-4.2315**	I1 (ADF)
Δfdi	-6.8214***	0.0000	3	2.0424	-5.6990***	
ltro	-1.5621	*0.4945*	4	0.4512	-2.3218	I1 (ADF)
Δltro	-6.1954***	*0.0000*	4	0.6350	-2.2079	
lrnw	-1.4528	*0.5494*	4	0.6849	-1.7238	I1 (ADF)
Δlrnw	-8.4039***	*0.0000*	1	2.7257	-4.9466***	
lurb	-11.9686***	*0.0000*	2	28.9751***	-20.3252***	I0 (FADF)
Δlurb	-2.7972*	*0.0675*	3	0.3015	-2.78692*	

Note: *, **, *** denotes 10 %, 5 % and 1 % significancy, respectively.

Given that there is no stationary series in the second difference among the variables in the model, it is possible to analyze the long-run relationship between the variables using the NARDL bounds test method. First, we estimate the short- and long-run asymmetry test of fdi. The results of the panel A test indicate that asymmetry is significant in the long run but insignificant in the short run. The F test statistic calculated in the bounds test results in Panel B is greater than the 1 % critical value, which implies that there is a cointegration relationship between the variables. In the next stage, long-run coefficients are estimated. Positive shocks of *fdi* have a positive and significant effect on CO2 emissions, while negative shocks are positive but insignificant. An increase of 1 % in *fdi* increases CO2 emissions by 0.058 %. The effects of trade openness and renewable energy on CO2 emissions are also positive. The polluting effect of urbanization is insignificant.

If the effects of both positive and negative shocks have the same direction in the long-run coefficients, the effect sizes are compared and evaluated in absolute value (Zheng et al., 2024). The positive shock has a greater effect than the negative shock in the NARDL analysis. Therefore, it can be concluded that the inflow of FDI into the country has a detrimental impact on the quality of the environment.

The coefficient of the error correction term in Panel D, which depicts the velocity of adjustment of short-run deviations, is negative and statistically significant in accordance with expectations. Consequently, 80 % of the short-run deviations are rectified in the subsequent period and converge towards the long-run equilibrium value. The short-run effects of the variables are insignificant. Finally, diagnostic tests of the model are analyzed. According to the findings, (Panel E) error terms are normally distributed and there is no autocorrelation, heteroscedasticity and model specification error.

Table 4: NARDL (1,1,3,3,3)

	Panel A: Asymmetry Test			
	Long Run		**Short Run**	
Variable	**F-stat**	**Prob.**	**F-stat**	**Prob.**
fdi	18.7192	0.0001	0.1908	0.6652
	Panel B: Long Run Relationship			
F-stat	**k**	**Critical Values**	**I0**	**I1**
6.8235***	5	**0.10**	2.435	3.600
		0.05	2.900	4.218
		0.01	3.955	5.583
	Panel C: Long Run Coefficient			
Variable	**Coefficient**	**Std. Error**	**t-Statistic**	**Prob.**
fdi^{+}	0.0586	0.0158	3.7120	0.0006
fdi^{-}	0.0125	0.0185	0.6728	0.5045
ltro	0.2261	0.0864	2.6180	0.0120
lrnw	0.0712	0.0393	1.8116	0.0767
lurb	0.4357	0.2705	1.6110	0.1142
	Panel D: Short Run Dynamics			
Variable	**Coefficient**	**Std. Error**	**t-Statistic**	**Prob.**
ect(-1)	-0.8063	0.1172	-6.8803	0.0000
Δfdi^{+}	-0.0063	0.0139	-0.4529	0.6533
Δfdi^{-}	0.0105	0.0205	0.5108	0.6125
$\Delta ltro$	0.0231	0.0362	0.6382	0.5273
$\Delta ltro(-1)$	-0.1517	0.0512	-2.9648	0.0053
$\Delta ltro(-2)$	-0.1371	0.0549	-2.4958	0.0172
$\Delta lrnw$	-0.0016	0.0237	-0.0657	0.9479
$\Delta lrnw(-1)$	-0.0235	0.0251	-0.9385	0.3541
$\Delta lrnw(-2)$	0.0464	0.0241	1.9256	0.0619
$\Delta lurb$	-1.4346	1.8918	-0.7583	0.4531
$\Delta lurb(-1)$	2.8283	2.9475	0.9596	0.3435
$\Delta lurb(-2)$	-3.0123	1.8677	-1.6128	0.1153
c	-1.8054	0.2632	-6.8604	0.0000

(*continued on next page*)

Table 4: Continued

Panel E: Diagnostic Tests		
Test	**Stat.**	**Prob.**
Jarque-Bera Normality	0.0015	0.9992
Breusch-Godfrey LM Test	1.0077	0.6042
BPG Heteroscedasticity Test	15.2532	0.5773
Ramsey RESET Test	0.0029	0.9577

The outcomes of the CUSUM and CUSUMQ assessments, which indicate the presence of systematic and sudden alterations in the long-term coefficients of the NARDL (1,1,3,3,3) model, are illustrated in Figure 1. The error term curves are situated within the confidence interval, suggesting that the coefficients are stable.

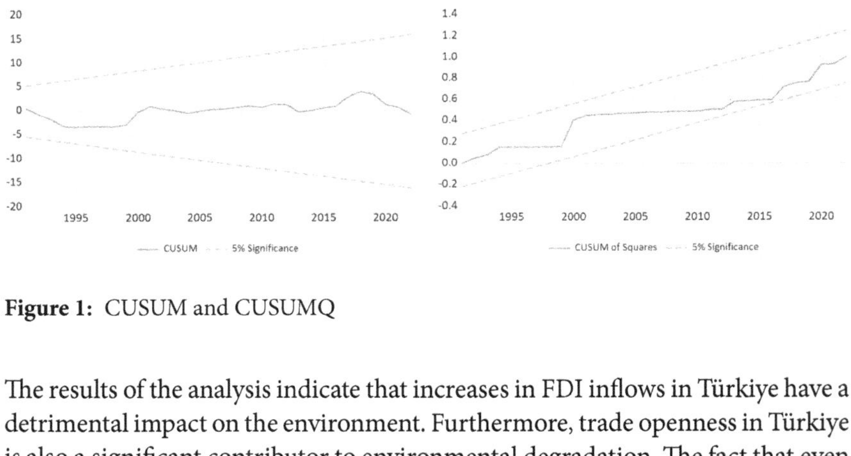

Figure 1: CUSUM and CUSUMQ

The results of the analysis indicate that increases in FDI inflows in Türkiye have a detrimental impact on the environment. Furthermore, trade openness in Türkiye is also a significant contributor to environmental degradation. The fact that even the effect of renewable energy in the model is pollution-enhancing demonstrates that Türkiye is a pollution haven. These results are consistent with those of Ur Rahman et al. (2019), Rana and Sharma (2019), Gyamfi et al. (2021), Bulus and Koc (2021), Balsalobre-Lorente et al. (2022), and Temurlenk and Lögün (2022).

5 Conclusion

This study evaluates whether FDI inflows into Türkiye have resulted in the country becoming a pollution haven. The NARDL method is employed in the analysis, with the assumption that the effect of FDI may be asymmetric. The results of the model, which includes trade openness, renewable energy, and urbanization, demonstrate that the pollution haven hypothesis is valid for Türkiye.

While the long-run results indicate that FDI flows have a negative impact on the environment in Türkiye, they do not demonstrate the environmental benefits of a reduction in FDI flows. This suggests that incoming investments result in long-term pollution. Considering that the foreign trade regime is also polluting, it makes Türkiye a destination country for the dirty industries of multinational companies. The polluting effect of renewable energy may be due to the data used. The polluting impact of renewable energy may be due to the data used. The data included in the model includes hydroelectric power, which may have a negative impact on the environment in the long run and has a high share for renewable energies in Türkiye. In addition, the installation of power plants used in renewable energy generation requires the use of fossil fuel-intensive energy. The environmental degradation of Türkiye is evident from the occurrence of recent environmental disasters. The most tangible example is the cyanide used by multinational companies in the extraction of gold mines in the country, which has resulted in the reduction of biodiversity and forest areas.

Policymakers need to take certain measures regarding the nature of FDI, which may have beneficial effects such as technology transfer, competition and efficiency, but may also have environmental damages in the long run. The most crucial initial step is to implement legal frameworks for environmental protection. Incentives, tax reductions, and subsidies for clean industries and additional taxes for dirty industries can be introduced.

References

Abbasi, M. A., Nosheen, M., & Rahman, H. U. (2023). An approach to the pollution haven and pollution halo hypotheses in Asian countries. *Environmental Science and Pollution Research, 30*(17), 49270–49289.

Abid, M., & Sekrafi, H. (2021). Pollution haven or halo effect? A comparative analysis of developing and developed countries. *Energy Reports*, *7*, 4862–4871.

Aydemir, C., Arslan, İ., & Uncu, F. (2012). Doğrudan yabancı yatırımların Dünya'daki ve Türkiye'deki gelişimi. *Kocaeli Üniversitesi Sosyal Bilimler Dergisi, 23*, 69–104.

Balsalobre-Lorente, D., Gokmenoglu, K. K., Taspinar, N., & Cantos-Cantos, J. M. (2019). An approach to the pollution haven and pollution halo hypotheses in MINT countries. *Environmental Science and Pollution Research, 26*, 23010–23026.

Balsalobre-Lorente, D., Ibáñez-Luzón, L., Usman, M., & Shahbaz, M. (2022). The environmental Kuznets curve, based on the economic complexity, and the pollution haven hypothesis in PIIGS countries. *Renewable Energy, 185*, 1441–1455.

Birdsall, N., & Wheeler, D. (1993). Trade policy and industrial pollution in Latin America: Where are the pollution havens?. *The Journal of Environment & Development, 2*(1), 137–149.

Bulus, G. C., & Koc, S. (2021). The effects of FDI and government expenditures on environmental pollution in Korea: The pollution haven hypothesis revisited. *Environmental Science and Pollution Research, 28*(28), 38238–38253.

Copeland, B. R., & Taylor, M. S. (1994). North-South trade and the environment. *The Quarterly Journal of Economics, 109*(3), 755–787.

De Mello Jr, L. R. (1999). Foreign direct investment-led growth: Evidence from time series and panel data. *Oxford Economic Papers, 51*(1), 133–151.

Dickey, D. A., & Fuller, W. A. (1979). Distribution of the estimators for autoregressive time series with a Unit Root. *Journal of the American Statistical Association, 74*(366a), 427–431.

Dickey, D. A., & Fuller, W. A. (1981). Likelihood Ratio Statistics for autoregressive time series with a Unit Root. *Econometrica, 49*(4), 1057–1072.

Dinda, S. (2006). Globalization and environment: Can pollution haven hypothesis alone explain the impact of globalization on environment?. Munich personal repec archive, 50590. MPRA Paper. https://mpra.ub.uni-muenchen.de/50590/1/MPRA_paper_50590.pdf

Enders, W., & Lee, J. (2012). The flexible Fourier form and Dickey–Fuller type unit root tests. *Economics Letters, 117*(1), 196–199.

Gill, F. L., Viswanathan, K. K., & Karim, M. Z. A. (2018). The critical review of the pollution haven hypothesis. *International Journal of Energy Economics and Policy, 8*(1), 167–174.

Gray, H. P. (2002). International trade and economic development or growth: A qualification. *The Journal of World Investment & Trade, 3*(1), 65–75.

Gyamfi, B. A., Bein, M. A., Udemba, E. N., & Bekun, F. V. (2021). Investigating the pollution haven hypothesis in oil and non-oil sub-Saharan Africa countries: Evidence from quantile regression technique. *Resources Policy, 73*, 102119.

Kim, S. E., & Seok, J. H. (2023). The impact of foreign direct investment on CO2 emissions considering economic development: Evidence from South Korea. *The Journal of International Trade & Economic Development, 32*(4), 537–552.

Kisswani, K. M., & Zaitouni, M. (2023). Does FDI affect environmental degradation? Examining pollution haven and pollution halo hypotheses using ARDL modelling. *Journal of the Asia Pacific Economy*, *28*(4), 1406–1432.

Mert, M., & Caglar, A. E. (2020). Testing pollution haven and pollution halo hypotheses for Turkey: A new perspective. *Environmental Science and Pollution Research, 27*, 32933–32943.

Pesaran, M. H., Shin, Y., & Smith, R. J. (2001). Bounds testing approaches to the analysis of level relationships. *Journal of Applied Econometrics, 16*(3), 289–326.

Pethig, R. (1976). Pollution, welfare, and environmental policy in the theory of Comparative Advantage. *Journal of Environmental Economics and Management, 2*(3), 160–169.

Porter, M. E., & Linde, C. V. D. (1995). Toward a new conception of the environment-competitiveness relationship. *Journal of Economic Perspectives*, *9*(4), 97–118.

Rana, R., & Sharma, M. (2019). Dynamic causality testing for EKC hypothesis, pollution haven hypothesis and international trade in India. *The Journal of International Trade & Economic Development*, *28*(3), 348–364.

Sadorsky, P. (2014). The effect of urbanization on CO2 emissions in emerging economies. *Energy Economics*, *41*, 147–153.

Shin, Y., Yu, B., & Greenwood-Nimmo, M. (2014). Modelling Asymmetric Cointegration and Dynamic Multipliers in a Nonlinear ARDL framework. *Festschrift in honor of Peter Schmidt: Econometric Methods and Applications,* 281–314.

Temurlenk, M. S., & Lögün, A. (2022). An analysis of the Pollution Haven Hypothesis in the context of Turkey: A nonlinear approach. *Economics and Business Review*, *8*(1), 5–23.

Ur Rahman, Z., Chongbo, W., & Ahmad, M. (2019). An (a) symmetric analysis of the pollution haven hypothesis in the context of Pakistan: a non-linear approach. *Carbon Management*, *10*(3), 227–239.

Yilanci, V., Cutcu, I., Cayir, B., & Saglam, M. S. (2023). Pollution haven or pollution halo in the fishing footprint: Evidence from Indonesia. *Marine Pollution Bulletin, 188*, 114626.

Zheng, J., Assad, U., Kamal, M. A., & Wang, H. (2024). Foreign direct investment and carbon emissions in China:"Pollution Haven" or "Pollution Halo"? Evidence from the NARDL model. *Journal of Environmental Planning and Management*, *67*(3), 662–687.

Yeşim Kubar[1] & Seyit Taha Ketenci[2] & İkranur Yılmaz[3]

Validity of Beveridge Curve: Türkiye Application with Non-Linear Cointegration Analysis

1 Introduction

The Beveridge Curve is used to measure the relationship between job vacancies and unemployment and provides important information about job vacancies and registered unemployment. In addition, it offers the opportunity to monitor the developments in the labor market. The Beveridge Curve shows the negative relationship between the job vacancy rate or number and the unemployment rate or number. This curve entered the literature with the Beveridge Report prepared by W. Henry Beveridge for the British economy in 1944. In this report, Beveridge examined the relationship between the job vacancy rate and the unemployment rate and explained the existence of a negative relationship between both variables.

The Beveridge Curve, which expresses the existence of an inverse and stable relationship between the job vacancy rate in the labor market and the unemployment rate, also shows the contraction and expansion tendencies in the labor market. Current job vacancies and unemployment rates may vary depending on the economic conjuncture. The fact that the number of vacancy is the same as the number of unemployment (exact and simultaneous match) indicates full employment. In this context, periods when the job vacancy rate is greater than the unemployment rate are called expansion periods, while periods when it is lower are called constriction periods. Through the Beveridge Curve it seems possible to follow business cycles in the economy by monitoring the changes in job vacancies and unemployment rates in the labor market in an economy

1 Ph.D., Fırat University, Faculty of Economics and Administrative Sciences Department of Economics, ykubar@fırat.edu.tr, ORCID:0000-0002-3439-9430.
2 Ph.D.,Selcuk University, Faculty of Economics and Administrative Sciences Department of Economics, stk@selcuk.edu.tr, ORCID:0000-0003-1230-5795.
3 Graduate Students, Fırat University, Social Sciences Institute, Branch of Economics, 231219109@firat.edu.tr, ORCID:0009-0003-4831-4471.

over time. Beveridge curves show marked differences between countries and the curves change over time. Although downward sloping in theory, some Beveridge curves have different shapes, indicating that economies have very heterogeneous levels of mismatches (Bova etc., 2016: 6).

It is possible to divide the economic literature examining the theoretical background of the Beveridge curve into two. The first group of studies that emerged in the late 1950s; Dow and Dicks-Mireaux (1958), Lipsey (1960), Hansen (1970) consider the Beveridge curve as a concept of supply-demand inequality within the framework of real wage rigidity. The second approach is Beveridge Curve analysis, which analyzes the structural dynamics of the labor market and the effects of business cycles. In studies carried out after 1980; Pissarides (1985); Börsch-Supan (1991); Mortensen and Pissarides (1994); Bleakley and Fuhrer (1997); Pissarides (2000); Petrongolo and Pissarides (2001) this approach is used.

The rising unemployment rate in Türkiye has become a serious problem that has been waiting to be solved for a long time. The unemployment problem brings with it social or economic difficulties and unless precautions and measures are taken to address this problem, long-term structural deteriorations and imbalances occur. What developing countries need to do in order to reduce the unemployment problem to reasonable levels is to examine the labor market correctly and carefully.

In this study, the relationship between the number of vacant jobs and registered unemployment in Türkiye for the period 2006:01-2023:12 was analyzed using the nonlinear time series method. In this context, the nonlinear Hepsağ (2021) cointegration analysis was used. As a result of the analysis, it was concluded that there was a symmetric ESTAR type cointegration. When the long-term coefficient was examined, the determination of a positive relationship between the number of unemployment and the number of vacant jobs did not confirm the existence of the Beveridge Curve in Türkiye.

2 Literature Review

The literature on the Beveridge Curve, which began to be discussed in the late 1950s, began to expand in the late 1980s; it accelerated especially in the 2000s when Diamond, Mortensen and Pissarides received the Nobel Prize for their study on this subject, and an important literature on the Beveridge Curve began to compose. Studies on the Beveridge Curve yield results that are consistent with theoretical and empirical literature. Some selected international studies on the estimation of the Beveridge Curve are given in Table 1.

Table 1: Literature Review

Author	Period	Method	Results
Yılancı 2009	1923–2007	Perron, ZivotAndrews, Lumsdaine – Papell, LM unit root tests	Shocks to the unemployment rate divert unemployment from its natural rate and this effect is permanent.
Güloğlu and İspir 2011	1988–2008	Panel unit root	The results were found to support the argument that unemployment is the natural rate of unemployment
Tokatlıoğlu, Öztürk and Ardor 2014	1988–2011	Ratchet model	Considering the entire population in these countries, the effect of unemployment hysteria is low.
Bayrak and Tatlı 2016	1988–2014	ADF and ARDL limit test	It has been concluded that PPI has a statistically significant and negative impact on youth unemployment, while economic growth has a statistically insignificant and negative impact on youth unemployment.
Özçelik and Uslu 2017	2007–2014	ARDL	It was found that a 1 unit change in the output gap creates a change of 0.0185 % in the unemployment gap in the short term and 0.2182 % in the long term.
Bilen etc. 2018	1961–2013	VAR, VECM	The job vacant is affected by both its own and its matching performance two years ago. Unemployment is affected only by its own past performance.
Çağlarırmak Uslu etc. 2019	2005–2017	ARDL and Hatemi-J Asymmetric Causality	A long-term relationship was found between the job vacancy rate and the unemployment rate. In the long term, job vacancies had a negative effect on unemployment, but in the short term, no statistically significant relationship was found.

(*continued on next page*)

Table 1: Continued

Author	Period	Method	Results
Özçelik and Akıncı 2020	2006–2020	Cointegration	They showed that there is a cointegration relationship between the unemployment rate and the job vacancy rate in the long run and that a 1 % increase in the job vacancy rate reduces the unemployment rate by -0.15 % in the long run.
Abraham and Katz 1986	1986	OLS	According to the findings, in an economy where sectors have different cyclical sensitivities, aggregate demand movements affect both the distribution of employment growth rate and the unemployment rate, creating a positive relationship between the two.
Evans 1989	1950–1985	Granger Causality,VAR	With the Granger causality test, a bidirectional causality was found between unemployment and economic growth, and with the VAR model, a negative relationship was found between the growth rate and the unemployment rate.
Teo etc. 2005	1987–2003	Regression Analysis	The matching efficiency for the Singapore economy is found to be higher than that of Japan and Korea.
Destefonis and Fonseca 2006	1992–2001	GLS method	In Italy, the Beveridge curve was found to differ across regions and in the Southern Italian region, unskilled workers (those with education levels below a university or college degree) reduced matching efficiency.
Choudhry etc. 2010	1980–2005	GMM panel regression	They proved that the crisis variable has a positive and statistically significant effect on youth unemployment according to the GMM panel regression model estimation results.
Kasrin and Lang 2010	2004–2010	Time Series	A negative relationship was found between private sector job vacancy rate and unemployment rates.

Table 2: Continued

Author	Period	Method	Results
Bonthuis etc. 2013	1991–2012	OLS	It has been found that the global financial crisis shifted the Beveridge curve of Spain and France to the right and the Beveridge curve of Germany to the left.
Tagkalakis 2016	2000 –2015	SVAR	He concluded that there is an inverse relationship between unemployment and job vacancies, resulting from developments after the 2008 crisis.
Pater 2017	1955–2016	B-VAR	The shock that creates a positive cooperation between job vacant and unemployment is due to the matching job and productivity decline. This positive cooperation occurs after recessions and in the long run.

Although the literature on the analysis of the Beveridge Curve in Table-1 has different calculation methods, different methodologies, different periods and economies, it is seen that the results obtained are compatible with the relevant theoretical literature.

3 Dataset and Method

The study uses an annual data set covering the period 2006:01-2023:12. The data for the variables used in the analysis are taken from the statistical yearbooks prepared by İŞKUR for each year, representing the entire labor market. Explanatory information about the data set is given in Table 2.

Table 2: Explanatory Information on Variables

Abbreviations	Variable	Data Source
aio	Job vacancies	İŞKUR
io	Registered unemployment	İŞKUR

Various economic events occurring in the Turkish economy may affect the number of job vacancies, job placements and unemployment, and may cause breaks in the variables in question. The basic hypothesis of this study is that there may be a negative and statistically significant long-term relationship between job vacancies and registered unemployed people for the period 2006:01-2023:12 and therefore the existence of the Beveridge Curve in Türkiye.

In this context, the hypothesis tested in the study is given below.

H_1: In the long run, there is a negative and statistically significant relationship between the number of vacant jobs and the number of registered unemployed. In the long run, vacant jobs affect unemployment.

For the econometric application of the study, the regression model determined for Türkiye by taking the logarithms of the vacant jobs and registered unemployed variables is shown equation (1):

$$\text{lio}_t = a_0 + a_1 laio_t + \varepsilon_t \qquad (1)$$

The laio variable given in equation (1) represents the open jobs and the lio variable represents the registered unemployed. The a_0 coefficient represents the constant term, a_1 coefficient represents the slope coefficient and ε is the error term.

In this study, Augmented Dickey Fuller (ADF) unit root test were applied to determine whether the series were stationary. Unit root tests are applied to determine the level at which the data is stationary. When interpreting the ADF unit root test results, the t-statistic values are compared with the critical values (1 %, 5 %, 10 %). If the t-statistic value is less than the critical value, the series is stationary.

Null and alternative hypotheses for ADF Unit root test is expressed as follows:

H_0: Series is not stationary (series contains unit root).
H_1: Series is stationary (series does not contain unit root)

The ADF unit root test results are given in Table 3.

Table 3: ADF Unit Root Test Results

ADF			**ADF**		
lio	constant	With constant and trend	dlio	constant	With constant and trend
t-statistic	-0.735684	-0.735684	t-statistic	-17.16808	-17.13508
1 %	-3.460739	-4.001311	1 %	-3.460884	-4.001516
5 %	-2.874804	-3.430864	5 %	-2.874868	-3.430963
10 %	-2.573917	-3.139056	10 %	-2.573951	-3.139114

Table 4: Continued

ADF			ADF		
laio	constant	With constant and trend	dlaio	constant	With constant and trend
t-statistic	-1.77048	-2.265598	t- statistic	-14.20804	-14.21364
1 %	-3.461030	-4.001722	1 %	-3.461030	-4.001722
5 %	-2.874932	-3.431062	5 %	-2.874932	-3.431062
10 %	-2.573985	-3.139173	10 %	-2.573985	-3.139173

According to Table 3, since the variables (-0.73) with the value of lio and (-1.77) with the value of laio are smaller than the critical values in absolute values at 1 %, 5 % and 10 % significance levels, it was concluded that the Ho hypothesis could not be rejected and the series was not stationary. When the ADF test statistic was examined in terms of first differences, the ADF test statistic value of (-17.16) with the value of lio and (-14.20) with the value of laio were found to be greater than the critical values in absolute value at the 1 %, 5 % and 10 % significance levels, so the Ho hypothesis was rejected and the series was found to be stationary. In the ADF unit root tests of the interest and inflation variables performed for the trend and fixed model with appropriate specification, it was determined that both variables were 1st degree stationary I(1).

If the trends of economic time series over time show non-linear properties, the stationarity condition of the time series can be investigated with non-linear unit root tests such as ESTAR (Exponential Smooth Transition Autoregressive) type KSS, SLS, KRS and Fourier Type FKPSS, FADF, FGLS. The results of Kruse (2011) unit root test, one of the non-linear unit root tests, are given in Table 4.

Table 4: Kruse (2011) Unit Root Test Results

Variables	$d_t = 0$	$d_t = 1$	$d_t = [1t]'$
laio t-ist	5.70501	6.88717	20.41402***
lio t-ist	9.32295	8.23071	10.90556***
1 %	13.15	13.75	17.10
5 %	9.53	10.17	12.87
10 %	7.85	8.60	11.10

*, ** and *** symbols are expressed as significant at 1 %, 5 % and 10 % levels, respectively, raw data is expressed as $d_t = 0$, reduced data is expressed as $d_t = 1$ and detrended data is expressed as d_t= [1 t]'.

As a result of the Kruse (2011) unit root test, it was determined that the variables contained a unit root and that the lio and laio variables were stationary variables in the first difference according to the ADF test results from traditional unit root tests. For this reason, it was concluded that the Hepsağ (2021) cointegration test could be used.

Hepsağ (2021) Cointegration test results are presented in Table 5.

Table 5: Hepsağ(2021) Cointegration Test Results

	Test statistic	**%1**	**%5**	**%10**
$lio_t = a_0 + a_1 laio_t + \varepsilon_t$				
FANEC-raw	5.33101**	6.494	4.536	3.678
FANEG-raw	5.80924**	6.767	4.953	4.088
FANEC-demeaned	5.56166***	8.112	5.972	5.009
FANEC-demeaned	5.75166***	8.433	6.325	5.373
FANEC-detrended	5.43219	9.669	7.351	6.303
FENEC-detrended	5.54401	10.082	7.741	6.655

H_0= There is no cointegrated relationship between variables.
H_1= There is a cointegrated relationship between the variables.

According to the results of the Hepsağ (2021) cointegration analysis, the null hypothesis is rejected when the calculated test statistic value is greater than the table value. According to Table 5, since the test statistic is greater than the table value in all models, it has been determined that there is a cointegrated relationship between the variables. In other words, it has been found that there is a cointegrated relationship between vacant jobs and registered unemployed in Türkiye. In the asymmetric nonlinear cointegration test developed by Hepsağ (2021), if the cointegration relationship between the variables is determined, it is possible to test symmetric ESTAR cointegration against asymmetric ESTAR cointegration (Hepsağ, 2022). In order to test the validity of symmetric ESTAR cointegration and asymmetric ESTAR cointegration, the following hypotheses are tested by taking into account the above two test regressions.

H_0: $\theta 2 = 0$
H_1: $\theta 2 \neq 0$

In the case where the F test statistic is greater than the critical values of the standard F distribution, the asymmetric ESTAR cointegration situation is valid. Accordingly, positive and negative shocks that will occur in the short term will have an asymmetric effect in the long-term equilibrium process. In order to apply this test, the coefficient of the cubic expression in the test statistics, the parameter ∅1, must have a negative value.

Table 6 shows the long-term models of the cointegrated relationship between vacant jobs and registered unemployed.

Table 6: The Results of long run and short run(AESTAR-ECM) Models

Long Run Estimates		
Dependent Variable: io_t		
Variables	coefficients	t-statistic
aio_t	1.0883060774	55.52579
Symmetric ESTAR Cointegration vs. Asymmetric ESTAR Cointegretion		
t(209)= -0.441224 or F(1.209)= 0.194679	Level=0.65950681	Symmetric ESTAR Cointegration
Dependent variable: Δio		
$\breve{U}^3_{t-1}$	-0.006842997	-1.94251
$\breve{U}^4_{t-1}$	-0.000307540	-0.44122
Δaio_t	0.103461022	0.24837
$\Delta \text{io}\{1\}$	-0.114668639	-1.66800
$\Delta\text{aio1}\{1\}$	-0.047582596	-0.11531

According to Table 6, since the F_{ANEC} test statistic calculated for the raw data set is greater than the critical value, the H0 hypothesis, which states that there is no cointegration relationship, is rejected. In addition, since the $\breve{U}^3_{t-1}$ (UCUBED) value is negative, the existence of a cointegration relationship is accepted. In this way, it is possible to apply the F test to determine the existence of symmetric ESTAR cointegration or asymmetric ESTAR cointegration between variables. The F test statistic calculated for this test can be compared with the standard F table at a significance level of 5 %. In addition, the existence of symmetric ESTAR cointegration or asymmetric ESTAR cointegration can be decided according to whether the level value is greater or less than 0.05. If the calculated level value is less than 0.05, there is an asymmetric ESTAR cointegration. If the calculated level value is greater than 0.05, there is a symmetric ESTAR cointegration. If there is a symmetric ESTAR cointegration, shocks that occur in the short and long term will have the same effect on the long-term equilibrium process.

In the study where the validity of the Beveridge curve which shows the negative relationship between the unemployment rate and the job vacancy rate was tested in Türkiye. According to the findings obtained with the econometric analysis with the Hepsağ cointegration method a positive relationship between the variables "io"-"aio" in the long-term coefficient statistics were calculated This situtaion shows that the Beveridge curve is not valid in Türkiye.

According to Blanchard and Diamond, who examined the matching process in the labor market and the nature of the shocks affecting this market, since the course of unemployment and job vacancies are given together in the Beveridge curve, an interpretation can be made about whether the shocks affecting the labor market are due to changes in aggregate demand, matching problems, or labor. If unemployment and job vacancies are moving in the opposite direction, this movement reflects the changes in labor demand seen in economic fluctuations. If unemployment and job vacancies increase at the same time, it reflects a matching problem in the labor market. The result obtained in the study shows that unemployment and job vacancies increased simultaneously for Türkiye in the period 2006:01-2023:12.

4 Conclusion

Employment and unemployment are concepts that concern both individuals and the vast majority of society. In order to ensure the level of economic and social welfare in a country, it is important to analyze these two concepts and the vacant jobs that affect these concepts in detail.

While employment is an important element for growth and development, unemployment is a phenomenon that causes individual and social problems and must be solved. Job vacancy is an indicator of growth potential for the economy. The existence of unemployment and the resulting underemployment in economy as a result of a portion of the workforce not participating physically and mentally in the production process brings many costs for both the unemployed and the country's economy as a whole. The steps that need to be taken and the policies that need to be implemented in order to combat this problem and reduce unemployment should be aimed at the sources that cause unemployment. Therefore, movements in the labor market should be examined by taking into account labor demand and supply. The analysis of the relationship between job vacancies constituting the demand side of the labour market and unemployment constituting the supply side is important in this respect. The Beveridge curve, which is used as a tool to measure this relationship, provides important information about the job vacancy and unemployment in the labor market. The Beveridge curve provides important information about job vacant rate and unemployment in the labor market. The Beveridge curve reveals the level of matching and compatibility in the labor market with the inverse relationship between job vacant rates and unemployment rates. In addition, it is possible to analyze the type and source of existing unemployment in the labor market by using the Beveridge curve. In this context, the Beveridge curve is important in terms of revealing the structure of the labor market from a different perspective and enabling the observation of the causes of the unemployment problem. The Beveridge curve is a method used in labor market analyses in developed countries and is an important instrument preferred in labor market comparisons between countries.

The solution to natural unemployment in Türkiye is to revitalize the economy and create new employment areas by developing solution proposals that take into account the reasons of unemployment. Periods when the negative relationship between the number of unemployed and job vacancies deteriorates indicate that matching in the labor market is weak. Matching shocks, which can be considered as an expression of the mismatch between the qualifications required for vacant jobs and the qualifications of job seekers, can lead to long-term unemployment, as in Türkiye, and to the unemployment rate stabilizing at a higher level even if aggregate demand recovers after economic crises, thus reversing the expected negative relationship between unemployment and vacant jobs in the Beveridge curve.

Matching shocks can be considered as an expression of the mismatch between the qualifications required for vacant jobs and the qualifications of job seekers. This mismatch can lead to long-term unemployment to the unemployment rate stabilizing at a higher level even if aggregate demand recovers after economic crises as in Türkiye. This can eventually lead to a reversal of the expected negative relationship between unemployment and vacant jobs in the Beveridge curve. For this reason, it is recommended that all actors in the labor market for the Turkish economy act with a sense of shared responsibility, contribute to employment by using resources effectively and developing and supporting policies that will reduce unemployment.

References

Abraham, K. G., & Katz, L. F. (1986). "Cyclical unemployment: Sectoral shifts or aggregate disturbances?". *Journal of Political Economy*, *94*(3-1), 507–522.

Bayrak, R., & Tatlı H. (2016). "Short and long term analysis of some factors effecting youth unemployment İn Turkey", *Theoretical and Applied Economics, 3*(608), 229–242.

Bilen, Ö., Bilen Kazancık, L., & Doğan, İ. (2018). "Türkiye İşgücü Piyasasında Arama Ve Eşleştirme Modeli", *Sayıştay Dergisi, 108*, 73–97.

Bonthuis, B., Jarvis, V., & Vanhala, J. (2013). "What's going on behind the Euro Area Beveridge Curve(S)?". *Working Paper Series*, 1586, 1–44.

Börsch- Supan. (1991). "Panel data analysis of the beveridge curve: Is there a macroeconomic relation between the rate of unemployment and the vacancy rate?". *Economica, 58*(231), 279–297.

Bova, E., Tovar Jalles, J., & Kolerus, C. (2016) "Shifting the Beveridge curve: What affects labour market matching?". *International Labour Review, 157*(2), 267306.

Bleakley, H., & Fuhrer, J. C. (1997). "Shifts in the Beveridge Curve, job matching and labor market dynamics", *New England Economic Review*, 5–19.

Choudhry, M. T., Marelli, E., & Signorelli, M. (2010). "Youth Unemployment and the Impact of Financial Crises", *This Paper Were Presented at the 11th biannual EACES Conference, University of Tartu*, Estonia, August 26–28: 1–15.

Çağlarırmak Uslu, N., Karahan-Dursun, P., & Çer, Ş. (2019). "Türkiye'de Beveridge Eğrisinin Geçerliliğine İlişkin Ampirik Bir Analiz". *Aksaray Üniversitesi İktisadi Ve İdari Bilimler Fakültesi Dergisi, 11*(1), 43–52.

Destefanis, S., & Fonseca, R. (2006). "Labour-Market Reforms And The Beveridge Curve Some Macro Evidence For Italy". *RAND Labor and Population Working Paper Series*, 1–32.

Dow, J. R., & Dicks-Mireaux, L. A. (1958). "The excess demand for labour. A study of conditions in Great Britain, 1946–56", *Oxford Economic Papers, 10*(1), 1–33.

Evans, G. (1989). "Output and unemployment dynamics İn The United States: 1950–1985", *Journal of Applied Econometrics, 4*(3), 213–237.

Güloğlu, B., & İspir, M. S. (2011). "Doğal İşsizlik Oranı mı? İşsizlik Histerisi mi? Türkiye için Sektörel Panel Birim Kök Sınaması Analizi". *Ege Akademik Bakış, 11*(2), 205–2015.

Hansen, B. (1970). "Excess demand, unemployment, vacancies, and wages". *The Quarterly Journal of Economics, 84*(1), 1–23.

Hepsağ, A. (2021). Testing for Cointegration in Nonlinear Asymmetric Smooth Transition Error Correction Models. Communications in Statistics-Simulation and Computation, 50(2), 400–412.

Hepsağ, A. (2022). "Ekonometrik Zaman Serileri Analizlerinde Güncel Yöntemler". İstanbul: Der yayınları.

Kasrin, Z., & Lang G. (2010). "Estimating the Beveridge Curve of Egypt: An econometric study for the period 2004 to 2010". German University İn Cairo. *Working Paper, 21*, 1–24.

Kruse, R. (2011) A New Unit Root Test against ESTAR-Based on a Class of Modified Statistics. Statistical Papers, 52, 71–85.

Lipsey, R. (1960). "The Relation Between Unemployment And The Rate Of Change Of Money Wage Rates İn The United Kingdom, 1862_1957: A Further Analysis". *Economica 27*, 1–31.

Mortansen, D. T., & Pissarides, C. (1994). "Job creation and job destruction in the theory of unemployment". *Review of Economic Studies*, *61*, 397–415.

Özçelik, Ö., & Akıncı, A. (2020). "İşgücü Piyasasında Beveridge Eğrisi'nin Geçerliliği: Türkiye Uygulaması". *Balkan Sosyal Bilimler Dergisi*, *9*(17), 152–161.

Pater, R. (2017). "Is there a Beveridge Curve in the short and the long run?". *Journal of Applied Economics, XX* (2), 283–303.

Petrongolo, B. & Pissarides A. C. (2001). "Looking into the Black Box: A Survey of the Matching Function." *Journal of Economic Literature*, 39 (2): 390–431.

Pissarides, C. A. (1985). "Short-run equilibrium dynamics of unemployment vacancies and real wages". *The American Economic Review*, *75*(4), 676–690.

Pissarides, C. A. (2000). "Equilibrium unemployment theory". Cambridge, MA: The MIT Press Second Edition. London, England.

Tagkalakis, A. O. (2016). "Unemployment dynamics and the Beveridge Curve İn Greece". *Journal of European Labor Studies*, *5*(13), 1–34.

Teo, E., Thangavelu, S. M., & Quah, E. (2005). "Singapore's Beveridge Curve: A comparative study of the unemployment and vacancy relationship for selected East Asian Countries". *SCAPE Working Paper*, 2005/08, 1–17.

Tokatlıoğlu, İbrahim, Öztürk, F., & Ardor, H. N. (2014). "AB Ülkeleri ve Türkiye İşgücü Piyasasında Histeri Etkisi: Ratchet Modeli Analizi". *Sosyo-Ekonomi*, *22*(22), 297–320.

Yılancı, Veli (2009), "Yapısal Kırılmalar Altında Türkiye İçin İşsizlik Histerisinin Sınanması". *Doğuş Üniversitesi Dergisi*, *10*(2), 324–335.

Ali Rauf Karataş[1] & Murat Ergül[2]

Unemployment and Youth Unemployment Hysteresis in Countries of Different Income Groups: Evidence from Fourier Unit Root Tests

1 Introduction

In the economic literature, the starting point of theoretical discussions on unemployment is often referred to the Great Depression years. In this context, the 1930s are still recognized as a historical threshold point at which macroeconomic imbalances are measured. In the nearly 100 years since the Great Depression, unemployment has remained an important macroeconomic problem across countries, both in terms of its incidence and its multidimensional consequences.

As the problem of unemployment has become central to macroeconomics, economists have focused on the causes, determinants, different dimensions and severity of unemployment. As a result of the approaches developed under the umbrella of Keynesian economics and the Neoclassical Synthesis, two theoretical approaches have gained weight in explaining the unemployment phenomenon. Accordingly, economies are assumed to have either unemployment hysteresis or natural rate of unemployment (NRU) or nonaccelerating inflation rate of unemployment (NAIRU).

As is well known, standard macroeconomic models are based on a sharp distinction between equilibrium and actual unemployment rates. Accordingly, the equilibrium unemployment rate depends on the institutional characteristics of the labor market. Its movements are moderate and not significantly affected by the actual unemployment rate. Therefore, unexpected and structural shocks at the supply and demand levels only cause deviations in the actual unemployment rate. Deviations in the actual unemployment rate trigger inflation through the mechanism of flexible prices and wages, and ultimately bring the actual

1 Ph.D., Asst. Prof., Karabük University, Faculty of Economics and Administrative Sciences Department of Economics, alıraufkaratas@karabuk.edu.tr, ORCID: 0000-0003-1031-6722.

2 Ph.D., Asst. Prof., Karabük University, Faculty of Economics and Administrative Sciences Department of Economics, muratergul@karabuk.edu.tr, ORCID: 0000-0003-2117-7561.

unemployment rate back to equilibrium (Tekin, 2018: 98). In this context, the natural rate hypothesis Friedman (1968) and Phelps (1967, 1968, 1994) formed the framework of his work. Friedman defines the natural rate of unemployment as a given level of unemployment consistent with labor market equilibrium. It is unemployment at a moment of full employment when there is price stability in the economy and is determined by real factors that affect the level of incidental (frictional) and structural unemployment. As can be seen, the NRU is built on the assumption of perfect competition. The concept of NAIRU entered the macroeconomic literature in the 1970s, a period of rapidly rising inflation rates on a global scale, and it still persists. The essence of NAIRU, which is roughly synonymous with NRU, is the idea that changes in monetary policy and aggregate demand more generally push inflation and unemployment in opposite directions in the short run, but that unemployment will remain at a consistent level in the long run once price stability is achieved. The NAIRU also changes depending on the relationship between inflation, unemployment and productivity. The NAIRU rises when labor and productivity growth slows down (as in the 1970s) and falls when productivity growth picks up (as in the 1990s). (Ball & Mankiw, 2002: 115).

The inclusion of the hysteresis hypothesis in the economic literature is related to the global economic developments in the 1970s. Declining productivity growth and high oil prices as a result of the oil shocks in the 1970s increased the actual unemployment rate throughout the 1970s. The simultaneous rise in inflation and the actual unemployment rate (stagflation) caused the classical macroeconomic models to be discussed as in the Depression years. Especially in the 1980s, despite the cessation of price increases, the persistence of rising unemployment rates, in other words, the failure to return to the equilibrium rate before supply shocks, fueled these debates. In this context, Blanchard and Summers (1986) developed an alternative theory of unemployment based on the idea that the equilibrium unemployment rate may depend on the past values of the actual unemployment rate and integrated the concept of "hysteresis" into the unemployment phenomenon. They defined "unemployment hysteresis" as a situation in which actual unemployment, which rises as a result of shocks and/or crises, does not fall back to its previous level after the crisis period and equilibrium unemployment rises for this reason.

As can be seen, the emergence of the hysteresis hypothesis was realized in proportion to the effects of structural economic shocks, either supply or demand-side. World economies have faced many real economic shocks in the recent past. In addition to real economic shocks, the financial system, which has developed and gained momentum since the 1970s, reinforces the effects of shocks due to its procyclicality. In this sense, the Global Financial Crisis demonstrated the procyclicality of the financial system and how micro-scale crises in the financial system have macro-scale consequences that spill over into the real sector. In addition to rising unemployment rates, the Global Financial Crisis has also exposed youth unemployment as a global problem and demonstrated the importance of understanding labor markets. The Covid-19 pandemic that started in 2019, the European Energy Crisis of 2021 and the Russia-Ukraine War that started in February 2022 have brought new economic shocks with new and different dimensions. Based on the above-mentioned theoretical requirements and the possible global effects of these economic shocks on the labor market, this study was conducted to test the existence of unemployment and youth unemployment hysteresis in different income groups in the world. By simultaneously testing the hysteresis hypothesis in countries in different income groups, it will be seen whether the effect of hysteresis differs in terms of welfare level. As is known, the welfare state concept developed as a result of the economic collapse and high unemployment rates during the Great Depression. It can be expected that labor market institutions are more efficient and well-functioning especially in countries with higher income levels. For this reason, it is more likely than in low-income countries that unemployment rates that rise after an economic shock will return to the natural rate level after the effects of the shocks have passed. In this context, the study also claims to reveal whether this theoretical expectation is realized in reality. For these purposes, the study uses annual data for the period 1991–2023 to test the hysteresis hypothesis for four different income groups (upper middle income, middle income, lower middle income, low income) and for youth and general unemployment rates worldwide. All data used in the study are obtained from the World Bank database. The empirical findings are presented in the econometric analysis section.

2 Econometric Analysis

Since hysteresis is associated with non-stationary unemployment rates, unit root tests are widely used to test the validity of the unemployment hysteresis hypothesis. (Chang et al., 2005: 882). While the proof of a unit root supports the hysteresis hypothesis, the rejection of a unit root is seen as evidence for the natural rate hypothesis. (Chen at al., 2022).

2.1 Data

In this study, the unemployment hysteresis hypothesis is tested with the dimensions of unemployment rate and youth unemployment rate for different country groups. The data used for the purpose of this study were obtained annually (1991–2023) from the World Bank database. Accordingly, the time series graphs of the data included in the analysis are presented in Figures 1 and 2.

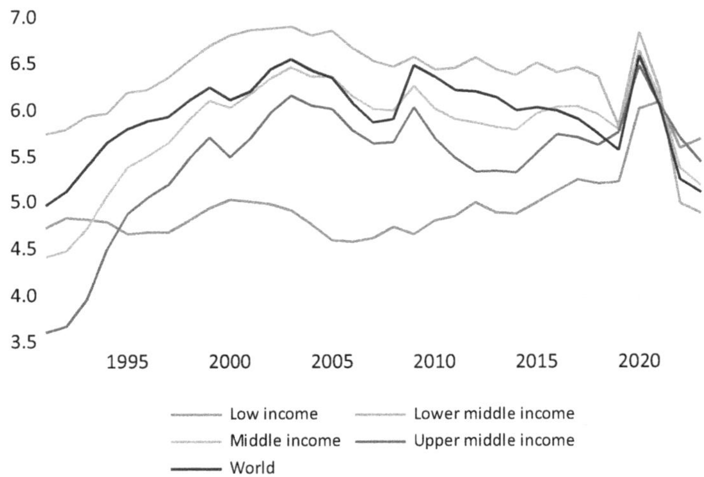

Figure 1: Unemployment, total (% of total labor force)
Source: (World Bank, 2024)

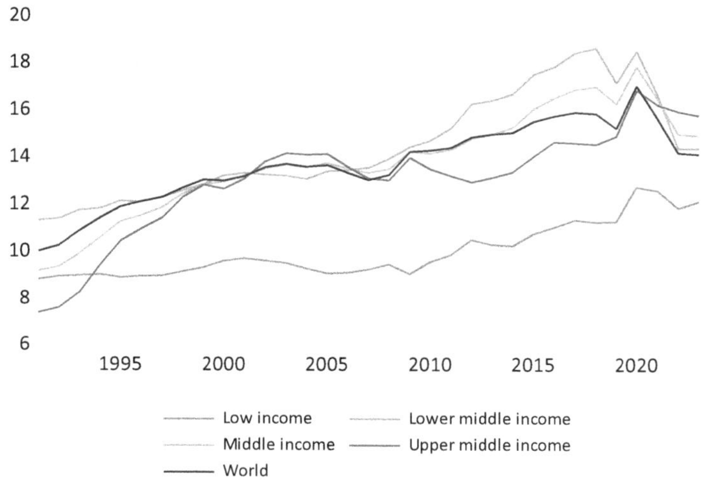

Figure 2: Unemployment, youth total (% of total labor force ages 15–24)
Source: (World Bank, 2024)

2.2 Unit Root Tests

At this stage, the quality of the unit root tests to be preferred for testing the hysteresis hypothesis becomes important. The existence of structural breaks in many macroeconomic variables such

as unemployment rate is known. For this reason, unit root tests that take structural breaks into account and have current values are used in this study. Accordingly, Table 1 shows the results of the Fourier Augmented Dickey-Fuller (Fourier ADF) unit root test for the model with constant and trend.

Table 1: Fourier ADF Unit Root Test Results

Unemployment, total (% of total labor force)				
Country group	**FADF stat.**	**Lag (l)**	**Fourier (k)**	**Conclusion**
Upper middle income	-3.4386	1	2	I(1)
Middle income	-2.1807	1	2	I(1)
Lower middle income	-1.5261	0	1	I(1)
Low income	-4.8750*	1	0	I(0)
World	-3.6712	1	1	I(1)

(continued on next page)

Table 2: Continued

Unemployment, youth total (% of total labor force ages 15-24)				
Country group	**FADF stat.**	**Lag (l)**	**Fourier (k)**	**Conclusion**
Upper middle income	-3.8467	1	2	I(1)
Middle income	-2.3024	1	2	I(1)
Lower middle income	0.1670	0	1	I(1)
Low income	-4.3371***	0	1	I(0)
World	-2.6778	1	2	I(1)

Notes: The critical values for unit root rejection according to Enders and Lee (2012b). "*", "**"and "***"indicate statistical significance at the 1 %, 5 %, and 10 % levels, respectively.

The Fourier ADF unit root test was developed by Enders and Lee (2012b) to determine the number and form of gradual breaks in the time series and their deterministic components. This unit root test uses the frequency component of the Fourier type function and tests for stationarity based on the ADF unit root test. (Çevikalp & Özdemir, 2021: 824). According to the results in Table 1, it is seen that the hysteresis hypothesis is valid in all definitions except for the low-income country groups in terms of both unemployment and youth unemployment rates. This result indicates that structural shocks in world economies lead to permanent increases in unemployment rates and that countries are mostly in an underemployment equilibrium. A comparative analysis of the Fourier ADF test statistic results reveals that the hysteresis effect is stronger on youth unemployment. It also implies that youth unemployment, which emerged as an important macroeconomic problem in continental Europe especially after the 2008 Global Financial Crisis, tends to turn into a structural problem.

In order to provide stronger econometric evidence, in addition to the Fourier ADF test, other unit root tests were included in the analysis. Accordingly, the results of the Fourier LM test, which is another unit root test developed by Enders and Lee (2012a), are presented in Table 2 for the constant and trend model.

Table 2: Fourier LM Unit Root Test Results

Unemployment, total (% of total labor force)				
Country group	**LM stat.**	**Lag (l)**	**Fourier (k)**	**Conclusion**
Upper middle income	-3.7010	1	1	I(1)
Middle income	-3.2526	1	1	I(1)
Lower middle income	-2.9182	0	1	I(1)

Table 2: Continued

Low income	-3.9021***	0	1	I(0)
World	-4.4204**	1	1	I(0)
Unemployment, youth total (% of total labor force ages 15-24)				
Country group	**LM stat.**	**Lag (l)**	**Fourier (k)**	**Conclusion**
Upper middle income	-3.4639	1	1	I(1)
Middle income	-2.7372	1	2	I(1)
Lower middle income	-1.3099	0	2	I(1)
Low income	-3.7146	0	1	I(1)
World	-2.9507	1	2	I(1)

Notes: The critical values for unit root rejection according to Enders and Lee (2012a). "*", "**"and "***"indicate statistical significance at the 1 %, 5 %, and 10 % levels, respectively.

According to the results of the Fourier LM unit root test reported in Table 2, the hysteresis hypothesis is valid in all series except for the lower income group and world unemployment in terms of unemployment rate. When the results are analyzed in the context of youth unemployment, it is seen that the hysteresis hypothesis is valid in all series included in the analysis and therefore in all country groups. Fourier Kwiatkowski-Phillips-Schmidt-Shin (Fourier KPSS) unit root test results are presented in Table 3 for the model with constant and trend.

Table 3: Fourier KPSS Unit Root Test Results

Unemployment, total (% of total labor force)			
Country group	**KPSS stat.**	**Fourier (k)**	**Conclusion**
Upper middle income	0.0947	1	I(1)
Middle income	0.1077	1	I(1)
Lower middle income	0.1059	1	I(1)
Low income	0.0555***	1	I(0)
World	0.0868	1	I(1)
Unemployment, youth total (% of total labor force ages 15-24)			
Country group	**KPSS stat.**	**Fourier (k)**	**Conclusion**
Upper middle income	0.0921	1	I(1)
Middle income	0.2847	2	I(1)
Lower middle income	0.1966	2	I(1)
Low income	0.0617***	1	I(0)
World	0.2755	2	I(1)

Notes: The critical values for unit root rejection according to Becker at al. (2006). "*", "**"and "***"indicate statistical significance at the 1 %, 5 %, and 10 % levels, respectively.

In the Fourier KPSS unit root test, the null hypothesis "there is no unit root in the series" is tested. The null hypothesis is accepted if the Fourier KPSS test statistic value calculated for each series is smaller than the relevant critical value (Becker et al., 2006). Accordingly, the Fourier KPSS unit root test results in Table 3 are similar to the Fourier ADF test results. Except for the low-income group, all series except for the low-income group have a unit root and hence the hysteresis hypothesis is valid. Finally, the results of the Fourier GLS unit root test developed by Rodrigues and Taylor (2012) for the model with constant and trend are presented in Table 4.

Table 4: Fourier GLS Unit Root Test Results

	Unemployment, total (% of total labor force)			
Country group	**GLS stat.**	**Lag (l)**	**Fourier (k)**	**Conclusion**
Upper middle income	-3.8263	1	1	I(1)
Middle income	-3.2826	1	1	I(1)
Lower middle income	-2.9953	0	1	I(1)
Low income	-4.8538*	0	1	I(0)
World	-4.3149**	0	1	I(0)
	Unemployment, youth total (% of total labor force ages 15-24)			
Country group	**GLS stat.**	**Lag (l)**	**Fourier (k)**	**Conclusion**
Upper middle income	-3.8371	1	1	I(1)
Middle income	-2.7005	1	2	I(1)
Lower middle income	-1.7500	1	2	I(1)
Low income	-4.4907**	0	1	I(0)
World	-3.0065	0	2	I(1)

Notes: The critical values for unit root rejection according to Rodrigues and Taylor (2012). "*", "**"and "***"indicate statistical significance at the 1 %, 5 %, and 10 % levels, respectively.

The null hypothesis of the Fourier GLS unit root test is "There is no unit root in the series". Accordingly, if the GLS test statistic obtained as a result of the test is smaller than the critical table values in absolute value, the null hypothesis is accepted. In this context, according to the results in Table 4, there is no unit root in the lower income group and world unemployment series in terms of unemployment rate. Therefore, while the hysteresis hypothesis is not valid in the definition of these two country groups, hysteresis exists in other series. In the youth unemployment rates, in line with the results of Fourier ADF and Fourier KPSS unit root tests, only the lower income group does not have a hysteresis effect, while all other series have a unit root and hysteresis effect. Similarly, in the Fourier GLS test, the hysteresis effect is stronger in youth unemployment rates.

4 Conclusion

In this study, the phenomenon of unemployment, which has been an important macroeconomic problem in most countries for many years regardless of income and welfare levels, has been tested on a world scale in the context of the hysteresis hypothesis. In this sense, the aim of the study is to determine whether the hysteresis effect is differentiated in countries with different income levels. For this purpose, unemployment and youth unemployment values of country groups with different income levels for the period 1991–2023 are analyzed with four different unit root tests extended with Fourier functions. As a result of the econometric analysis, two important conclusions were reached. Accordingly, it is found that both unemployment and youth unemployment hysteresis are valid in all country definitions except for low-income countries. The second important result is that the hysteresis effect is higher for youth unemployment at all income levels. The findings of the study suggest that supply and/or demand-side structural shocks have global effects and lead to permanent increases in unemployment rates. These findings also show that the youth unemployment problem, which became evident after the Global Financial Crisis, has turned into a permanent and structural phenomenon. The absence of the hysteresis effect in lower-income countries may be related to the concentration of employment in the agricultural sector in these countries. In this context, it is possible to say that the negative effects of economic shocks on the labor market stand out in the industry and services sectors. All these results emphasize the importance of better understanding the labor market. For this reason, determining the micro and macroeconomic foundations of unemployment and especially youth unemployment will be important for economies to reach full employment equilibrium.

References

Ball, Laurence, ve N. Gregory Mankiw. 2002. "The NAIRU in Theory and Practice". *Journal of economic Perspectives* 16(4): 115–136.

Becker, Ralf, Walter Enders, ve Junsoo Lee. 2006. "A Stationarity Test in the Presence of an Unknown Number of Smooth Breaks". *Journal of Time Series Analysis* 27(3): 381–409. doi: 10.1111/j.1467-9892.2006.00478.x.

Blanchard, O. J., & Summers, L. H. (1986). Hysteresis in unemployment. In *Economic models of Trade unions* (pp. 235-242). Dordrecht: Springer Netherlands.

Chang, Tsangyao, Kuei-Chiu Lee, Chien-Chung Nieh, ve Ching-Chun Wei. 2005. "An Empirical Note on Testing Hysteresis in Unemployment for Ten European Countries: Panel SURADF Approach". *Applied Economics Letters* 12(14): 881–886. doi: 10.1080/13504850500365871.

Chen, Hong, Baljeet Singh, ve Wesley Steve Aru. 2022. "Relationship between Government Expenditure and Economic Growth: Evidence from Vanuatu". *Journal of the Asia Pacific Economy* 27(4): 640–659. doi: 10.1080/13547860.2020.1844610.

Çevikalp, Sedef, ve Onur Özdemir. 2021. "FourierBirim Kök Testleri Temelinde Ekolojik Açık, Ekonomik Büyüme ve Özel Sektör Kredilerinin Analizi: Türkiye Örneği". *Turk Turizm Arastirmalari Dergisi* 4(9): 815–832. doi: 10.26677/TR1010.2021.800.

Enders, Walter, ve Junsoo Lee. 2012a. "A Unit Root Test Using a Fourier Series to Approximate Smooth Breaks*". *Oxford Bulletin of Economics and Statistics* 74(4): 574–599. doi: 10.1111/j.1468-0084.2011.00662.x.

Enders, Walter, ve Junsoo Lee. 2012b. "The Flexible Fourier Form and Dickey–Fuller Type Unit Root Tests". *Economics Letters* 117(1): 196–199. doi: 10.1016/j.econlet.2012.04.081.

Friedman, Milton. 1968. "The Role of Monetary Policy". 58(1–17).

Phelps, Edmund S. 1967. "Phillips Curves, Expectations of Inflation and Optimal Unemployment Over Time". *Economica* 254–281.

Phelps, Edmund S. 1968. "Money-Wage Dynamics and Labor-Market Equilibrium". *Journal of Political Economy* 76(4, Part 2): 678–711. doi: 10.1086/259438.

Phelps, Edmund S. 1994. *Structural Slumps: The Modern Equilibrium Theory of Unemployment, Interest, and Assets*. Harvard University Press.

Rodrigues, Paulo M. M., ve A. M. Robert Taylor. 2012. "The Flexible Fourier Form and Local Generalised Least Squares De-trended Unit Root Tests *". *Oxford Bulletin of Economics and Statistics* 74(5): 736–759. doi: 10.1111/j.1468-0084.2011.00665.x.

Tekin, İpek. 2018. "Türkiye'de işsizlik histerisi: Fourier fonksiyonlu durağanlık sınamaları". *Dokuz Eylül Üniversitesi İktisadi İdari Bilimler Fakültesi Dergisi* 33(1): 97–127.

The World Bank. (2024). World Bank Open Data. 13.06.2024, https://data.worldbank.org/.

Ogün Baykuş[1] & Selahattin Bektaş[2]

Examining the Sustainability of Current Account Deficits in D8 Countries Using Panel Unit-Root Tests: The Example of the 1981–2022 Period

1 Introduction

The current account (CA) is one of the four sub-components of the balance of payments, generally representing the record of monetary transactions a country's economy engages in with other economies. If the mathematical sum of the CA components is negative, it indicates a current account deficit (CAD), while a positive sum indicates a surplus (Bektaş, 2017: 52; Yardımcı, 2023: 1).

The CA is a significant macroeconomic indicator impacting economic growth, stable exchange rates, and the overall competitiveness of the economy. Therefore, deviations and imbalances in the CA attract the attention of policymakers. The size and persistence of these imbalances have raised the fundamental question of sustainability (Shastri et al., 2018: 2).

The CA is not a policy variable like money supply or interest rates. Additionally, it is not a target variable like inflation or unemployment. However, the CA provides information about the current state of macroeconomic policies and the behavior of economic units. At the same time, the CA reflects the optimal decisions made over time by rational borrowers and lenders in the global economy. A country's CA does not run a continuous deficit because the borrowed funds need to be repaid. Creditors' confidence depends on a country's ability to meet its financial obligations. In other words, a sustainable CA deficit determines a country's credibility (Hassan et al., 2015: 886).

The sustainability of the CAD has been the subject of numerous studies and debates in the literature. Studies on the dynamic characteristics of CA indicate that if stationarity is not adhered to, the CA of economies will not revert to the

1 Ph.D., Independent Researcher, ogunbaykus@gmail.com, ORCID: 0000-0003-4905-4715.
2 Ph.D., Asst. Prof., KTO Karatay University, Faculty of Economics, Administrative and Social Sciences, Department of Islamic Economics and Finance., selahattin.bektas@karatay.edu.tr, ORCID: 0000-0001-6285-8318.

mean. If this tendency not to revert to the mean persists, meaning an unexpected shock does not bring the CA balance, the economy will be driven to bankruptcy, and foreign investments will exit the market (Wu, 2000: 216).

Measuring the sustainability of the CA deficit is a contentious issue. Advances in time series techniques, particularly stationarity tests, allow for econometric testing of the sustainability hypothesis. The stationarity of the CA is significant for two reasons. First, a stationary CA is consistent with the sustainability of the deficit. Second, the stationarity of the CA aligns with the application of the modern intertemporal approach to the deficit, thus supporting the validity of this approach (Baharumshah et al., 2005: 515–516).

The sustainability of the CAD is interpreted in various ways. The most general interpretation is that a CAD is sustainable if an economy can meet the present value of its intertemporal budget constraint, meaning it has the capacity to pay its debts. In other words, the economy must generate enough trade surplus in the future to repay its debts. Therefore, the sustainability of the CAD is crucial for the survival of both developed and developing economies (Shastri et al., 2018: 2).

This study examines the sustainability of the CAD of D8 countries for the period 1981–2022 using panel unit-root tests. The introduction provides theoretical information about the CA and its sustainability. The following section summarizes the relevant literature review in a table. The methodology section then presents the analysis of the sustainability of the CAD of D8 countries for the period 1981–2022. The conclusion section offers a general assessment based on the findings.

2 Literature Review

Studies on the sustainability of the CAD began with Hakkio and Rush's (1991) examination of the long-term relationship between exports and imports in the U.S. economy. This research was further developed by Husted (1992), who made it testable econometrically.

The literature on the sustainability of CAD is divided into two main categories. The first category, led by Trehan and Walsh (1991), includes studies arguing that the stationarity of the CA is sufficient to meet the intertemporal budget constraint. The second category, led by Husted (1992), includes studies asserting that the existence of cointegration between imports and exports is a prerequisite for fulfilling the intertemporal budget constraint. Table 1 presents a summary of the literature under both categories.

Table 1: Literature Review

Author(s)	**Period**	**Country(ies)**	**Method**	**Result**
Husted (1992)	1967–1989	USA	Cointegration Test	Unsustainable
Wu (2000)	1977–1997	10 OECD Countries	Panel Unit Root, Panel Cointegration	Sustainable
Baharumshah et al. (2005)	1970–2000	8 East Asian Countries	Panel Unit Root, Panel Cointegration	Sustainable
Cunado et al. (2010)	1960–2005	16 EU Countries	Ng Perron Unit Root Test	Unsustainable
Chen (2010)	1960–2008	USA, UK, Canada, France	Traditional Unit Root	Unsustainable
Hassan et al. (2015)	1970–2010	Indonesia, Malaysia, Philippines, Thailand	Panel Error Correction and DOLS	Sustainable
Bektaş (2017)	1970–2014	Selected 5 Emerging Economies	Panel Unit Root, Panel Cointegration	Sustainable
Singh (2017)	1970–2006	24 OECD Countries	Panel FMOLS-DOLS	Sustainable
Öztürk & Çoltu (2018)	1995–2015	15 OECD Countries	Panel Unit Root, Panel Cointegration	Sustainable
Shastri et al. (2018)	1985–2016	India, Pakistan, Bangladesh, Sri Lanka, Nepal	Panel Cointegration	Sustainable
Uz et al. (2018)	1990–2014	Fragile Five Countries	Panel Unit Root	Sustainable (Except Brazil)
Demir (2019)	1998Q1-2018Q1	Turkey	Structural Break Unit Root	Unsustainable

3 Data and Methodology

In this study, to analyze the sustainability of the CAD, the ratio of the CAD to GDP for D8 countries (Bangladesh, Turkey, Nigeria, Malaysia, Egypt, Indonesia, and Iran) was used for the period 1981–2022. The data set was obtained from the World Bank database. Due to the unavailability of data for Iran, Iran was excluded from the analysis, and the analysis continued with the remaining seven countries. The cross-sectional dimension of the analysis comprises 7 units, and the time dimension covers a 41-year period. Therefore, a panel data set consisting of 287 observations was used in the analysis.

The sustainability of the CAD for the D8 countries will be examined using panel unit-root tests on the "CAD/GDP" variable. Before conducting the unit-root test analysis, a series of specification tests must be performed to determine which generation of unit-root tests (first generation, second generation) will be used to measure the stationarity of the "CAD/GDP" panel data. In this regard, cross-sectional dependence was first examined using the Breusch-Pagan LM (1980), CD Pesaran (2004), LM adj Adjusted LM Pesaran et al. (2008), CDNT Pesaran (2015) CD tests, and homogeneity was examined using Swamy's (1970) S test. This approach helps determine whether the panel data is heterogeneous or homogeneous and whether there is cross-sectional dependence. The analyses in the study were conducted using STATA and EViews software packages. Table 2 shows the summary statistics of the data set.

Table 2: Summary Statistics of "CAD/GDP" Panel Data Set

Variable	Mean	Median	Max.	Min.	Std. Dev.	Skewness	Kurtosis	Obs
Panel	-0.218111	-0.964847	20.794	-13.375	4.8604	1.3017	5.9536	294
Bangladesh	-0.565305	-0.344059	3.4701	-5.0204	1.8049	-0.3170	2.7783	42
Turkey	-2.411508	-2.401004	2.0137	-8.8702	2.4595	-0.1585	2.7314	42
Nigeria	2.490496	1.197095	20.794	-5.1003	5.3863	1.3791	5.1143	42
Egypt	-1.086329	-1.786476	9.0109	-9.6442	3.8434	0.6664	3.5744	42
Indonesia	-0.830510	-1.586510	4.8430	-7.8196	2.9163	0.2246	2.7352	42
Malaysia	3.376423	3.485015	16.859	-13.375	8.3267	-0.1460	2.1231	42
Pakistan	-2.500044	-2.704151	3.9357	-7.7419	2.3664	0.4729	3.8028	42

Source: Authors' Calculations

According to Table 2, the panel data consists of 249 observations. The mean of the data is -0.218, with a standard deviation of 4.86. The minimum value is -13.375, the maximum value is 20.794, the skewness is 1.3017, and the kurtosis is 5.9536.

Table 3: Horizontal Cross-Section Dependence Test Results

Cross-Section Tests	Statistics	d.f.	p- values
Breusch-Pagan LM	97.34510	21	0,0000
Pesaran scaled LM	11.78030		0,0000
Bias-corrected scaled LM	11.69494		0,0000
Pesaran CD	5.265782		0,0000

Source: Authors' Calculations

Table 3 presents the results of the cross-sectional dependence tests. According to the test results, the p-values are found to be less than 0.05, indicating that the panel data exhibits cross-sectional dependence.

Table 4: Unit Homogeneity Test Results (Swammy S)

Variable	chi2(6)	Prob > chi
Cari Açık/GSYH	61,17	0,0000

Source: Authors' Calculations

Table 4 presents the results of the Swamy S (homogeneity) test. According to the Swamy S test, the p-value is found to be less than 0.05, indicating that the panel data has a heterogeneous structure.

According to cross-sectional dependence and homogeneity tests, the "CAD/GDP" panel data is heterogeneous and exhibits cross-sectional dependence. Therefore, it is appropriate to use second-generation unit-root tests in the analysis. In this study, Pesaran's CIPS and PESCADF (2007), IPS (2003), and Breitung and Das (2005)'s (Robust Cross-Sectional) unit-root tests will be used to investigate sustainability.

Table 5: Results of Unit-Root Tests (CADF, CIPS and IPS (Robust)), Breitung (Robust)

(PESCADF) Test Result					
t-bar	**Cv10**	Cv5	Cv1	Z[t-bar]	p-value
-3,450	-2,210	-2,330	-2,550	-4,705	0,0000
(CIPS) Test Result					
CIPS Statistics	**Critical Values**		10 %	5 %	1 %
-3,514			-2,21	-2,33	-2,55

(*continued on next page*)

Table 5: Continued

IPS (Robust Cross-Sectional) Test Result				
Lag Length	**Information Criterion**	W-t-bar	Statistics	p-value
0,57	Akaike (AIC)		-4,476	0,0000
Breitung (Robust Cross-Sectional) Test Result				
Lambda	**Statistics**		p-value	
	-3,6205		0,0001	

Source: Authors' Calculations

Table 5 presents the results of the unit-root tests used in the analysis of the sustainability of CAD for D8 countries. Four unit-root tests were applied to the "CAD/GDP" panel data. The "CAD/GDP" panel data was included in the analysis at its level value.

Upon evaluating the obtained findings, it is observed that the t-bar statistic value according to the PESCADF test is -3.450. Since this calculated value is absolutely greater than the critical values, it is concluded that the "CAD/GDP" panel data is stationary at the level value. Additionally, the p-value of the Z[t-bar] statistic supporting the t-bar statistic is $p < 0.05$. Therefore, it is understood that the "CAD/GDP" panel data is stationary. Based on these results, it can be interpreted that the CAD is sustainable in the D8 countries over the examined periods according to the PESCADF test results.

CIPS test results, the CIPS statistic value is calculated as -3.514. Since the calculated value is absolutely greater than the critical values, it is determined that the "CAD/GDP" panel data is stationary at the level value, meaning it does not contain a unit-root. According to the results obtained from this test, it can also be stated that the CAD is sustainable for the D8 countries from 1981 to 2022.

IPS Robust test, the calculated value of the IPS test is -4.476. The p-value of the test statistic is $p < 0.05$. The results of the IPS test also indicate that the "CAD/GDP" panel data is stationary at the level value, meaning the panel data does not contain a unit-root. According to the results of the IPS test, it is also determined that the CAD is sustainable for the D8 countries during the examined years.

Lastly, Breitung Robust test, the calculated value of the IPS test is -3.6205. The p-value of the test statistic is $p < 0.05$. The results of the Breitung test also indicate that the "CAD/GDP" panel data is stationary at the level value. According to the results of the Breitung test, it is observed that the CAD is sustainable for the D8 countries during the examined periods.

In this study, conducted to determine the sustainability of the CAD of D8 countries for the period 1981–2022, all four unit-root tests consistently found that the "CAD/GDP" panel data is stationary at the level value. Therefore, it is possible to conclude that the CAD is sustainable for this group of countries during the examined period.

4 Conclusion

The balance of payments is one of the important macroeconomic variables that determine an economy's credibility and indicate the national savings balance. In an economy, if incomes exceed expenditures, it results in a CA surplus, whereas if expenditures exceed incomes, it results in a CAD. Increases in the CAD can stem from increased investments or decreased savings. Temporary CAD, as indicated here, are not necessarily detrimental because they reflect the reallocation of capital to countries where it is more productive. However, persistent CAD can have serious implications: Interest rates may be increased to attract foreign portfolio investments. Additionally, the accumulation of external debt from increased interest payments can burden future generations with excessive debt. Therefore, the sustainability of the CA balance is a crucial policy consideration for national economies.

In this study, the sustainability of the CAD in D8 countries, predominantly consisting of developing economies, will be examined using panel unit root tests from 1981 to 2022. In this context, panel data set homogeneity and cross-sectional dependence were initially tested to determine the appropriate generation of unit root tests. Based on the results of these tests, it was determined that second-generation unit root tests are suitable for the panel data set.

The second-generation unit-root tests used in this study include the CIPS and PESCADF tests by Pesaran (2007), as well as the IPS (2003) and Breitung and Das (2005) (Robust Cross-Sectional) tests. The results from all mentioned unit-root tests show that the calculated values are greater than the critical values, and the probability values are significant. This indicates that the "CAD/GDP" panel data set does not contain a unit-root. In other words, the panel data set is stationary. Consequently, it is concluded that the CAD is sustainable for the D8 countries over the examined periods and that there are no significant imbalances.

References

Baharumshah, A. Z., Lau, E., & Fountas, S. (2005). "Current Account Deficit Sustainability: A Panel Approach". *Journal of Economic Integration*, 20(3): 514–529.

Bektaş, V. (2017). "Gelişmekte Olan Ülkelerde Cari Açıkların Sürdürülebilirliği: Bir Panel Veri Analizi". *AİBÜ Sosyal Bilimler Enstitüsü Dergisi*, 17(1): 51–66.

Breitung, J, & Das, S. (2005). "Panel Unit Root Tests under Cross-Sectional Dependence". *Statistica Neerlandica*, 59: 414–433.

Breusch, T., & Pagan, A. (1980). "The Lagrange Multiplier Test and Its Applications to Model Specification in Econometrics". *Review of Economic Studies*, 47: 239–253.

Chen, S. W. (2010). "Testing for the Sustainability of the Current Account Deficit in Four Industrial Countries: A Revisitation". *Economics Bulletin*, 30(2): 1474–1495.

Cunado, J., Gil-Alana, L. A., & Gracia, F. P. D. (2010). "European Current Account Sustainability: New Evidence Based On Unit Roots and Fractional Integration". *Eastern Economic Journal*, 36(2): 177–187.

Demir, M. A. (2019). "Yapısal Kırılmalı Birim Kök Testleri Çerçevesinde Türkiye İçin Cari AçıkSürdürülebilirliliği". *Üçüncü Sektör Sosyal Ekonomi Dergisi*, 54(1): 261–273.

Hakkio, C.S. & Rush, M. (1991). "Is the Budget Deficıt "Too Large?". *Economic Inquiry*. 19: 429–445.

Hassan, K., Hoque, A., & Rao, A. (2015). "Sustainability of current account balance in ASEAN countries: Evidence from a panel error correction model". *The Journal of Developing Areas*, 49(6): 189–204.

Husted, S. (1992). "The Emerging U.S. Current Account Deficit in the 1980s: A Cointegration Analysis". *The Review of Economics and Statistics*, 74(1): 159.

Im, K. S., Pesaran, M. H. V & Shin, Y. (2003). "Testing for Unit Roots in Heterogeneous Panels". *Journal of Econometrics*, 115: 53–74.

Öztürk, S., & Çoltu, S. (2018). "Seçilmiş Oecd Ülkelerinde Cari Açığın Sürdürülebilirliği Analizi (1995–2015)". *Bingöl Üniversitesi Sosyal Bilimler Enstitüsü Dergisi*, 8(16): 37–48.

Pesaran, M. H. (2004). "General Diagnostic Tests for Cross Section Dependence in Panels", University of Cambridge, Faculty of Economics, Cambridge WP 0435 in Economics.

Pesaran, M. H. (2007). "A Simple Panel Unit Root Test in the Presence of Cross Section Dependence". *Journal of Applied Econometrics*, 22 (2): 265–312.

Pesaran, M. H. (2015). "Testing Weak CrossSectional Dependence in Large Panels". *Econometric Reviews*, 34: 1089–1117.

Pesaran, M. H., Ullah A., & Yamagata T. (2008). "A Bias-Adjusted LM Test of Error Cross Section Independence". *Econometrics Journal*, 11: 105–127.

Shastri, S., Giri, A. K., & Mohapatra, G. (2018). "Testing the Sustainability of Current Accounts for Major South Asian Economies: A Panel Data Approach". *South Asia Economic Journal*, 19(1): 1–21.

Singh, T. (2017). "Sustainability of Current Account Deficits in the OECD Countries: Evidence from Panel Data Estimators". *Global Economy Journal*, 17(4): 1–16.

Swamy, P. A. (1970). "Efficient inference in a random coefficient regression model". *Econometrica: Journal of the Econometric Society*, 38(2): 311–323.

Trehan, B. & Walsh C.E. (1991). " Testing Intertemporal Budget Constraints: Theory and Applications to U. S. Federal Budget and Current Account Deficits". *Journal of Money, Credit and Banking*. 23(2): 206–223.

Uz, İ., Çevİş, İ., & Ceylan, R. (2018). "Kırılgan Beşlide Cari Açıkların Sürdürülebilirliği: Doğrusal Olmayan Birim Kök Testleri İle Kanıtlar". Ege Akademik Bakis (Ege Academic Review), 18(1): 121–134.

Yardimci, C. (2023). "Cari İşlemler Açığının Sürdürülebilirliği: Türkiye Örneği (2003–2023)". *Ulusal Ceteris Paribus Ekonomi Araştırmaları Dergisi*, 1(1): 1–10.

Wu, J.-L. (2000). "Mean reversion of the current account: Evidence from the panel data unit-root test". *Economics Letters*, 66(2): 215–222.

Ömer Faruk Özyalçın[1]

The Effects of Uncertainties and Risks in Türkiye on Türkiye and United States of America 10-Year Bond Yield Spread: A Long Run Empirical Analysis

1 Introduction

Just as in real life, economics is based on uncertainties and risk. While uncertainty states that people or agents do not know the possible results of an attempt, people or agents know and are ready to face up to probabilities of possible outcomes at risk (Leroy & Singell, 1987, pp. 394–395). In other words, both the range of results and the result are not known in an uncertain event, but the range of results is known, but the result is not known in a risky event (Anderson et al., 2009, p. 234). Therefore, risk is a subset of uncertainty and comes out of uncertainty.

The U.S. (United States of America) 10-year treasury yield is treated as the risk-free discounted rate for stocks (Zhang & Cui, 2023, p. 503). Risks and Türkiye and U.S. 10-year bond yield spread are closely related to each other. The more risks for Türkiye increase, the more the bond yield spread rises.

The purpose of this study is to analyze the uncertainty and risk indices that affect 10-year bond yield spread will be analyzed by the Autoregressive Distributed Lag Model (ARDL) and Error Correction Model (ECM) and will be made a robustness check by Fully-Modified Ordinary Least Square (FMOLS), Dynamic Ordinary Least Square (DOLS) and Canonical Cointegrating Regression (CCR) models for the long run. In the literature, several studies focus on treasury bonds with different maturity for several countries. However, there is no study on how different types of risks affect Türkiye and the U.S. 10-year yield spread. Consequently, this study is expected to fill a void in the literature. In this manner, the factors that increase the risks in Türkiye are evaluated with respect to the uncertainty index, country-specific geopolitical risk index, and credit default risk, and how these risks affect the course of Türkiye and U.S. 10-year bond yield spread are examined both theoretically and empirically.

1 Ph.D., Asst. Prof., Karabuk University, Faculty of Economics and Administrative Sciences Department of Economics, omerfozyalcin@karabuk.edu.tr, ORCID: 000-0001-8542-3165.

2 Uncertainty and Risk Measures for Türkiye

2.1 Uncertainty Index

Between 2019 and 2024, both Türkiye and the world experienced several uncertain times. For example, the COVID-19 pandemic was the most notable uncertain period. Additionally, the variants of COVID-19, the Russia-Ukraine conflict, turmoil in Syria and the earthquake in Kahramanmaraş on February 6, 2023, were the events that led to uncertainty for Türkiye. The effects of these events are observed in Graph 1, which shows the uncertainty index for Türkiye on the left and Türkiye and U.S. 10-year bond yield spread on the right between August 2019 and May 2024. The prospect is that as the uncertainty index for Türkiye rises, Türkiye and U.S. 10-year bond yield spread increases as well.

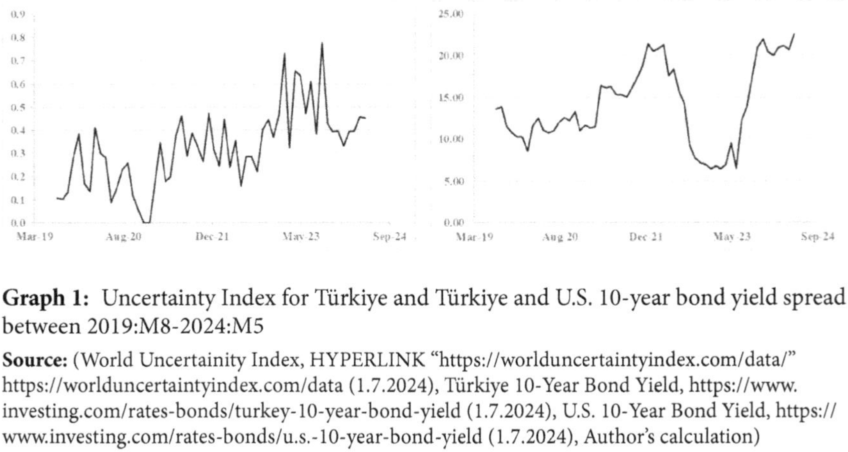

Graph 1: Uncertainty Index for Türkiye and Türkiye and U.S. 10-year bond yield spread between 2019:M8-2024:M5

Source: (World Uncertainity Index, HYPERLINK "https://worlduncertaintyindex.com/data/" https://worlduncertaintyindex.com/data (1.7.2024), Türkiye 10-Year Bond Yield, https://www.investing.com/rates-bonds/turkey-10-year-bond-yield (1.7.2024), U.S. 10-Year Bond Yield, https://www.investing.com/rates-bonds/u.s.-10-year-bond-yield (1.7.2024), Author's calculation)

2.2 Country Specific Geopolitical Risk Index

Geopolitical risk is defined as any bad events breaking out, such as war, terrorism, interior and exterior conflicts that affect the tranquility and peace of both the country itself and the world together. In order to measure the geopolitical risk of countries, an algorithm computes the share of articles that mention bad geopolitical events in the famous newspapers of the United States, United Kingdom, and Canada (Caldara & Iacoviello, 2022, p. 1195). Graph 2 shows Türkiye's geopolitical risk index on the left and Türkiye and U.S. 10-year bond yield spread on the right between August 2019 and May 2024. The expectation is that Türkiye and U.S. 10-year bond yield spread is in direct proportion to Türkiye's geopolitical risk index.

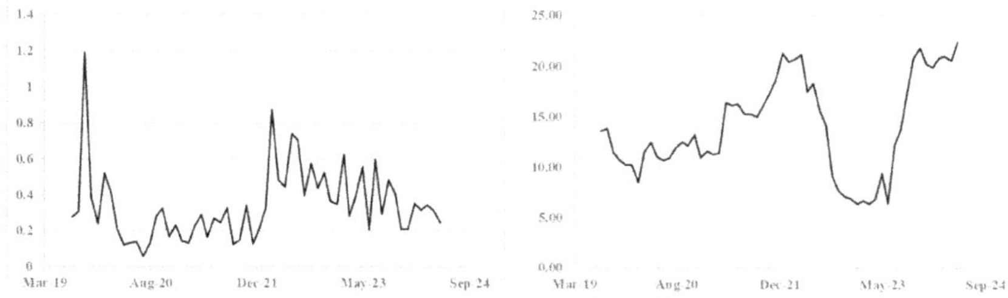

Graph 2: Geopolitical Risk Index for Türkiye and Türkiye and U.S. 10-year bond yield spread between 2019:M8-2024:M5

Source: (Country-Specific Geopolitical Risk Index, https://www.matteoiacoviello.com/gpr.htm (1.7.2024), Author's calculation)

2.3 Credit Default Swap

Credit default swap is a type of insurance that protects the default risk of a specific bond or loan. When a person or an institution buys this insurance, the person or institution makes an amount of payment until the agreement, in other words, a swap contract between them, matures. This agreement is valid even for bad days such as bankruptcy or failure to pay. If the worst-case scenario comes true, the one who sells this insurance either buys the defaulted bond or loan at its par value. The price of the credit default swap serves as a measure of risk (Greatrex, 2008, pp. 1–2). Graph 3 depicts Türkiye and U.S. 10-year bond yield spread on the left and Türkiye's 10-year credit default swap on the right between August 2019 and May 2024. It is expected that Türkiye's 10-year bond yield spread increase as the credit default swap of Türkiye rises.

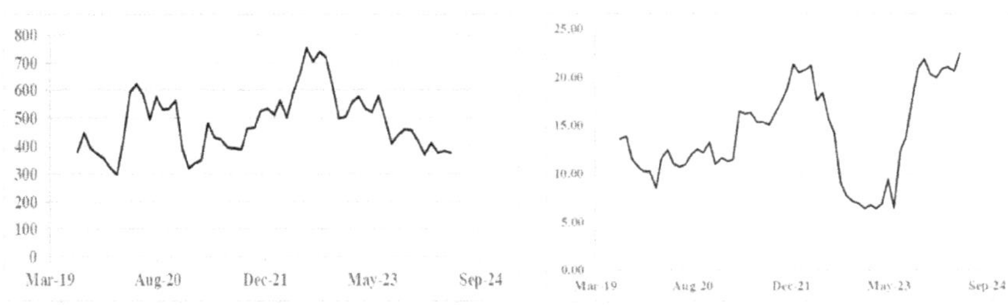

Graph 3: 10-Year Credit Default Swap of Türkiye and Türkiye and U.S. 10-year bond yield spread between 2019:M8-2024:M5

Source: Credit Default Swap, https://tr.investing.com/rates-bonds/turkey-cds-10-years-usd-historical-data (1.7.2024), Author's calculation.

3 Methodology

3.1 Data Set and Model

In the empirical part of this study, it will be analyzed the effects of Türkiye's uncertainty index, geopolitical risk index, and 10-year credit default swaps on Türkiye and U.S. 10-year yield spread between August 2019 and May 2024 by using monthly data. Table 1 gives abbreviations and definitions of variables used in the model and the data sources.

Table 1: Abbreviations (Abbr.), Description of Variables and Sources

ABBR.	VARIABLE	SOURCE
intsa	The difference between Türkiye 10-year Treasury yield and U.S. 10-year Treasury Yield	https://tr.investing.com and Author's calculation
uncsa	Türkiye's Uncertainty Index	https://worlduncertaintyindex.com/data/
geosa	Türkiye's Geopolitical Risk Index	https://www.matteoiacoviello.com/gpr.htm
cdssa	Türkiye's 10-years Credit Default Swap	https://tr.investing.com/rates-bonds/turkey-cds-10-years-usd

After defining the variables used in the model, the lag lengths which are used while applying Dickey-Fuller (DF) and Augmented Dickey-Fuller (ADF) unit root tests to each variable are determined. Proper lag lengths with the lowest values of Akaike and Schwarz information criteria and do not have autocorrelation problems are selected, and these lag lengths are given in parenthesis under the DF and ADF unit root test results. After finding proper lag lengths, the variables are subjected to the conditional hypothesis test, and proper models of variables are determined (D. A. Dickey & Fuller, 1979; D. Dickey & Fuller, 1981). While applying the Phillips Perron (PP) unit root test, the lag length is determined with the formula $T^{1/3}$. Since there are 58 observations in this analysis, $58^{1/3} = 3.871$, and 3 is selected (Phillips & Perron, 1988). DF, ADF and PP unit root test results are shown in Table 2.

Table 2: Results of DF, ADF and PP Unit Root Tests

VARIABLE	LEVEL	MODEL	DF-ADF	PP
intsa	at level	intercept	-2.609664 (3)	-1.528063 (3)
uncsa	at level	trend and intercept	-5.452275* (0)	-5.595890* (3)
geosa	at level	no trend and intercept	-1.724962 (2)	-1.950612** (3)
cdssa	at level	no trend and intercept	-0.467205 (0)	-0.428124 (3)

Note: * and ** show that there is not unit root at 1 % and 5 % significance level, respectively. The values in parenthesis represent the lag lengths.

According to the unit root test results, intsa and cdssa are nonstationary at level. Since there is a conflict between ADF and PP about the stationarity of geosa, it is also accepted as nonstationary at the level. To specify whether the reason for the nonstationarity of variables comes from structural breaks, they are firstly subjected to Zivot-Andrews one break structural break unit root test (Zivot & Andrews, 1992). Maximum lag length is determined according to the Schwert criteria formula. That is $12\ (T/100)^{1/4}$. T is an observation number, $12\ (58/100)^{1/4} = 10.472$, and 10 is selected for maximum lag. Test results are given in Table 3.

Table 3: Results of Zivot-Andrews Unit Root Test with One Structural Break

VARIABLE	MODEL	BREAK DATE	t-STAT	MAXIMUM LAG	CHOSEN LAG LENGTH
intsa	Both	2022M08	-4.411038	10	2
geosa	Both	2022M03	-5.018927	10	2
cdssa	Both	2022M05	-3.802542	10	0

According to the results of Zivot-Andrew unit root test with one structural break, all variables are found as nonstationary at all significance levels. Therefore, in the next step, Lee-Strazicich unit root test with multiple breaks is applied to test the probability that the nonstationarity can be due to two structural breaks (Lee & Strazicich, 2001, 2003). Lee-Strazicich unit root test results can be seen in Table 4.

Table 4: Results of Lee-Strazicich Unit Root Test with Two Structural Breaks

VARIABLE	MODEL	BREAK DATES	λ	t-STAT	CRIT. VALUE (5 %)	θ	Υ	Ω	Ψ	k
intsa	CC	2021:M10 2023:M05	λ_1:0.4 λ_2:0.7	-5.575	-5.65	-1.081	-4.225	3.751	4.291	8
geosa	CC	2022:M02 2023:M05	λ_1:0.5 λ_2:0.7	-8.198*	-5.73	9.080	-7.167	-5.528	7.842	8
cdssa	CC	2020:M10 2022:M05	λ_1:0.2 λ_2:0.5	-4.968	-5.74	2.479	-0.539	-3.051	3.640	8

Note: * and ** show that there is no unit root at 1 % and 5 % significance level, respectively. λ represents the break location and is calculated with the observation number until the break date is divided by the total observation number. θ and Υ are the dummy variables that show level break magnitudes, Ω and Ψ are the dummy variables that show trend break magnitudes. k indicates the lag length.

In Table 4, Model CC considers the changes both in level and trend. If θ, Υ, Ω and Ψ dummy variables' t statistics are greater than 1.645, they are statistically significant. In order to take into consideration test results, all of the dummy variables must be statistically significant. In Table 4, all dummy variables of geosa are statistically significant, and so t-statistic can be considered. Since the t- statistic is significant at 5 % significance level, geosa is stationary with two structural breaks.

Since intsa and cdssa are found nonstationary, the first differenced (I(1)) unit root test results of intsa and cdssa according to DF, ADF and PP unit root tests are given in Table 5.

Table 5: Results of DF, ADF and PP Unit Root Tests of intsa and cdssa at First Difference

VARIABLE	LEVEL	MODEL	DF-ADF	PP
dintsa	the first difference	intercept	-3.003269**(3)	-7.046537*(3)
dcdssa	the first difference	no trend and intercept	-7.758762* (0)	-7.798987* (3)

Note: * and ** show that there is not unit root at 1 % and 5 % significance level, respectively.

According to Table 5, intsa and cdssa are stationary at first difference. According to Narayan, the variables can be either I(0) and I(1) individually, or they are integrated at the same level (Narayan, 2004, pp. 6–7). Yılancı and Kırca propose that the dependent variable must be I(1) (Yılancı & Kırca, 2024, p. 412). The results from the unit root test results, while intsa, which is the dependent variable, and cdssa are found as I(1), uncsa and geosa are I(0). Consequently, the long run ARDL model can be set up as in Equation 1.

$$intsa_t = \alpha_0 + \alpha_1\, intsa_t + \alpha_2\, uncsa_t + \alpha_3\, geosa_t + \alpha_4\, cdssa_t + \epsilon_t \quad (1)$$

After setting up the ARDL model, the next stage is to determine the optimum lag length for the model. Since the maximum lag length is found as 10 according to Schwert criteria, lag lengths from 0 to 10 are tested. ARDL (1,0,4,4) model, which does not have any autocorrelation problem and its Akaike information criteria take its minimum is determined by applying trend specification as Restricted Trend. The results are shown in Table 6.

Table 6: Model Selection

MODEL	AKAIKE INFORMATION CRITERIA	SCHWARZ INFORMATION CRITERIA	HANNAN-QUINN INFORMATION CRITERIA	BREUSCH-GODFREY SERIAL CORRELATION LM TEST
ARDL (1,0,1,1)	3.985639	4.272383	4.097078	0.1687 (0.1251)
ARDL (1,0,2,1)	3.998005	4.323508	4.124202	0.2339 (0.1736)
ARDL (1,0,2,2)	3.998753	4.360423	4.138972	0.3570 (0.2778)
ARDL (3,3,3,2)	3.912930	4.496882	4.138749	0.6530 (0.5346)
ARDL (1,0,4,4)*	3.853630	4.369293	4.052501	0.5825 (0.4690)
ARDL (7,5,0,6)	3.868446	4.739662	4.201363	0.5051 (0.2711)
ARDL (7,5,2,1)	3.868735	4.626313	4.158228	0.5771 (0.3873)
ARDL (7,8,2,1)	3.882662	4.762192	4.217592	0.8237 (0.6805)
ARDL (7,7,9,9)	3.748214	5.176732	4.290192	0.5205 (0.0498)
ARDL (9,7,10,10)	3.604109	5.202427	4.208116	0.0937 (0.0000)

After deciding the model as ARDL (1,0,4,4), the model has been subjected to the normality test (Jarque-Bera), Heteroskedasticity test (Breusch-Pagan-Godfrey), Breusch-Godfrey Serial Correlation LM test, and Ramsey Regression Specification Error test (RESET). There is not normality, heteroskedasticity, autocorrelation and regression specification error problems according to 5 % significance level. Moreover, the ARDL (1,0,4,4) model is stable according to the

CUSUM and CUSUM of Squares tests. Then, the next step is to test the presence of a cointegration relationship between long-term coefficients, which are β_1, β_2, β_3 and β_4. The presence of cointegration is found with the help of F-test statistics (Pesaran et al., 2001).

$$intsa_t = \alpha_0 + \sum_{i=1}^{n} \alpha_{1i} intsa_{t-i} + \sum_{i=1}^{n} \alpha_{1i} uncsa_{t-i} + \sum_{i=1}^{n} \alpha_{1i} geosa_{t-i} + \sum_{i=1}^{n} \alpha_{1i} cdssa_{t-i} + \beta_1 intsa_{t-i}$$

$$+ \beta_2 uncsa_{t-i} + \beta_3 geosa_{t-i} + \beta_4 geosa_{t-i} + \epsilon_t \qquad (2)$$

Table 7: Bound Test Results

TEST STAT.	VALUE	SIGNIFICANCE	I(0)	I(1)
F-statistics	6.882145	10 %	2.97	3.74
k	3	5 %	3.38	4.23
		2.5 %	3.8	4.68
		1 %	4.3	5.23

According to bound test results in Table 7, the calculated F-statistics value is greater than the I(1) value, and it is statistically significant at 5 % significance level, so there is a cointegration relationship between long term coefficients. Table 9 presents the long run coefficients that are significant at 1 % and 5 % are given.

Table 8: Long Run Coefficients

VARIABLE	COEFFICIENT	t-STAT.
UNCSA	2.641166	0.371617
GEOSA	-9.868461	-1.446439
CDSSA	-0.000489	-4.633900*
@TREND	-4.633900	3.313819*

Note: * and ** show that there is not unit root at 1 % and 5 % significance level, respectively.

According to Table 8, cdssa is statistically significant and affects intsa negatively in the long run. In the third stage, short run coefficients (α_{1i}, α_{2i}, α_{3i}) and error correction coefficient (EC_{t-1}) are estimated. The model turns into Equation 3 as the following.

$$intsa_t = \alpha_0 + \sum_{i=1}^{n} \alpha_{1i} intsa_{t-i} + \sum_{i=1}^{n} \alpha_{2i} uncsa_{t-i} + \sum_{i=1}^{n} \alpha_{3i} geosa_{t-i} + \sum_{i=1}^{n} \alpha_{4i} cdssa_{t-i} + \lambda EC_{t-1} + \epsilon_t \quad (3)$$

The estimation results of the short run and error correction coefficient are given in Table 9.

Table 9: ARDL (1,0,4,4) Short Run and Error Correction Coefficients

VARIABLE	COEFFICIENT	t-STAT.
D(GEO_SA)	4.583608	3.379997*
D(GEO_SA(-1))	7.573920	4.029111*
D(GEO_SA(-2))	5.850116	3.546943*
D(GEO_SA(-3))	3.365661	2.829159*
D(CDS_SA(-1))	0.000144	3.482265*
D(CDS_SA(-3))	0.000103	2.804736*
CointEq(-1)	-0.300784	-6.152381*

Note: * and ** show that there is not unit root at 1 % and 5 % significance level, respectively.

According to Table 9, geosa and cdssa affect intsa positively in all lags in the short run. Moreover, the results that the error correction coefficient is estimated between -1 and 0, and the t statistics value is greater than 2 in absolute value fulfill the conditions for functioning the Error Correction mechanism properly. That also indicates the presence of long run relationship between variables (Dell'Anno & Halicioglu, 2010, p. 636). The shocks that occurred in Türkiye and U.S. 10-year bond yield spread are corrected by 0.03 of short-run shocks after a month, the effect of these shocks disappear in three months (1/0.300784) and then converge to long run equilibrium.

3.2 Robustness Check the Long Run Relationship with FMOLS, DOLS and CCR

FMOLS, DOLS, and CCR crosscheck the long-term coefficients of the ARDL model. The FMOLS test changes the Ordinary Least Square (OLS) method to clarify the cointegration relationship and its effects on serial correlation and endogeneity in the independent variables (Pedroni, 2000, 2001). The DOLS test corrects the bias of simultaneity of variables (Jackman & Lorde, 2010, p. 3). The CCR test works by correcting errors in the OLS method and using long run covariance matrix. Thus, this test abolishes the asymptotic internality (Zimon et al., 2023, p. 11). Equation 2 is estimated with FMOLS, DOLS and CCR tests by applying trend specification as Linear Trend, and the results are shown in the following Table 10. According to Table 10, uncsa and cdssa are statistically significant at 5 % and 1 %, respectively, and affect intsa negatively according to both FMOLS, DOLS and CCR tests in the long run. Furthermore, the results of cdssa in Table 10 correspond to Table 8.

Table 10: FMOLS, DOLS and CCR Tests Results

VARIABLE	FMOLS		DOLS		CCR	
	Coefficient	**t-stat**	**Coefficient**	**t-stat**	**Coefficient**	**t-stat**
uncsa	-14.25642	-2.330399**	-22.43610	-2.617060**	-17.64294	-2.291878**
geosa	0.124253	0.031435	2.299780	0.397219	0.605955	0.126237
cdssa	-0.000219	-2.763579*	-0.000265	-2.891563*	-0.000230	-2.841618*
C	22.78775	5.577544*	25.72426	5.782349*	23.51911	5.689608*
@TREND	0.231103	3.828986*	0.274134	3.726910*	0.256525	3.725522*
R^2	0.201665		0.609732		0.164814	

Note: * and ** show that there is not unit root at 1 % and 5 % significance level, respectively.

4 Conclusion

This study has analyzed the effects of Türkiye's uncertainty index, geopolitical risk index, and 10-year credit default swaps on Türkiye and U.S. 10-year yield spread between August 2019 and May 2024 by using monthly data with long run ARDL model. Moreover, it has robustness checked of the estimated results with FMOLS, DOLS and CCR models. According to the mutual results of all long run models, cdssa affects intsa negatively in the long run. In the period between 2019 and 2024, many uncertainties existed, and these uncertainties led to appear risks. Since Türkiye and the U.S. 10-year yield spread give remarkable signs about the way of the Turkish economy, policymakers should take precautions timely and try to develop effective strategies to avoid increasing risks. By doing so, risks are minimized, and economic indicators become steady.

References

Anderson, E. W., Ghysels, E., & Juergens, J. L. (2009). The impact of risk and uncertainty on expected returns. *Journal of Financial Economics, 94*, 233–263. https://doi.org/10.1016/j.jfineco.2008.11.001

Caldara, D., & Iacoviello, M. (2022). Measuring Geopolitical Risk. *American Economic Review, 112*(4), 1194–1225. https://doi.org/10.1257/aer.20191823

Country-Specific Geopolitical Risk Index. (2024). Retrieved July 1, 2024, from https://www.matteoiacoviello.com/gpr.htm

Credit Default Swap. (2024). Retrieved July 1, 2024, from https://tr.investing.com/rates-bonds/turkey-cds-10-years-usd-historical-data

Dell'Anno, R., & Halicioglu, F. (2010). An ARDL model of unrecorded and recorded economies in Turkey. *Journal of Economic Studies, 37*(6), 627–646. https://doi.org/10.1108/01443581011086666

Dickey, D. A., & Fuller, W. A. (1979). Distribution of the estimators for autoregressive time series with a Unit Root. *Journal of the American Statistical Association, 74*(366), 427. https://doi.org/10.2307/2286348

Dickey, D., & Fuller, W. (1981). Likelihood Ratio Statistics for autoregressive time series with a Unit Root. *Econometrica, 49*(4), 1057–1072. Retrieved from http://www.jstor.org/stable/1912517

Greatrex, C. A. (2008). The Credit Default Swap Market's Determinants. In *Fordham University Department of Economics Discussion Paper Series* (No. 2008–05). https://doi.org/10.2139/ssrn.1104887

Jackman, M., & Lorde, T. (2010). On the relationship between tourist flows and household expenditure in Barbados: A dynamic OLS approach. *Economics Bulletin, 30*(1), 472–481.

Lee, J., & Strazicich, M. (2001). Break point estimation and spurious rejections. *Oxford Bulletin of Economics and Statistics, 63*(5), 535–558.

Lee, J., & Strazicich, M. (2003). Minimum Lagrange Multiplier Unit Root Test with Two Structural Breaks. *Department of Economics, University of Central Florida Working Paper Series, 85*(4), 1082–1089. https://doi.org/10.1162/003465303772815961

Leroy, S. F., & Singell, L. D. (1987). Knight on Risk and Uncertainty. *The University of Chicago Press, 95*(2), 394–406. https://doi.org/10.1086/261461

Narayan, P. K. (2004). *Reformulating Critical Values for the Bounds F-statistics Approach to Cointegration: An Application to the Tourism Demand Model for Fiji*. Australia.

Pedroni, P. (2000). Fully modified OLS for heterogenous cointegrated panels. *Recent Developments in the Econometrics of Panel Data, 15*, 93–130. Retrieved from https://econpapers.repec.org/scripts/redir.pf?u=https%3A%2F%2Fweb.williams.edu%2FEconomics%2Fwp%2Fpedroniaie.pdf;h=repec:wil:wileco:2000-03

Pedroni, P. (2001). Purchasing Power Parity Tests in Cointegrated Panels. *The Review of Economics and Statistics*, *83*(4), 727–731. https://doi.org/https://doi.org/10.1162/003465301753237803

Pesaran, M. H., Shin, Y., & Smith, R. J. (2001). Bounds testing approaches to the analysis of level relationships. *Journal of Applied Econometrics*, *16*(3), 1–33.

Phillips, P. C. B., & Perron, P. (1988). Testing for a Unit Root in Time Series Regression. *Biometrika*, *75*(2), 335–346. https://doi.org/10.1093/biomet/75.2.335

Türkiye 10-Year Bond Yield. (2024). Retrieved July 1, 2024, from https://www.investing.com/rates-bonds/turkey-10-year-bond-yield

U.S. 10-Year Bond Yield. (2024). Retrieved July 1, 2024, from https://www.investing.com/rates-bonds/u.s.-10-year-bond-yield

World Uncertainity Index. (2024). Retrieved July 1, 2024, from https://worlduncertaintyindex.com/data/

Yılancı, V., & Kırca, M. (2024). Testing the relationship between employment and tourism: A fresh evidence from the ARDL Bounds Test with sharp and smooth breaks. *Journal of Hospitality and Tourism Insıghts*, *7*(1), 394–413. https://doi.org/10.1108/JHTI-10-2022-0509

Zhang, L., & Cui, H. (2023). Local Dependence Test between Random Vectors Based on the Robust Conditional Spearman's ρ and Kendall's τ. *Acta Mathematicae Applicatae Sinica, English Series*, *39*(3), 491–510. https://doi.org/https://doi.org/10.1007/s10255-023-1073-4

Zimon, G., Pattak, D. C., Voumik, L. C., Akter, S., Kaya, F., Walasek, R., & Kochanski, K. (2023). The impact of fossil fuels, renewable energy, and nuclear energy on South Korea's environment based on the STIRPAT model: ARDL, FMOLS, and CCR approaches. *Energies*, *16*(6198), 93–130. https://doi.org/htpps://doi.org/10.3390/en16176198

Zivot, E., & Andrews, D. W. K. (1992). Further evidence on the Great Cash, the Oil-Price Shock, and the Unit Root Hypothesis. *Journal of Business & Economic Statistics*, *10*(3), 251–270.

Tuğba Özyildiz[1] & Eda Dineri[2] & Zeynep Köse[3]

The Impact of Tourism and Environmental Technologies on Environmental Degradation: Countries with a High Income from Tourism

1 Introduction

The tourism sector, which acts as a locomotive in ensuring economic growth and development, offers economic, social and cultural contributions. The tourism sector has become the fastest growing sector in recent years and also contributes to regional development by providing employment opportunities to local people, earning foreign currency, contributing to the development of infrastructure, preserving historical buildings and monuments, and preserving and beautifying the region. As a result, it becomes an alternative source of growth, contributing to balance of payments and national income (Long, 2012:29; Ali et al., 2021: 1; Zheng et al., 2023: 2). In contrast, tourism is an integral part of the environment and climate as a rapidly growing and important part of the global economy. If tourism is not planned, managed and developed, its positive effects on the country's economy can be reversed and the natural environment can be damaged. For this reason, tourism activities play a crucial role in economic growth and development, environmental degradation, energy use and natural resources (Khan et al., 2020: 2; Ghosh, 2022:3886).

There are many pollutants that affect environmental quality, and CO_2 emissions, which increase under various conditions, are one of them. (Geng et al., 2018: 6162–6163). Environmental damage may occur due to the activities

1 Ph.D., Gaziantep University Islahiye Faculty of Economics and Administrative Sciences, Department of Economics, tyesilyurt@gantep.edu.tr, ORCID: 0000-0002-2101-2954.

2 Ph.D., Assoc. Prof., Hasan Kalyoncu University, Faculty of Economics, Administrative and Social Sciences, Department of Economics, Gaziantep, Turkey, eda.dineri@hku.edu.tr, ORCID: 0000-0002-5637-594X.

3 Ph.D., Assoc. Prof., Hasan Kalyoncu University, Faculty of Economics, Administrative and Social Sciences, Department of Economics, Gaziantep, Turkey, zeynep.kose@hku.edu.tr, ORCID: 0000-0002-9494-3098.

carried out in the tourism sector (Buckley, 2011: 402–403). The development of sustainable tourism is extremely important in terms of global environmental crises and climate changes, as well as economic challenges. The existence of decreasing resources around the world and the damage that people cause to the environment due to their lives make it necessary for societies to act more environmentally aware. Environmental degradation experienced by the increasing needs and expectations of the world population indicates that consumer and environmental awareness should be carried out together in order to achieve sustainable development in green/sustainable/environmental innovations (Hall et al., 2005: 186; Loureiro et al., 2022: 255) . Reducing the environmental impacts of energy production is one of the fundamental mechanisms for sustainability. Many leaders and businesses around the world advocate the need to switch to more diverse energy sources (Dong et al., 2017: 403).

The only action plan that can ensure the continuity of a country's environmental resources is possible by ensuring the evolution of that country from its development process to a sustainable development process through modern technologies. For this reason, environmental technologies continue to be a focus for countries to achieve green development (Pan et al., 2018: 457). Environmentally friendly innovations in technology is played a crucial role in sustainable development (Sun et al., 2008). The need for effective destinations to realize environmentally friendly tourism and carbon-free tourism is a priority for a sustainable environment in the globalized world. Environmental innovations refer to all processes, practices, systems and products that enhance environmental sustainability by providing environmental benefits (Oltra, 2008: 4).

Many governments are trying to address the threat of climate change by developing and implementing green policies within their borders. Governments plan to achieve a low-carbon economy with policies such as green taxes, green subsidies, green loans and green financing (Lee et al., 2022; 136; Centobelli et al., 2022: 455).

There is very extensive academic research about the connection between tourism sector and CO_2 emissions. The unique aspect of the study is the addition of the interaction of environmental technologies with the tourism sector to the model.

The study is organized into 4 sections. Information on the general outlook for the sector is provided by explaining the connection between the tourism sector, economic growth and environmental degradation in the first section. After the first section, previous research related to our study is included. The data set and model are explained in the third section. The results obtained are explained in relation to previous studies. Finally, the fourth section gives conclusions and proposals.

2 Literature Review

Findings examining the relationship between tourism, economic growth and CO_2 emissions vary across countries, periods and country groups. If the tourism sector development causes economic growth, it defines the tourism-led growth hypothesis. This growth spreads throughout the economy can be experienced through the government economic policies that encourage tourism development (Balsalobre-Lorente et al., 2023; Xia et al., 2022; Kyara et al. 2021: 2–4; Balsalobre-Lorente & Leitão 2020: 45883; Wang, 2019; Sokhanvar, et al., 2018: 98; Brida et al., 2014: 394; Bassil et al., 2015, 43).The second hypothesis suggests that economic development leads to tourism development. The third hypothesis is the feedback hypothesis (Jebli et al., 2019). Neutrality hypothesis defines the no causality between economic growth and tourism development (Sokhanvar et al., 2018: Kyara et al., 2021: 2–4).

Studies of the relationship between tourism and CO_2 emissions have come to two different conclusions. One finding is that the amount of emissions released into the air increases as the energy needs of tourism activities are met by fossil fuels (Razzaq et al., 2023; Koçak et al., 2020; Shakouri et al., 2017; Raza et a., 2017). Another finding is that as tourism developments decrease emissions as a result of investments in green energy, sustainable infrastructure and environmental protection. (Rahaman et al., 2022; Razzaq et al., 2023; Doğru et al., 2020, Paramati et al., 2018; Lee & Brahmasrene, 2013).

Academic studies often focus on the interactions between sustainable development and growth, the tourism sector and environmental degradation. In recent years researchers have begun to investigate the impact of renewable energy and environmental technologies. Table 1. lists some studies on renewable energy and technological developments.

Table 1: A Summary of the Studies in the Literature

Author(s)	Sample and Period	Method	Dependent Variable	Independent Variable	Results
Leitão and Balsalobre-Lorente (2020)	EU 28 Countries (1995–2014)	FMOLS DOLS GMM	CO_2	Tourism Renewable Energy Economic growth	Tourism-CO_2 (-) Renewable energy-CO_2 (-) Growth- CO_2 (+)
Erdoğan and friends (2022)	Most visited countries (1995–2018)	Panel Quantile Regression	CO_2	Tourism Innovation	Tourism- CO_2 (+) Innovation- CO_2 (-)
Ullah and friends (2023)	BRICS (1995–2018)	CS-ARDL	CO_2	Tourism, Technical innovation Economic Growth Natural Resource Rent	Tourism- CO_2 (-) Growth- CO_2 (+) Technical innovation-CO_2 (+) Nature – CO_2 (+)
Razzaq and friends (2023)	Top 10 GDP countries (1995–2018)	Panel co-integration and Panel Causality	CO_2	Tourism Development Index Green Technology Innovation Economic Growth	TDI- CO_2 (+) GTI- CO_2 (-) Growth- CO_2 (+)
Thi and friends (2023)	53 nations (1990–2019)	FMOLS DOLS GMM	CO_2	Tourism Renewable Energy Economic growth Innovation	Tourism- CO_2 (+) Renewable energy -CO_2(-) Growth- CO_2 (+) Innovation-CO_2(-)
Hossain and friends (2024)	Bangladesh (1995–2019)	ARDL technique	CO_2	Tourism, Renewable energy Innovation	Tourism- CO_2 (-) Renewable energy-CO_2 (-) Innovation- CO_2 (+)

3 Data and Model

The purpose of the research is to investigate the moderating effect of technological innovation on CO_2 emissions in 8 selected countries with the highest tourism income (America, Austria, Germany, Spain, Italy, Mexico, Turkey, United Kingdom, United States). The panel data set covers the annual period between 1995 and 2020. In the model followed in the study, the model in the study of Erdoğan et al. (2022) is taken as an example.

The model in which CO_2 emissions are the dependent variable is created as follows:

$$lnco_{2i,t} = \beta_0 + \beta_1 \, lngdp_{i,t} + \beta_2 \, lntou_{i,t} + \beta_3 \, lntech_{i,t} + \beta_4 \, lnpop_{i,t} + \beta_5 \, lnren_{i,t} + \beta_6 \, lntech{*}lntou_{i,t} + \varepsilon_{i,t}$$

β_0 is defined constant term; β_1,β_2 … are the estimation coefficients. $\varepsilon_{i,t}$ is defined an error term. All the series are expressed in the logarithmic form. Information about the series is given in Table 2.

Table 2: Disclosure of Series

Series	Abbreviations	Measurement	Source
CO_2 emissions	$lnco_2$	per capita metric tons	World bank (2024)
GDP per capita	lngdp	constant 2015US$	World Bank (2024)
Number of tourist arrivals	lntou	International tourism, number of arrivals	World Bank (2024)
Patents in environment-related technologies	lntech	Patents on environmental technologies, Percentage (%)	OECD Data (2024)
Population density	lnpop	People per sq. km of land area	World Bank (2024)
Renewable energy	lnren	% of total final energy consumption	World Bank (2023)

The expected sign of economic growth is positive while the renewable energy is negative. The expected impact of tourism on emissions is uncertain. Studies have found a positive effect of tourism on CO_2 emissions (Aini et al., 2024; Erdogan et al., 2022; Raihan & Kuspekova, 2022; Kumail et al., 2020). On the contrary, some academic studies have found a negative relationship between tourism and CO_2 emissions (El-Menyari, 2021; Leitão & Lorente, 2020). The population density expected sign can be either positive or negative.

There are also results where environmental technologies have a statistically insignificant, negative or positive effect on CO_2 emissions. Ullah et al. (2023) showed in their study that technological progress raises CO_2 emissions. Mongo et al. revealed that in the long run CO_2 emissions is negatively affected by technological innovation (Mongo et al., 2021). Finally, the sign of the lntech* lntou is negative. The use of eco- friendly technology in the tourism sector is expected to reduce environmental degradation.

4 Methodology and Findings

Firstly, the descriptive statistics of the variables are shown. Subsequently, CSD and slope homogeneity tests are performed to test whether there is dependence between countries. The stationarity of the series was tested using the second-generation unit root test. Durbin Hausman Test is performed for examining the cointegration in after the results of the CIPS test. After verifying the existence of cointegration, panel ARDL/PMG technique is applied to estimate the coefficient. The causality relationship between the series is examined by Dumitrescu and Hurlin Causality Test. Empirical analysis consists of 6 stages. The step of the analysis is given Figure 1.

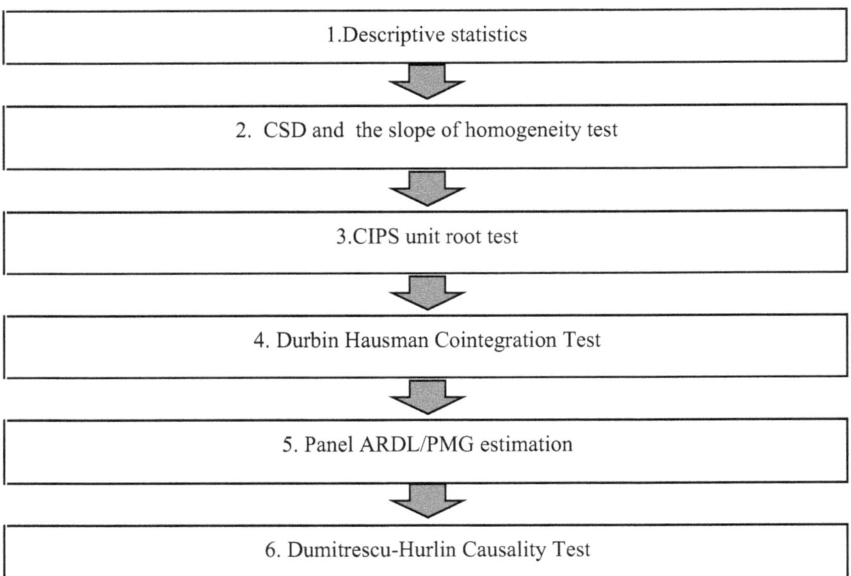

Figure 1: The Stage of the Analysis

Table 3 is shown the descriptive statistics. The maximum value of economic growth is 4.787678 and the minimum value is 3.756056 and economic growth has the highest mean value. The mean value is 7.657659 in tourism series and the value is high when we observe the minimum value 6.874308 and maximum value 8.262873.

Table 3: Descriptive of Statistics

Series	Observations	Mean	Median	Maximum	Minimum	Std. Deviation
lnco2	208	0.850588	0.870129	1.311114	0.452603	0.210404
lngdp	208	4.414369	4.519563	4.787678	3.756056	0.288745
lntou	208	7.657659	7.636472	8.262873	6.874308	0.330966
lntech	208	0.983538	1.006461	1.330414	0.623249	0.142637
lnpop	208	2.031481	1.992015	2.442911	1.463489	0.295204
lnren	208	0.976076	1.012837	1.553883	-0.096910	0.354501
lntech*lntou	208	7.532089	7.680115	9.729880	4.284407	1.143259

The second stage of the analysis reports the CSD and slope of homogeneity test. The CSD test result shows that any shock occurring in one country will also affect other countries. Since the probability value is $p < 0.01$, it is decided that there is CSD between the units. The Ho hypothesis, which states "There is no CSD", is rejected.

Slope homogeneity test developed by Pesaran and Yamagata (2008). The Ho hypothesis is defined as "The slope coefficients are homogeneous". The null hypothesis is rejected if $p < 0.01$.

Table 4: CSD and Slope of Homogeneity Test

Series	Breusch-Pagan LM	Pesaran scaled LM	Bias-corrected scaled LM	Pesaran CD
lnco2	407.1332	49.59476	49.43476	9.975663
lngdp	446.8089	54.89664	54.73664	19.36352
lntou	427.3021	52.28995	52.12995	18.82099
lntech	365.0057	43.96524	43.80524	14.96934
lnpop	498.9332	61.86205	61.70205	19.04266
lnren	512.0123	63.60982	63.44982	6.036318

(continued on next page)

Table 4: Continued

Series	**Breusch-Pagan LM**	**Pesaran scaled LM**	**Bias-corrected scaled LM**	**Pesaran CD**
lntechtou	370.7270*	44.72977*	44.56977*	16.72582*
Δ test	p- value	Δ_{adj} test	p- value	
11.834	0.000	13.843	0.000	

According to the CSD and slope of homogeneity test results, it is necessary to provide second generation unit root tests as the appropriate unit root test. The results of the CIPS unit root test, developed by Pesaran (2007) is given Table 3.

In the traditional method unit root tests used, each unit in the panel is independent and it is assumed that all other units are affected with the same severity by a shock occurring in any of the units. This situation is far from reflecting reality, and each unit is affected by a shock in the panel to a different degree. It was deemed appropriate to use the CIPS unit root test, which is one of the second-generation tests applied by taking into account the cross-sectional dependence and slope homogeneity test In CIPS statistics, the calculation is made by taking the average of the t statistics of each unit.

It is determined that the lnco2, lngdp, lntou and lnren series are not stationary at the level. The series are stationary in first-order difference. The series lntech, lntech*lntou and lnpop are stationary at level.

Table 5: The Results of the Unit Root Test

Series	**Inco2**	**Ingdp**	**Intou**	**ntech**	**Inpop**	**Inren**	**Intechtou**
CIPS I(0)	-1.004	-0.451	-1.379	-2.748*	-2.814*	-2.091	-2.773*
CIPSI(1)	-4.631*	-2.531*	-3.856*	-4.925*	-1.885**	-4.323*	-4.702*

*,**,*** *It indicates significance at 1 %, 5 % and 10 % levels.*

After testing stationarity, the existence of a long-term relationship between series is performed with the Durbin Hausman cointegration test developed by Westerlund (2008) in Table 6. In the test results, a significance level of 10 % is found in the panel test results and a significance level of 5 % in the group test results. The cointegration test reveal that the Ho hypothesis is rejected. The series have a cointegration relationship in the long run.

Table 6: Durbin Hausman Cointegration Test

	Constant		Constant + trend	
Tests	Statistics	p- value	Statistics	p- value
dh_group	-1.972	0.024	-1.763	0.039
dh_panel	-1.376	0.084	-1.351	0.088

Table 7 shows that in the short run tourism has no significant impact on CO_2 emissions. Tourism has significant positive CO_2 emissions for the long run. We can say that the number of tourist arrivals causes environmental pollution in the long run. These findings are in line with the Aini (2024), Erdoğan et al. (2022), Raihan and Tuspekova (2022), Kumail et al. (2020) is consistent with their studies.

A positive relationship is found among economic growth and CO_2 emissions in both the short and long run. Another variable that is statistically significant in both the short and long run is renewable energy. Renewable energy is found negative effect on CO_2 emissions. These results are similar to Voumik et al. (2024) study. Voumik et al. (2024) found that in 40 Asian countries.

economic development increases emissions and renewable energy reduces emissions.

While population density has no significant effect on CO_2 emissions in the short run, it has a negative effect on CO_2 emissions in the long run. We can say that population density contributes to the mitigation of environmental pollution in the long run. One of the supporting studies on the emission-reducing effect of population density was conducted by Chen et al. (2020) and concluded that population density reduces air pollution in China. Abdouli et al. (2017) revealed that inverted-U relationship between population density and CO_2 emissions for BRICTS countries between 1990 and 2014. The same result is obtained by Lantz and Feng (2006) for Canada between 1970 and 2000. The results suggest that, with technological developments, there will be more pressure on turning to environmentally friendly production techniques and adopting environmental standards (Abdouli et al. (2017); Lantz & Feng, 2006).

Environmental technologies have a positive impact on emissions in the long run. We can explain it as the interaction of environmental technologies with tourism has a reducing effect on emissions. In the short run, neither variable had a statistically significant effect on emissions. The error correction coefficient is to be negative and significant. It means that short-term deviations will be corrected in the long run.

Table 7: The Results of the Estimation

lnco$_2$= (lngdp, lntou,lntech, lnpop, lnren, lntech*lntou)							
Long run estimation	**coefficient**	**standard error**	**p-value**	**Short run estimation**	**coefficient**	**standard error**	**p-value**
lngdp	0.677442	0.146628	0.0000	lngdp	0.880182	0.131682	0.000
lntou	0.364221	0.112195	0.0015	lntou	-0.068828	0.253933	0.7867
lntech	5.995914	1.083129	0.0000	lntech	-1.066791	1.890370	0.5734
lnpop	-2.226242	0.253625	0.0000	lnpop	-0.445677	1.197227	0.7102
lnren	-0.237932	0.029628	0.0000	lnren	-0.239047	-0.098484	0.0164
lntechtou	-0.728349	0.135502	0.0000	lntechtou	0.130230	0.247069	0.5989
				ECTt-1	-0.135348	0.075150	0.0738

Note: *,**,*** *indicates significance at 1 %, 5 % and 10 % levels.*

Finally, the direction of the causality between series is analyzed causality test. Table 8. summarize the causality test results developed by Dumitrescu and Hurlin (2012).

Table 8: *The Panel Causality Test Results*

H_0	W-Stat	Z bar - Stat	p- value	H_0	W-Stat	Z bar - Stat	p-value
lnco$_2$ →lngdp	5.9025	9.8049	0.0000*	lnco$_2$ →lnpop	6.3484	10.6968	0.0000*
lngdp →lnco2	1.4068	0.8136	0.5909	lnpop→lnco$_2$	19.0563	36.1127	0.0000*
lnco$_2$ →lntou	3.5395	5.0791	0.0000*	lnco$_2$ →lnre	3.6145	5.2290	0.0000*
lntou →lnco$_2$	1.9405	1.8811	0.0600**	lnren→lnco$_2$	3.3059	4.6118	0.0000*
lnco$_2$ →lntech	3.1281	4.2563	0.0000*	lnco$_2$→lntech*tou	3.4736	4.9473	0.0000*
lntech→lnco$_2$	3.9141	5.8283	0.0000*	lntech*tou→lnco$_2$	6.2711	10.5421	0.0000*

Note: *Optimal lag length is determined as 1. ***p < 0.10, **p < 0.05, *p < 0.01*

Two-way causality has been detected between CO_2 emissions and tourism İn Table 8. The same results are verified for renewable energy, population density, and economic growth. A bilateral causality is detected between CO_2 emissions and lntech*lntou. It has also been demonstrated unidirectional causality from CO_2 emissions to economic growth.

5 Conclusion

Economic growth and development have become the main focus of every country. Sustainable development needs to be addressed in its economic, social and environmental dimensions and seeks solutions to problems in this direction. In the process of sustainable development, the use of resources in production must be determined in line with technological developments and the institutional changes must be carried out in line with the potential to meet the needs and expectations (Bebbington, 2000: 3). In the tourism sector, fossil energy resources are largely used to carry out transportation, accommodation and entertainment activities. Energy consumption in the tourism sector increases CO_2 emissions into the atmosphere (Mester et al., 2023).

it is important to meet the energy needs from renewable energy sources and use environmentally friendly technologies in tourism sector for the minimize the impact of environmental damage.

In the study we aimed to investigate the connection between tourism sector, environmental technologies, economic growth, renewable energy, population density and CO_2 emissions over the period 1995–2020 for the 8 selected countries with the highest tourism income (America, Austria, Germany, Spain, Italy, Mexico, Turkey, United Kingdom, United States) applying the panel ARDL/PMG method. Analysis results showed that tourism, environmental technologies and economic growth enhance CO_2 emissions while renewable energy consumption and,population density series mitigates CO_2 emissions in the long run. The new variable created by the interaction of tourism with environmental technology is found to have a negative effect on emissions.

We can say that in the study, environmentally friendly technologies are used to reduce emissions in the sector. It is also hoped that the following recommendations will shed light on future research and guide policy makers.

(i) Sustainable growth and sustainable tourism policies will contribute to reducing environmental pollution and increasing environmental sustainability.
(ii) Policy makers should encourage the use of renewable energy through practices aimed at increasing private sector investments.
(iii) It is important for countries to establish and support legal frameworks to reduce renewable energy production costs.
(iv) The use of environmental technologies should be promoted throughout the country and the public should be given easy access to environmental technologies. Policies need to be formulated to minimize the role of tourism in negative environmental externalities.

(v) Sector requires strong collaboration between policy makers, tourism operators, local communities, technology providers and various stakeholders. Sustainability can be achieved through these collaborations

(vi) Governments should initiate educational initiatives to raise public awareness in order to create a sustainable environment and a more responsible perception of tourism. In this way, local and foreign tourists should be motivated to adopt greener and more sustainable habits.

(vii) For a sustainable environment, government support should be provided for recycling, waste reduction efforts and composting processes.

(viii) Green tourism practices should be supported by following the United Nations Sustainable Development Strategies.

References

Abdouli, M., Kamoun, O., & Hamdi, B. (2017). The impact of economic growth, population density, and FDI inflows on CO2 emissions in BRICTS coun tries: Does the Kuznets curve exist?. Empirical Economics, 54(4), 1717–1742.

Aini, Y. N. (2024). Sustainable tourism in Southeast Asia: Balancing economic growth, employment, and carbon emissions through evidence- based strate gies. Jurnal Kepariwisataan Indonesia: Jurnal Penelitian dan Pengembangan Kepariwisataan Indonesia, 0000-0002-0887-0145

Ali, Q., Yaseen, M. R., Anwar, S., Makhdum, M. S. A., & Khan, M. T. I. (2021). The İmpact of tourism, renewable energy, and economic growth on ecological footprint and natural resources: A panel data analysis. *Resources Policy*, *74*, 102365.

Anser, M. K., Yousaf, Z., Awan, U., Nassani, A. A., Qazi Abro, M. M., & Zaman, K. (2020). Identifying the carbon emissions damage to International Tourism: Turn a blind eye. Sustainability, 12(5), 1937.

Balsalobre-Lorente, D., Luzon, L. I., Usman, M., & Jahanger, A. (2023). The relevance of International Tourism and Natural Resource rents in economic growth: Fresh evidence from MINT countries in the Digital Era. *Environmental Science and Pollution Research*, *30*(34), 81495–81512.

Balsalobre-Lorente, D., & Leitão, N. C. (2020). The role of tourism, trade, renewable energy use and carbon dioxide emissions on economic growth: evidence of tourism-led growth hypothesis in EU-28. *Environmental Science and Pollution Research*, *27*(36), 45883–45896. https://doi.org/10.1007/s11356-020-10375-1

Bassil, C., Hamadeh, M., & Samara, N. (2015). The tourism led growth hypothesis: the Lebanese case. *Tourism Review*, *70*(1), 43–55.

Bebbington, J. (2000). Sustainable Development: A Review of the International Development, Business and Accounting Literatüre, Aberdeen Papers in Accountancy, Finance & Management, Working Paper.

Breusch, T. S. & Pagan, A. R. (1980), "The Lagrange Multiplier Test and its applications to model specification tests in Econometrics". Review of Economic Studies, 47, 239–253.

Brida, J. G., Cortes- Jimenez, I., & Pulina, M. (2014). Has the tourism- led growth hypothesis been validated? A literature review. Current Issues in Tourism, 19(5), 394–430. https://doi.org/10.1080/13683500.2013.868414

Buckley, R. (2011). Tourism and environment. *Annual Review of Environment and Resources*, *36*(1), 397–416.

Centobelli, P., Cerchione, R., & Esposito, E. (2022) Environmental sustainability in the service industry of transportation and logistics service providers: Systematic literature review and research directions. Transport Research Part D Transport & Environment, 53, 454–470.

Chen, J., Wang, B., Huang, S., & Song, M. (2020). The influence of increased population density in China on air pollution. *Science of the Total Environment*, *735*, 139456. https://doi.org/10.1016/j.scitotenv.2020.139456

Dogru, T., Bulut, U., Kocak, E., Isik, C., Suess, C., & Sirakaya-Turk, E. (2020). The Nexus between tourism, economic growth, renewable energy consumption, and carbon dioxide emissions: Contemporary evidence from OECD countries. *Environmental Science and Pollution Research*, *27*, 40930–40948.

Dong, K., Sun, R., Hochman, G., Zeng, X., Li, H., & Jiang, H. (2017). Impact of natural gas consumption on CO2 emissions: Panel data evidence from China's provinces. Journal of Cleaner Production, 162, 400–410.

Dumitrescu, E. I., & Hurlin, C. (2012). Testing for Granger non-causality in heterogeneous panels. *Economic Modelling*, *29*(4), 1450–1460. https://doi.org/10.1016/j.econmod.2012.02.014

El Menyari, Y. (2021). The effects of international tourism, electricity consumption, and economic growth on CO2 emissions in North Africa. *Environmental Science and Pollution Research*, *28*(32), 44028–44038. https://doi.org/10.1007/s11356-021-13818-5

Erdoğan, S., Gedikli, A., Cevik, E. I., & Erdoğan, F. (2022). Eco-friendly technologies, international tourism and carbon emissions: Evidence from the most visited countries. *Technological Forecasting and Social Change*, *180*, 121705. https://doi.org/10.1016/j.techfore.2022.121705

Geng, Z., Kong, X., Chen, W., Su, H., Liu, Y., Cai, F., & Zeng, J. (2018). Oxygen vacancies in ZnO nanosheets enhance CO2 electrochemical reduction to CO. *Angewandte Chemie, 130*(21), 6162–6167. https://doi.org/10.1002/ange.201711255

Ghosh, S. (2022). Effects of tourism on carbon dioxide emissions, a panel causal ity analysis with new data sets. Environment, Development and Sustainability, 24(3), 668-021-01592-w

Hall, D.R., Kirkpatrick, I., & Mitchell, M. (2005). *Rural Tourism and Sustainable Business.* Channel View Publications, London.

Hossain, M. A., Acharjee, S. K., Urbee, A. J., Kibria, M. G., & Tahrim, F. (2024). The Intricate impression of tourism, patent, and renewable energy on environmental sustainability. *South Asian Journal of Social Sciences and Humanities, 5*(1), 1–22. https://doi.org/10.1002/sd.2139

Jebli, M. B., Youssef, S. B., & Apergis, N. (2019).The dynamic linkage between renewable energy, tourism, CO2 emissions, economic growth, Foreign Direct Investment, and Trade. *Latin American Economic Review, 28*(1), 1–19. https://doi.org/10.1186/s40503-019-0063-7

Khan, A., Chenggang, Y., Hussain, J., Bano, S., & Nawaz, A. (2020). Natural Resources, Tourism Development, and Energy-Growth-CO2 Emission Nexus: A Simultaneity Modeling Analysis of BRI.

Koçak, E., Ulucak, R., & Ulucak, Z. Ş. (2020). The impact of tourism developments on CO2 emissions: An advanced panel data estimation. *Tourism Management Perspectives,* 33, 100611. https://doi.org/10.1016/j.tmp.2019.100611

Kyara, V. C., Rahman, M. M., & Khanam, R. (2021). Tourism expansion and economic growth in Tanzania: A Causality Analysis. *Heliyon, 7*(5).

Kumail, T., Ali, W., Sadiq, F., Wu, D., & Aburumman, A. (2020). Dynamic linkages between tourism, technology and CO2 emissions in Pakistan. *Anatolia,* 1–13.

Lantz, V., & Feng, Q. (2006). Assessing income, population, and technology impacts on CO_2 emissions in Canada: where's the EKC?. *Ecological Economics, 57*(2), 229–238. https://doi.org/10.1016/j.ecolecon.2005.04.006

Leitão, N. C., & Lorente, D. B. (2020). The linkage between economic growth, renewable energy, tourism, CO2 emissions, and international trade: The evidence for the European Union. *Energies, 13*(18), 4838. https://doi.org/10.3390/en13184838

Lee, C., Wang, C., & Ho, S. (2022). The dimension of Green Economy: Culture viewpoint. *Economic Analiz Policy, 74,* 122–138.

Lee, J. W., & Brahmasrene, T. (2013). Investigating the influence of tourism on economic growth and carbon emissions: Evidence from panel analysis of the

European Union. *Tourism Management*, *38*, 69–76. https://doi.org/10.1016/j.tourman.2013.02.016

Leitão, N. C., & Lorente, D. B. (2020). The linkage between economic growth, renewable energy, tourism, CO2 emissions, and international trade: The evidence for the European Union. *Energies*, *13*(18), 4838. https://doi.org/10.3390/en13184838

Loureiro, S. M. C., Guerreiro, J., & Han, H. (2022). Past, present, and future of pro-environmental behavior in tourism and hospitality: A text-mining approach. *Journal of Sustainable Tourism*, *30*(1), 258–278.

Long, H. P. (2012). "Tourism impacts and support for Tourism Development in Ha Long Bay, Vietnam: An examination of residents' perceptions". Asian Social Science, 8(8), s. 28–39. http://www.ccsenet.org/ass

Mester, I., Simuţ, R., Meşter, L., & Bâc, D. (2023). An investigation of tourism, economic growth, CO2 emissions, trade openness and Energy Intensity Index Nexus: Evidence for the European Union. Energies, 16(11), 4308.

Mongo, M., Belaïd, F., & Ramdani, B. (2021). The effects of environmental innovations on CO2 emissions: Empirical evidence from Europe. *Environmental Science & Policy*, *118*, 1–9. https://doi.org/10.1016/j.envsci.2020.12.004

Oltra, V. (2008). Environmental Innovation and Industrial Dynamics: The Contributions of Evolutionary Economics. DIME Working Papers. HYPERLINK "https://econpapers.repec.org/scripts/redir.pf?u=http%3A%2F%2Fcahiersdugretha.u-bordeaux.fr%2F2008%2F2008-28.pdf;h=repec:grt:wpegrt:2008-28" http://cahiersdugretha.u-bordeaux.fr/2008/2008-28.pdf

Pan, S. Y., Gao, M., Kim, H., Shah, K. J., Pei, S. L., & Chiang, P. C. (2018). Advances and challenges in Sustainable Tourism toward a Green Economy. *Science of the Total Environment*, *635*, 452–469.

Paramati, S. R., Alam, M. S., & Lau, C. K. M. (2018). The effect of Tourism Investment on Tourism Development and CO2 emissions: Empirical evidence from the EU Nations. *Journal of Sustainable Tourism*, *26*(9), 1587–1607. https://doi.org/10.1080/09669582.2018.1489398

Pesaran, M. H. (2007). "A simple panel Unit Root Test in the presence of Cross-Section Dependence". *Journal of Applied Econometrics*, *22*(2), 265–312.

Pesaran, M. H., ve Yamagata, T. (2008). "Testing slope homogeneity in large panels". *Journal of Econometrics*, *142*(1), 50–93.

Rahaman, M. A., Hossain, M. A., & Chen, S. (2022). The impact of foreign direct investment, tourism, electricity consumption, and economic development on CO2 emissions in Bangladesh. *Environmental science and pollution research*, *29*(25), 37344–37358. https://doi.org/10.1007/s11356-021-18061-6

Raihan, A., & Tuspekova, A. (2022). The nexus between economic growth, energy use, urbanization, tourism, and carbon dioxide emissions: New insights from Singapore. Sustainability Analytics and Modeling, 2, 100009. https://doi.org/10.1016/j.samod.2022.100009

Raza, S. A., Sharif, A., Wong, W. K., & Karim, M. Z. A. (2017). Tourism Development and Environmental degradation in the United States: Evidence from Wavelet-Based Analysis. *Current Issues in Tourism*, *20*(16), 1768–1790. https://doi.org/10.1080/13683500.2016.1192587

Razzaq, A., Fatima, T., & Murshed, M. (2023). Asymmetric effects of Tourism Development and Green İnnovation on economic growth and carbon emissions in top 10 GDP countries. *Journal of Environmental Planning and Management*, *66*(3), 471–500. https://doi.org/10.1080/09640568.2021.1990029

Shakouri, B., Yazdi, K. S., & Ghorchebigi, E. (2017). Does Tourism Development promote CO2 emissions?. *Anatolia*, *28*(3), 444–452. https://doi.org/10.1007/s11356-023-26089-z

Sokhanvar, A., Çiftçioğlu, S., & Javid, E. (2018). Another look at Tourism-Economic Development Nexus. Tourism Management Perspectives, 26, 97–106.

Sun, Y., Lu, Y., Wang, T., Ma, H., & He, G. (2008). Pattern of patent-based environmental technology innovation in China. *Technological Forecasting and Social Change*, *75*(7), 1032–1042. https://doi.org/10.1016/j.techfore.2007.09.004

Thi, D., Tran, V. Q., & Nguyen, D. T. (2023). The relationship between renewable energy consumption, international tourism, trade openness, innovation and carbon dioxide emissions: International evidence. *International Journal of Sustainable Energy*, *42*(1), 397–416. https://doi.org/10.1080/14786451.2023.2192827

This is how travel and tourism can reach its net zero goals | World Economic Forum. Accessed date: 09.07.2024.

Ullah, A., Raza, K., & Mehmood, U. (2023). The impact of Economic Growth, Tourism, Natural Resources, Technological Innovation on carbon dioxide emission: Evidence from BRICS countries. *Environmental Science and Pollution Research*, *30*(32), 78825–78838. https://doi.org/10.1007/s11356-023-27903-4

Voumik, L. C., Islam, M. A., & Nafi, S. M. (2024). Does tourism have an impact on carbon emissions in Asia? An application of Fresh Panel Methodology. *Environment, Development and Sustainability*, *26*(4), 9481–9499. https://doi.org/10.1007/s10668-023-03104-4

Wang, Z. (2018). Dynamic relationship between tourism, economic growth, and environmental quality. Journal of Sustainable Tourism, 26(11), 1928–1943.

Westerlund, J. (2008). Panel cointegration tests of the Fisher effect. Journal of Applied Econometrics, 23(2), 193–233. DOI: 10.1002/jae.967

Xia, W., Doğan, B., Shahzad, U., Adedoyin, F. F., Popoola, A., & Bashir, M. A. (2022). An empirical investigation of tourism-led growth hypothesis in the European countries: evidence from augmented mean group estimator. *Portuguese Economic Journal*, 1–28. https://doi.org/10.1007/s10258-021-00193-9

Zheng, C., Wu, S., Teng, Y. P., Wu, S., & Wang, Z. (2023). Natural resources, tourism resources and economic growth: A new direction to natural resources perspective and investment. *Resources Policy*, *86*, 104134.

Süleyman Koç[1]

Investigating the Stationarity Properties of the Average Propensity to Consume in the Turkish Economy

1 Introduction

Households and Non-profit institutions serving households consumption spending are the largest and most critical component of aggregate expenditures. Savings and consumption behavior have a central role in many macroeconomic models. Thus, the stationarity properties of the average propensity to consume (APC), have attracted considerable scholarly interest and examination over many years by academic inquires and policy-oriented discussions in macroeconomics. In economies that are subject to continuous shocks, it is crucial to determine whether consumption expenditures are permanently affected by these shocks. Should shocks have a permanent impact on the APC with no tendency towards mean reversion, in other words exhibiting non-stationary behavior, this would validate several prominent consumption theories. Specifically, it would lend support to Keynes' absolute income hypothesis (1936) which posits that consumption levels are primarily determined by current disposable income. Additionally, it would align with Deaton's involuntary savings theory (1977) as well as Marxist underconsumption theory. Conversely, a finding that the APC demonstrates stationary qualities would corroborate alternative hypotheses. Namely, it would corroborate Duesenberry's (1949) relative income hypothesis that consumption is influenced more by relative than absolute income levels. It would also validate Friedman's permanent income hypothesis (1957) asserting lifetime resources determine spending. Similarly, it would concord with Ando and Modigliani's (1963) life cycle hypothesis positing consumption smoothing over one's lifetime. All of these latter hypotheses imply stable equilibrium will eventually be reached between consumption and income in the long-run, necessitating the APC converge to a long-run equilibrium level.

1 Ph.D. Res. Asst. Aksaray University, FEAS Department of Economics, suleymankoc@aksaray.edu.tr, ORCID: 0000-0002-6976-045X.

Turkey is a dynamic economy in the high middle-income group, which has approached the middle-income trap and has undergone a serious transformation in terms of urbanization and opening up to foreign trade. Moreover, as a founding member of the OECD and G20, it is among the major economies of the world and its region. Furthermore, as an economy that has been frequently exposed to shocks such as monetary and fiscal policy shocks as well as political, geopolitical, and natural disasters, the Turkish economy is a very suitable sample to test whether the effects of these shocks on the stochastic properties of the APC are persistent.

This chapter sets out to conduct a comprehensive empirical investigation and contribute fresh evidence to the related literature by testing the stationarity of the APC over the datasets compiled for the long run (annually from 1960 to 2023). We will test for stationarity in both the short run (quarterly from 1998 to 2024) and the long run (annually from 1960 to 2023) using a range of tests, including those that consider sharp structural breaks and unit root tests, as well as those that consider gradual breaks extended with Fourier functions and conventional unit root tests. The rest of the chapter continues with data and methodology. After a brief literature review, empirical findings are reported, and the findings obtained in the discussion section are compared with the Turkish economy and relevant literature. The chapter is completed with concluding remarks.

2 Brief Overview of the Literature

Theories of consumption are divided on whether the average propensity to consume reverts to a mean value over time. To resolve this debate and uncover the true nature of the phenomenon, researchers have examined the stability of APC using time series and panel data unit root or stationarity procedures. Baykara and Telatar (2012), as well as Cerrato et al. (2013), provide a conceptual overview of differing consumption hypotheses and their empirical implications regarding the stationarity of APC. As empirical works on this topic have utilized the cutting-edge econometric techniques available at the time they were conducted, studies investigating the stationarity of APC have similarly embraced advances in statistical methodology. Empirical investigations into the unit root properties of the APC have produced diverse findings influenced by evolving econometric approaches to unit roots or stationarity testing. The literature examining the stationarity of the APC for Turkey via stationarity and unit root tests is summarized in Table 1. As can be gleaned from Table 1, conventional tests as well as those accounting for sharp structural breaks and various nonlinear unit root tests have been applied to investigate the stationarity of the APC in Turkey's economy.

Meanwhile, no prior studies have benefited from Fourier or fractional Fourier form unit root or stationarity tests on Turkey's income-consumption data. This study represents one of the inaugural endeavors to deploy Fourier and Fractional Flexible Fourier unit root and stationarity procedures for the purpose of evaluating the stationarity of APC in the Turkish economy.

Table 1: Empirical Literature on the Testing Stationarity of APC in Turkish Economy

Author(s) (Year)	Period	Method	Result
Arı and Özcan (2015)	1955–2010	ZA(1992), LP(1997) LS(2003;2004) Kapetanios et al. (2005) KSS (2003)	Non-stationarity APC; AIH is valid
Sivri and Seven (2017)	1987Q1-2017Q1 1987Q1-2016Q2	ADF(1981), PP(1987), KPSS(1992), ERS(1996), Ng-Perron(2001); LS (2003;2004)	Stationarity; PIH, RIH, LCH are valid.
Ceylan and Karaağaç (2019)	1950–2014	ZA(1992), LP(1997) KSS(2003), LNV(1998), LNV Sollis (2004), Sollis [AESTAR] (2009)	Stationarity; PIH, RIH, LCH are valid.
Uğur et al. (2020)	1960–2015	ADF(1981), PP(1987), ZA(1992)	Non-stationarity APC; AIH is valid
Yiğit (2020)	2006Q1-2019Q4	ADF(1981), PP(1987), KPSS(1992), ERS(1996), Ng-Perron(2001); ZA(1992); LS (2003;2004)	Non-stationarity APC, AIH is valid

Some of the recent multi-country studies testing the stationarity of the APC are summarized in Table 2. It is understood from the table that findings were obtained both for and against stationarity. It can be argued that the method, sample and period used also contributed to the difference in the results.

Table 2: Selected Literature on the Testing Stationarity of APC in Multi-country Samples

Author(s)	Sample	Period	Method(s)	Results
Elmi and Ranjbar (2013)	16 OECD countries	1960–2010	Becker et al. (2006) Fourier KPSS	APC of 12 of 16 OECD countries is stationarity
Cerrato et al. (2009)	24 countries from the OECD and 33 non-OECD countries	1951–2003	Cerrato et al. (2009) unit root test, Pesaran (2007) CIPS	The results indicated non-stationary behavior for 83 % of the OECD country sample and 74 % of the non-OECD country sample.
Gözgör (2013)	European (CEE) countries	1997–2012	Maddala – Wu (1999) and Pesaran (2007) CIPS	Stationarity for 9 of the sample
Yılancı et al. (2013)	9 Southeast Asia countries	1970–2010	IM et al. (2005) Panel LM Unit Root Test	Non-stationarity is rejected
Chen and Xie (2015a)	20 OECD countries	1950–2010	Chen and Xie (2015b)	Accounting for asymmetry and non-linear trends over the long run, the data exhibit stationary behavior.
Solarin (2017)	10 ASEAN Countries	1970–2015	Maddala – Wu (1999), Breuer et al. (2001), Kapetanios et al. (2003), Becker et al. (2006), Pesaran (2007), Cerato et al. (2009), Sollis (2009) and Kruse (2011) tests	The results show that APC generally does not demonstrate stationary properties across most of the sample time period analyzed.
Solarin et al. (2018)	50 African Countries	1970–2014	RALS-LM Unit root test	The analysis finds that APC exhibits stationary behavior after accounting for structural breaks in the majority (44 out of 50) of African countries examined.

Table 3: Continued

Author(s)	Sample	Period	Method(s)	Results
Eryer and Konuk (2022)	BRICS-TM countries	1990–2020	Fisher ADF, Fisher PP	Non-stationarity APC
Aksu et al. (2024)	D-8 Countries	1982–2020	Panel Panic and Fourier LM unit root tests	Non-stationarity APC

2 Data and Methodology

2.1 Data

Two consumption-income series with different frequencies have been compiled for econometric analyses. The first APC series is derived from current national accounts, and it's at a quarterly frequency. This data set is mainly for to understand recent shocks effect on APC. The second series is derived from the World Bank and is mainly for long-term shocks' effects on APC. Basic information of APC series is summarized in Table 3.

Table 3: Basic Information of the Data Set Used in Econometric Analyses

Variable	Period	Description	Source
APC1	1998Q1-2024Q1	Households and Non-profit institutions serving households final consumption expenditure (% of GDP) Quarterly data	CBRT
APC2	1960–2023	Households and Non-profit institutions serving households final consumption expenditure (% of GDP) Annual data	World Bank

Note: The APC1 series is calculated using nominal variables (consumption and GDP). With the new 2009-based series, TURKSTAT has made a very important change by starting to calculate GDP as a Fisher-type (chain-weighted) index instead of a Laspeyres-type index. One of the main features of Fisher-type indices is that the ratio calculation based on real variables does not make sense. See Sivri and Seven (2017) and Whelan (2002). CBRT refers Central Bank of Republic of Turkiye.

The basic descriptive statistics of the APC series are summarized in Table 4. The Jarque-Bera test indicated that both time series exhibited normal distributions. While series have different frequencies, their mean and median are so close to each other. However, the long-term annual series' standard deviation is greater than the short-term series. This statistic may imply that long-term series might be affected by more powerful shocks or gradual transformations than short-term series.

Table 4: Basic Descriptive Stats of APC Series

	APC1	APC2
Mean	61.1578	60.4137
Median	61.3844	60.1947
Maximum	68.4495	70.6013
Minimum	54.1944	51.1111
Std. Dev.	3.3429	5.2748
Skewness	-0.1379	0.1510
Kurtosis	2.2265	2.1941
Jarque-Bera Test Stat.	2.9506	1.9751
Prob. of Jarque-Bera Test	0.2287	0.3725
Sum	6421.5663	3866.4749
Sum Sq.	393890.9296	235340.8464
Sum Sq. Dev.	1162.2278	1752.9125
Observations	105	64

2.2 Methodology

Unit root and stationarity testing, which arose from (Dickey & Fuller, 1979; 1981) began to attract more attention in research after Nelson and Plosser (1982) continued to progress with a series of tests. The literature, which became richer after Perron (1989), started to take structural breaks into consideration. Unit root tests have also been developed for nonlinear frameworks (Leybourne et al., 1998; Kapetanios et al., 2003). Tests with structural breaks continue to be developed, with tests that take into account gradual breaks as well as tests that take into account sharp breaks (Becker, Enders & Lee 2006; Enders & Lee, 2012; Omay, 2015). In this study, in order to carry out a comprehensive investigation, tests that take into account both sharp breaks and gradual breaks are used in addition to conventional tests.

3 Empirical Results

Findings of conventional unit root tests are summarized in Table 5. APC1, which is a short-term series, appears to be stationary in almost all series in trend and intercept specifications. APC2, the long-term series, does not appear to be stationary at all.

Table 5: Conventional Unit Root and Stationarity Tests

		ADF	DF-GLS	PP	KPSS	NP-MZ_{α}	NP-MZ_{t}	NP-MSB	NP-MPT	ERS-PO
APC1	*Intercept*	-1.378	-1.095	-1.002	1.11899	-3.261	-1.07	0.328	7.33982	8.1477
APC2	*Intercept*	-2.121	-1.917	-1.958	0.374***	-6.563*	-1.809	0.2756	3.74185*	4.17342
APC1	*Intercept and Trend*	-4.39***	-3.251**	-4.41***	0.065***	-17.26*	-2.935**	0.1701*	5.2931**	6.5935*
APC2	*Intercept and Trend*	-0.955	-2.1807	-1.9714	0.2199	-8.583	-2.0353	0.2371	10.7445	28.0867

Note: The *, **, and *** denote stationarity tests fulfill the stationarity requirement. Namely, ADF refers to the Augmented Dickey-Fuller test, DF-GLS to the Dickey-Fuller GLS detrended, PP to the Phillips-Perron test, KPSS to the Kwiatkowski-Phillips-Schmidt-Shin test, ERS-PO to the Elliot-Rothenberg-Stock point-optimal test, NP-MZα to the Ng-Perron Zα test, NP-MZt to the Ng-Perron Zt test, NP-MSB to the Ng-Perron MSB test, and NP-MPT to the Ng-Perron MPT test. The lag lengths are calculated according to the Bayesian Information Criterion (BIC).

Table 6 reports one- and two-break unit root test results taking into account sharp structural breaks. Notable findings: The APC1 series, which was identified as stationary in all conventional unit root tests in the trend and intercept specification, appears to be stationary only in Narayan-Popp (2010) and Carrion-i-Silvestre and Sansó (2007). On the other hand, the APC2 series, which is not stationary in conventional tests, appears to be stationary in all sharp-break tests, especially in the Break in level and trend specification.

Table 6: Unit Root and Stationarity Tests with Sharp Structural Breaks

		APC_1	APC_2	APC_1	APC_2
		Break in Level		**Break in level and trend**	
Zivot & Andrews (1992)	ZA Test Stat	-3.466	-4.454	-3.990	-6.576***
	Break Date	2021Q3	1985	2017Q3	1985
	Fraction(λ)	0.905	0.406	0.752	0.406

(continued on next page)

Table 7: Continued

		APC_1	APC_2	APC_1	APC_2
		Break in Level		**Break in level and trend**	
Lee & Strazicich (2013)	LS	-3.362*	-2.594	-3.674	-5.249***
	Break Date	2016Q1	1979	2001Q3	1990
	Fraction(λ)	0.695	0.313	0.143	0.484
Narayan & Popp (2010)	NP Test Stat	-4.730**	-7.095	-5.583***	-8.445***
	Break Date1	2009Q3	1971	2009Q3	1982
	Break Date2	2021Q2	1985	2020Q2	1985
	Fraction (λ_1, λ_2)	0.448 / 0.895	0.188 / 0.406	0.448 / 0.857	0.359 / 0.406
Lee & Strazicich (2003)	LS Test Stat	-3.788	-2.924	-4.989	-6.447**
	Break Date1	2003Q1	1979	2005Q4	1987
	Break Date2	2015Q2	1989	2010Q1	1996
	Fraction (λ_1, λ_2)	0.200 / 0.667	0.313 / 0.469	0.305 / 0.467	0.438 / 0.578
Carrion-i-Silvestre & Sansó (2007)	CS Test Stat	0.081***	0.059***	0.030***	0.026***
	Break Date1	2009Q4	1971	2009Q4	1971
	Break Date2	2022Q3	1986	2021Q4	1986
	Fraction (λ_1, λ_2)	0.457 / 0.943	0.188 / 0.422	0.457 / 0.914	0.172 / 0.422

Note: The *, **, and *** denote varying levels of statistical significance at 10 %, 5 %, and 1 %, respectively, demonstrating that the associated stationarity tests fulfill the stationarity requirement. ZA, LS, NP, CS refer to the Zivot & Andrews, Lee & Strazicich, Narayan & Popp, Carrion-i-Silvestre & Sansó, respectively. The lag lengths are calculated according to the Akaike Information Criterion (AIC).

Table 7 summarizes the findings of the unit root procedures that considers gradual structural breaks by employing fourier functions. When gradual breaks are taken into account, APC1 is stationary in the trend and intercept specification, while APC2 is stationary only in the intercept specification.

Table 7: Fourier Unit Root and Stationarity Tests for Gradual Structural Breaks

		Intercept		**Trend and Intercept**	
		APC1	**APC2**	**APC1**	**APC2**
Fourier ADF (Enders & Lee, 2012)	Fourier	1	1	2	1
	FADF	-2.5040	-3.0030	-4.4584***	-3.1992

Table 7: Continued

		Intercept		Trend and Intercept	
		APC1	APC2	APC1	APC2
Fourier GLS (Rodrigues & Taylor, 2012)	Fourier	2	1	2	1
	FGLS	0.0465	-2.8593*	-4.5797***	-3.0926
Fourier KPSS (Becker, Enders & Lee, 2006)	Fourier	1	1	2	1
	FKPSS	1.2678	0.1145***	0.0554***	0.0904

Note: The *, **, and *** symbols denote statistical significance at 10 %, 5 %, and 1 % levels, respectively

Table 8 reports the results of the fractional frequency fourier ADF test. employed procedure is the extended version of Bozoklu et al. (2020) to allow fractional frequencies between 0 and 5. In the Fractional Frequency Fourier ADF test, if the frequency is an integer, it means that the shocks are temporary, while fractional means that they are permanent. On the other hand, before analyzing the FADF test statistics, when the F-test statistics, which express whether the Fourier terms are significant or not, are examined, significance is found only in the Intercept and Trend specification for the APC2 series. The FADF test statistic also indicates weak stationarity at the 10 % significance level. On the other hand, the frequency of this series is fractional at the 1.2 level. This implies the persistence of shocks. It should be emphasized that when we transformed the Fourier unit root test into fractional flexible form and applied the test, no statistically significant stationarity was detected for both series at a significance level of 5 % and below in any specification.

Table 8: Fractional Frequency Fourier ADF Test Results for Gradual Breaks

	Variable	Frequency	Min. SSR	F Test Stat	Opt. Lag	FADF Test Stat
Intercept	APC1	0.2	95.82462	1.940540	8	-2.301399
	APC2	1.0	315.7774	6.866861	4	-3.771126*
Intercept and Trend	APC1	2.2	85.7791	5.527288	8	-3.33936
	APC2	1.2	297.3366	9.561854**	4	-4.314295*

Note: The symbols *, **, and *** denote statistical significance at the 10 %, 5 %, and 1 % levels respectively.

Table 9 reports the results of Harvey and Leybourne (2007) and Harvey et al. (2008) non-linearity tests for the APC1 and APC2 series. The series are found to be linear, according to Harvey and Leybourne (2007) furthermore according to Harvey et al. (2008) test findings, both series are linear at the 5 % level. Therefore, nonlinear unit root tests could not be employed when testing the stationarity of the APC series.

Table 9: Non-Linearity Tests

Panel I: Harvey and Leybourne (2007) Test					
Test Stat			Table Critical Values		
%1	%5	%10	%1	%5	%10
APC1: 1.18	1.16	1.15	13.27	9.48	7.77
APC2: 7.38	7.28	7.22			
Panel II: Harvey Leybourne and Xiao (2008) Test					
Test Stat			Table Critical Values		
			%1	%5	%10
APC1: 4.75*			9.21	5.99	4.60
APC2: 4.16					

Note: The *symbol denote statistical significance at 10 % level.

4 Discussion

The results of the conventional, sharp break, and gradual break unit root tests employed in the empirical analysis phase yielded mixed results in line with the literature. Different specifications and unit root test procedures may lead to different findings in both the long-run APC series and the relatively short-run APC series in the Turkish economy. The break dates obtained from the unit root tests that account for sharp structural breaks and the APC series used in the analyses are plotted together in Figures 1 and 2. In Figure 1, development of the APC1 series is illustrated with structural breaks. It is clearly seen that the 2000–2001 crisis, the 2008–2009 great recession, the political crisis stemming from the failed coup attempt in 2016, the COVID-19 pandemic, and the subsequent expansionary policy shocks are also detected by the alternative sharp structural break unit root procedures.

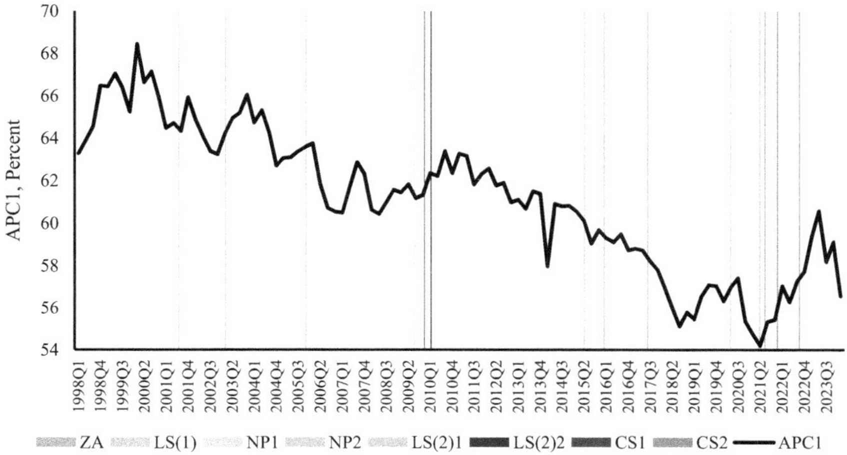

Figure 1: Development of Quarterly Consumption-Income Ratio in Turkish Economy with Structural Breaks

Source: Data from CBRT and Author's calculations

In Figure 2, the development of the APC of the Turkish economy over a period of more than half a century is depicted together with the break dates obtained from alternative sharp structural break unit root tests. The striking point in this graph is that the sudden break dates indicated by the alternative methods have accumulated between 1971 and 1996. Turkey is an economy that has experienced intensive and significant transformations in this period, such as urbanization with rural-urban migration in the 1970s, opening out after 1980, financial deregulation after 1989, and joining the customs union in 1996. The methods used in the study find the shocks in this period more significant than the previous and subsequent shocks. The stationarity of the average propensity to consume not only provides important information on the dynamics of consumption and saving in an economy but also provides insight into the dynamics of saving and investment. It is also known that one of the main drivers of Turkey's chronic current account deficit is the investment-savings gap. In this context, in light of the empirical results in this section, we can argue that the mixed findings obtained regarding the stability of the average propensity to consume are affected by the same factors as the volatile structure we observe in the pace of economic advancement of the Turkish economy both in the short and long term. In light of the findings of this study, it would not be correct to argue that the APC is clearly stationary or not. Additionally, through graphical examination and empirical evidence, we are able to validate the influence of these shocks on Turkish economy, which faces numerous shock types. This validation occurs when analyzing both short- and long-term APC data, capturing impacts over differing time horizons.

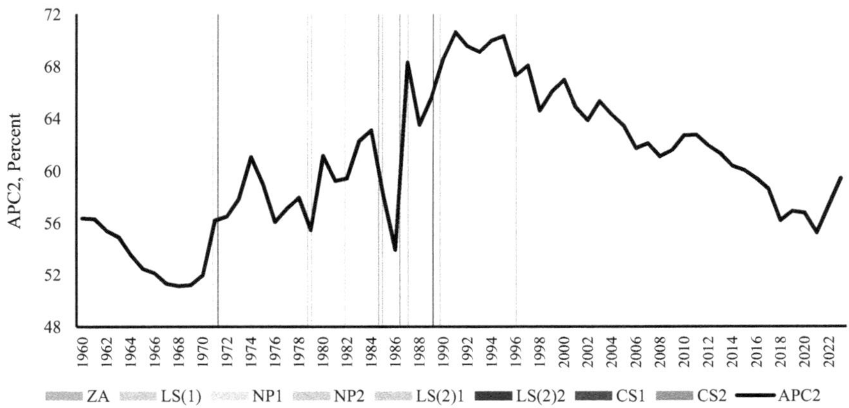

Figure 2: Development of Annual Consumption-Income Ratio in Turkish Economy with Structural Breaks

Source: Data from World Bank and Author's calculations

When Figures 1 and 2 are evaluated together, it is noteworthy that APC2, which reached its historical high value of 70.6 in 1991, decreased with a strong trend until 2021 (55.2). We can also observe this decline in the quarterly data from 2000Q1 (68.4) to 2021Q2 (54.2). GDP growth rates will be more volatile in periods of decreasing APC. Components that determine aggregate demand, such as investment, exports, and imports, may have a more determinant impact on cyclical fluctuations. The automatic stabilizing feature of consumption expenditures will weaken (Abeysinghe & Choy, 2004). The APC in an economy is positively related to the long-run growth rate in that economy (Baykara & Telatar, 2012), so the APC will be stable in the long run as long as the long-run growth rate does not change. Accordingly, the declining APC implied in the absolute income hypothesis requires interventionist fiscal policies. Incentives to stimulate private consumption expenditures, such as tax cuts, can be implemented by increasing disposable income. In this case, monetary policy instruments such as interest rates and money supply may not be needed. On the other hand, as Brady (2008) points out, if hypotheses such as PIH, RIH, and LCH, which assume that consumers smooth their intertemporal incomes, are valid, interventionist demand management may not be effective enough. Since the empirical findings of this study imply the existence of mixed evidence for the validity of both types of hypotheses, we should say that a systematic policy mix is more appropriate for the Turkish economy, although it may have discretionary elements.

5 Concluding Remarks

In this section, Turkey's consumption income ratios are analyzed using both short-run and long-run data using conventional, sudden structural break, and gradual structural break unit root tests. Since the findings from 15 different test procedures provide evidence of both stationarity and non-stationarity, it can be said that mixed findings have been reached in line with the previous literature. Some unit root procedures support the AIH hypothesis, while others support the RIH, PIH, and LCH hypotheses. Although there is empirical evidence in favor of both stationarity and non-stationarity, and these findings imply the need to focus on different types of policies, it cannot be denied that both sudden and gradual breaks that the Turkish economy and thus the APC is exposed to should be taken into account in policy design. Analyses based on the consumption-income ratio suggest that the Turkish economy needs a systematic combination of both fiscal and monetary policies to stabilize the APC in the face of shocks. For future research it is suggested that the application of residual augmented unit root tests and tests that take into account smooth structural breaks with asymmetric adjustments together can contribute to the literature.

References

Abeysinghe, T., & Choy, K. M. (2004). The aggregate consumption puzzle in Singapore. *Journal of Asian Economics, 15*(3), 563–578.

Aksu, T., Uğur, B., & Dallı, T. (2024). D-8 Ülkelerinde Ortalama Tüketim Eğiliminin Ampirik Analizi. *Equinox Journal of Economics Business and Political Studies, 11*(1), 38–52.

Ando, A., & Modigliani, F. (1963). The life-cycle hypothesis of saving: Aggregate implications and tests. *American Economic Review, 53*, 55–84.

Arı, A., & Özcan, B. (2015). Tüketim-Gelir Oranının Durağanlığı: Türkiye Örneği. *Hacettepe University Journal of Economics & Administrative Sciences/ Hacettepe Üniversitesi Iktisadi ve Idari Bilimler Fakültesi Dergisi, 33*(3).

Baykara, S., & Telatar, E. (2012). The stationarity of consumption-income ratios with nonlinear and asymmetric unit root tests: Evidence from fourteen transition economies. *Hacettepe University Department of Economics Working Papers, 20129*, 1–41.

Becker, R., Enders, W., & Lee, J. (2006). A stationarity test in the presence of an unknown number of smooth breaks. *Journal of Time Series Analysis, 27*(3), 381–409.

Bozoklu, S., Yilanci, V., & Gorus, M. S. (2020). Persistence in per capita energy consumption: A fractional integration approach with a Fourier function. *Energy Economics, 91*, 104926.

Brady, R. R. (2008). Structural breaks and consumer credit: Is consumption smoothing finally a reality?. *Journal of macroeconomics, 30*(3), 1246–1268.

Carrion-i-Silvestre, J. L., & Sansó, A. (2007). The KPSS test with two structural breaks. *Spanish Economic Review, 9*, 105–127.

Cerrato, M., De Peretti, C., & Stewart, C. (2013). Is the Consumption–Income Ratio Stationary? Evidence from Linear and Non-Linear Panel Unit Root Tests for OECD and Non-OECD Countries. *The Manchester School, 81*(1), 102–120.

Ceylan, R., & Karaağaç, G. E. (2019). Türkiye'de sürekli gelir hipotezinin test edilmesi: doğrusal olmayan birim kök testlerinden kanıtlar. *Ufuk Üniversitesi Sosyal Bilimler Enstitüsü Dergisi, 8*(15), 219–237.

Chen, S. W., & Xie, Z. (2015a). Nonlinear mean reversion in the consumption-income ratio: New evidence from the OECD countries. *International Review of Accounting, Banking & Finance, 7.*

Chen, S. W., & Xie, Z. (2015b). Testing for current account sustainability under assumptions of smooth break and nonlinearity. *International Review of Economics & Finance, 38*, 142–156.

Deaton, A. (1977). Involuntary saving through unanticipated inflation. *The American Economic Review, 67*(5), 899–910.

Dickey, D. A., & Fuller, W. A. (1979). Distribution of the estimators for autoregressive time series with a unit root. *Journal of the American statistical association, 74*(366a), 427–431.

Dickey, D. A., & Fuller, W. A. (1981). Likelihood ratio statistics for autoregressive time series with a unit root. *Econometrica: Journal of the Econometric Society*, 1057–1072.

Duesenberry, J. S. (1949). *Income, Saving, and the Theory of Consumption Behavior*. Cambridge, MA: Harvard University Press.

Elliott, G., Rothenberg, T. J., & Stock, J. H. (1996). Efficient Tests for an Autoregressive Unit Root. *Econometrica: Journal of the Econometric Society*, 813–836.

Elmi, Z. M., & Ranjbar, O. (2013). Nonlinear adjustment to the mean reversion of consumption–income ratio. *Economic Modelling*, *35*, 477–480.

Enders, W., & Lee, J. (2012). A unit root test using a Fourier series to approximate smooth breaks. *Oxford bulletin of Economics and Statistics*, *74*(4), 574–599.

Eryer, A., & Konuk, T. (2022). BRICS-TM ÜLKELERİNDE ORTALAMA TÜKETİM EĞİLİMİNİN GEÇERLİLİĞİ. *Uluslararası Muhasebe ve Finans Araştırmaları Dergisi*, *4*(2), 142–152.

Friedman, M. (1957). *A Theory of the Consumption Function*. Princeton, NJ: Princeton University Press.

Gozgor, G. (2013). Stochastic properties of the consumption-income ratios in central and eastern European countries. *Zbornik radova Ekonomskog fakulteta u Rijeci: časopis za ekonomsku teoriju i praksu*, *31*(2), 193–207.

Harvey, D. I., & Leybourne, S. J. (2007). Testing for time series linearity. *The Econometrics Journal*, *10*(1), 149–165.

Harvey, D. I., Leybourne, S. J., & Xiao, B. (2008). A powerful test for linearity when the order of integration is unknown. *Studies in Nonlinear Dynamics & Econometrics*, *12*(3).

Kapetanios, G. (2005). Unit-root testing against the alternative hypothesis of up to m structural breaks. *Journal of Time Series Analysis*, *26*(1), 123–133.

Kapetanios, G., Shin, Y., & Snell, A. (2003). Testing for a unit root in the nonlinear STAR framework. *Journal of Econometrics*, *112*(2), 359–379.

Keynes, J. M. (1936). *The General Theory of Employment Interest and Money*. New York: Macmillan and Company.

Lee, J., & Strazicich, M. C. (2003). Minimum Lagrange multiplier unit root test with two structural breaks. *Review of Economics and Statistics*, *85*(4), 1082–1089.

Lee, J., & Strazicich, M. C. (2013). Minimum LM unit root test with one structural break. *Economics Bulletin*, *33*(4), 2483–2492.

Leybourne, Stephen, Paul Newbold, & Dimitrios Vougas. "Unit roots and smooth transitions." *Journal of Time Series Analysis* 19.1 (1998): 83–97.

Lumsdaine, R. L., & Papell, D. H. (1997). Multiple trend breaks and the unit-root hypothesis. *Review of Economics and Statistics*, *79*(2), 212–218.

Narayan, P. K., & Popp, S. (2010). A new unit root test with two structural breaks in level and slope at unknown time. *Journal of Applied Statistics*, *37*(9), 1425–1438.

Nelson, C. R., & Plosser, C. R. (1982). Trends and random walks in macroeconmic time series: Some evidence and implications. *Journal of Monetary Economics*, *10*(2), 139–162.

Ng, S., & Perron, P. (1995). Unit root tests in ARMA models with data-dependent methods for the selection of the truncation lag. *Journal of the American Statistical Association*, 90(429), 268–281.

Omay, T. (2015). Fractional frequency flexible Fourier form to approximate smooth breaks in unit root testing. Economics letters, 134, 123–126.

Perron, P. (1989). The great crash, the oil price shock, and the unit root hypothesis. *Econometrica: journal of the Econometric Society*, 1361–1401.

Phillips, P. C., & Perron, P. (1988). Testing for a unit root in time series regression. *biometrika*, *75*(2), 335–346.

Rodrigues, P. M., & Robert Taylor, A. M. (2012). The flexible fourier form and local generalised least squares de-trended unit root tests. *Oxford Bulletin of Economics and Statistics*, *74*(5), 736–759.

Sivri, U., & Seven, B. (2017). Ortalama tüketim eğilimi durağan mıdır? Türkiye ekonomisi için bir zaman serisi analizi. *Anadolu İktisat ve İşletme Dergisi*, *1*(1), 50–65.

Solarin, S. A. (2017). the stationarity of consumption-income ratios: Nonlinear evidence in ASEAN countries. *Romanian Journal of Economic Forecasting*, *20*(2), 109–123.

Solarin, S. A., Shahbaz, M., & Stewart, C. (2018). Is the consumption-income ratio stationary in African countries? Evidence from new time series tests that allow for structural breaks. *Applied Economics*, *50*(38), 4122–4136.

Uğur, B., Özbek, S., & Türkmen, S. (2020). Ortalama Tüketim Eğiliminin Durağanlığı: Türkiye Örneği. *The Journal of Social Sciences*, *45*, 294–307.

Whelan, K. (2002). A guide to US chain aggregated NIPA data. *Review of income and wealth*, 48(2), 217–233.

Yılancı, V., Zeren, F., & Arı, A. (2013). Tüketim-Gelir Oranı Güneydoğu Asya Ülkelerinde Durağan Mı?: Panel Birim Kök Testi. *Journal of Management and Economics Research*, *11*(21), 130–139.

Yiğit, M. (2020). Testing the validity of absolute income hypothesis through average propensity to consume for turkish economy. in *Theoretical and Applied Studies on Turkish Economy*, Vol. I, 105.

Zivot, E., & Andrews, D. W. (1992). Further Evidence on the Great Crash, the Oil-Price Shock, and the Unit-Root Hypothesis. Journal of Business & Economic Statistics, 10(3), 251–270.

Zivot, E., & Andrews, D. W. K. (2002). Further evidence on the great crash, the oil-price shock, and the unit-root hypothesis. *Journal of Business & Economic Statistics*, *20*(1), 25–44.

www.ingramcontent.com/pod-product-compliance
Ingram Content Group UK Ltd.
Pitfield, Milton Keynes, MK11 3LW, UK
UKHW020154290926
15471UKWH00005B/866

9 783631 927557